Computational Anatomy Based on Whole Body Imaging

Hidefumi Kobatake • Yoshitaka Masutani
Editors

Computational Anatomy Based on Whole Body Imaging

Basic Principles of Computer-Assisted Diagnosis and Therapy

Springer

Editors
Hidefumi Kobatake
Professor Emeritus
Tokyo University of Agriculture
and Technology
Tokyo, Japan

Yoshitaka Masutani
Hiroshima City University
Hiroshima, Japan

ISBN 978-4-431-55974-0 ISBN 978-4-431-55976-4 (eBook)
DOI 10.1007/978-4-431-55976-4

Library of Congress Control Number: 2017934506

Printed on acid-free paper

This Springer imprint is published by Springer Nature
The registered company is Springer Japan KK
The registered company address is: Chiyoda First Bldg. East, 3-8-1 Nishi-Kanda, Chiyoda-ku, Tokyo 101-0065, Japan

Preface

The origin of research on computational anatomy can be traced back approximately a century. The famous book *On Shape and Growth* by D'Arcy Wentworth Thompson was published in 1917. Its central theme was to reveal the importance of physical laws and mechanics as the fundamental determinants of the form and structure of living organisms. In Chapter XVII of the book, "The Comparison of Related Forms," he showed that the differences in forms of related animals could be explained by relatively simple mathematical transformations.

Modern computational anatomy is emerging as a discipline focused on the quantitative analysis of variabilities in organ shape and the application of this analysis to computer-aided diagnosis (CAD) and computer-aided surgery. The spectrum of topics in computational anatomy has expanded to encompass all aspects of intelligent segmentation, modeling, recognition and understanding of complex three-dimensional (3D) objects, man–machine interface technologies, and other applications. Reflecting these developments, advanced computational anatomy provides a technical platform for a better understanding of anatomic variability, an aid in the diagnosis of disease, and a means to simulate surgical interventions.

October 2003 saw the start of an extensive research project on CAD in medical imaging, "Intelligent Assistance in Diagnosis of Multi-Dimensional Medical Images," in Japan. It was a 4-year research project supported by a Grant-in-Aid for Scientific Research on Priority Areas from the Ministry of Education, Culture, Sports, Science and Technology (MEXT). The state-of-the-art CAD system at that time was quite limited in its capabilities. The objective of this research project was to develop a multi-organ, multi-disease CAD system that made full use of human anatomical data and diagnostic knowledge of multiple diseases. The research organization consisted of nine planned research groups and ten research groups selected from publicly offered research plans. Almost all researchers in the area of medical image processing in Japan joined this project. Typical conventional CAD systems adopted two processing steps to detect specific abnormal regions on medical images. The first one is the processing to detect candidates of suspicious regions and the second one is to identify whether they are truly abnormal. Image features such as film densities, shapes, textural characteristics, and so on, which are distinctive to a specific disease on medical images, are used to detect suspicious regions. It was *the abnormality-dependent approach.* That is, image characteristics used in the first processing step were dependent on the kinds of diseases. However, we felt that the first processing step of future CAD systems whose targets are multiple diseases of multiple organs should be to understand the normal structure of the patient from input images, with potential ability to detect abnormal regions as regions having structures and/or characteristics that are different from normal ones. This fundamentally different approach was called *the normality-dependent approach.* That is, the CAD system depended on the understanding of normal organ structure and departures from those normal structures. The development of two databases, a digital atlas of human anatomy and a digital representation of chest and abdominal abnormalities, were set as primary targets. This project ended in March, 2007. The resulting technologies for analysis of medical images were improved, but had not achieved the level of sophistication of CAD technologies for the brain, which had been a priority for a longer period of time and involved an organ with less inter-individual variability. This provided strong motivation for us to organize a new research project. The 5-year research project on "Computational Anatomy for Computer-Aided Diagnosis and Therapy: Frontiers of Medical Image Sciences" (CA project) was initiated in October, 2009. It was supported by a Grant-in-Aid for Scientific Research on Innovative Areas from MEXT, Japan. The CA project employed a mathematical approach to provide a computational framework to deal with human anatomy in the chest and abdomen. The challenges consisted of (1) development of models for representation of anatomy that cover inter-individual variability in shape and topology and their construction through statistical analysis of population data, (2) investigation of methodologies for precise and robust retrieval of anatomical information from medical images, virtually equivalent to real human body dissection, and (3) development of innovative technologies to assist medical diagnosis and interventions based on computational anatomy. Details of the CA project are shown on the website: http://www.comp-anatomy.org/

The CA project completed its work at the end of March 2014. It greatly contributed to the development of advanced computational anatomy for the chest and abdomen and its applications to CAD and surgery. The purpose of this book is to introduce the basics and the state-of-the-art of this technology and its applications in the chest and abdomen. It contains not only the cutting-edge technologies produced by the CA project but also the basic mathematics and fundamentals. This book will be helpful and informative for researchers wishing to systematically survey the state-of-the-art in computational anatomy. We still have a long way to the final goal, that is, to realize perfect understanding of anatomical structures of patients from medical imageanalysis and intelligent assistance in medical diagnosis and interventions. I believe that this book will help to accelerate computational anatomy research.

Tokyo, Japan
July 31, 2015

Hidefumi Kobatake

Acknowledgments

This book is a review of the state of the art of the basics of computational anatomy focused primarily on the trunk area and its application to computer-aided diagnosis and surgery. As I mentioned in the preface, computational anatomy primarily focused on the brain came first and developed rapidly. The research project "Computational Anatomy for Computer-Aided Diagnosis and Therapy: Frontiers of Medical Image Sciences," which was supported by the Ministry of Education, Culture, Sports, Science and Technology, Japan (MEXT, Japan), was a challenge to establish computational anatomy on the modeling and analysis of the biological variability of organs in the chest and abdomen. About half the contents of this book are a compilation of the results of that project. As leader of the project, I am grateful to MEXT, Japan for the special support of our research. We will be very happy if this book is used as a standard textbook on computational anatomy for young researchers and as a result it contributes to accelerating the advance of computational anatomy not only on the trunk area but also on the whole body.

On behalf of the organizing committee of the research project, I would like to express sincere thanks to all the contributors, especially to the leaders of eight primary planned research groups who promoted the project successfully. And I also would like to express my great appreciation to the advisory committee members:

Dr. Karl Heinz Hoehne, Professor Emeritus, University of Hamburg, Germany,
Dr. Nicholas Ayache, Research Director, INRIA, France,
Dr. Takahiro Kozuka, Professor Emeritus, Osaka University, Japan,
Dr. Takeshi Iinuma, Honorary Research Fellow, National Institute of Radiological Sciences, Japan,
Dr. Koichiro Deguchi, Professor, Tohoku University, Japan,
and
Dr. Yoshimasa Kurumi, Professor, Shiga University of Medical Science, Japan.

They provided helpful directions for our research policy and gave us invaluable advice on our research activities. The role they played was essential for the success of our project.

The final goal of computational anatomy is the perfect understanding of human anatomy through the analysis of medical images and its application to intelligent assistance in diagnosis and surgery. There is still a long way to that goal, but I hope that this book will serve as a guideline for further work in this field and will mark a new era of computational anatomy research.

Tokyo University of Agriculture and Technology
Tokyo, Japan

Hidefumi Kobatake

Contents

Chapter 1
Introduction

Yoshitaka Masutani, Sakon Noriki, Shoji Kido, Hidetaka Arimura, Morimasa Tomikawa, Hidekata Hontani, and Yoshinobu Sato

Abstract This chapter presents the background and purpose of the computational anatomy research field from medical (needs) and technical (seeds) perspectives. We begin with a historical overview of the emergence of the discipline of computational anatomy (Sect. 1.1). Then, overviews of existing fields and the potential impact of computational anatomy on them are described (Sect. 1.2). To clarify the value of computational anatomy from the clinical viewpoint, medical education, diagnostic imaging, surgery, and radiation therapy are discussed, including situations that

---

Y. Masutani
Faculty of Information Science, Hiroshima City University, Hiroshima, 731-3194, Japan
e-mail: masutani@hiroshima-cu.ac.jp

S. Noriki
Faculty of Medical Sciences, University of Fukui, Eiheiji-cho, Fukui, 910-1193, Japan
e-mail: noriki@u-fukui.ac.jp

S. Kido
Department of Information Science and Engineering, Yamaguchi University, Ube-shi, Yamaguchi, 755-8611, Japan
e-mail: kido@ai.csse.yamaguchi-u.ac.jp

H. Arimura
Department of Health Sciences, Kyushu University, Fukuoka, 812-8582, Japan
e-mail: arimurah@med.kyushu-u.ac.jp

M. Tomikawa
Department of Surgery, Fukuoka City Hospital, Fukuoka, 812-0046, Japan
e-mail: mtomikaw@surg2.med.kyushu-u.ac.jp

H. Hontani
Department of Computer Science, Nagoya Institute of Technology, Aichi, 466-8555, Japan
e-mail: hontani@nitech.ac.jp

Y. Sato (✉)
Graduate School of Information Science, Nara Institute of Science and Technology, Ikoma-shi, Nara, 630-0192, Japan
e-mail: yoshi@is.naist.jp

H. Kobatake, Y. Masutani (eds.), *Computational Anatomy Based on Whole Body Imaging*, DOI 10.1007/978-4-431-55976-4_1

motivated the emergence of computational anatomy (Sect. 1.2.1). Similarly, from the technical (computer science) viewpoint, important technological developments providing the theoretical and algorithmic basis of computational anatomy are explored (Sect. 1.2.2). This book mainly addresses the development of whole-body computational anatomy, which is supported by the rapid progress of whole-body 3D imaging technologies. Thus, the impact of whole-body imaging (Sect. 1.3.1) and its utilization (Sect. 1.3.2) are discussed. Finally, the structure of this book is outlined (Sect. 1.4).

Keywords Medical education • Autopsy imaging • Computer-aided diagnosis • Radiation therapy • Computer-assisted surgery • Computer vision • Whole-body imaging

1.1 What Is Computational Anatomy?

Yoshitaka Masutani
Faculty of Information Science, Hiroshima City University, Hiroshima, 731-3194, Japan
e-mail: masutani@hiroshima-cu.ac.jp

Computational anatomy is an emerging discipline deriving from medical anatomy and several other sciences and technologies, including medical imaging, computer vision, and applied mathematics. The main focus of the discipline covers the quantitative analysis and modeling of variability of biological shapes in human anatomy in health and disease.

Beyond just adding numerical and quantitative information to the conventional anatomy describing human body structures, a wide spectrum of research topics are involved including simulation of average anatomies and normal variations, discovery of structural differences between healthy and diseased populations, detection and classification of pathologies from structural anomalies, and so on.

Disciplines such as morphometrics and anthropology have long been involved with analyzing biological shapes. Among them, the book *On Shape and Growth* by Thompson published a century ago [1] is regarded as the Bible for morphometrics. It focuses on the importance of the roles of physical laws and mechanics as the fundamental determinants of the form and structure of living organisms.

From the technical viewpoint, statistical analysis of shapes in pattern recognition [2], computer vision [3], and artificial intelligence [4] can be regarded as one of the origins. In the development of medical imaging research, digitizing data, including spatial and functional relationships of anatomical structures based on high-resolution images of a cadaver [5], was an essential step. It allowed creation of a digital atlas, which has been widely used in medical education. In the medical imaging research field, one of the pioneer works for computational anatomy was initiated by a so-called digital atlas of human anatomy created from high-resolution optical cross-sectional images of a cadaver [5], in which systematically organized knowledge was implemented. The digital atlas is now widely used in medical education.

An important role of computational anatomy in the clinical setting is in computer-assisted diagnosis (CAD) and computer-assisted surgery (CAS). In such application-oriented aspects of the discipline, one of the key demands is computational understanding of medical images with high accuracy and robustness. In other words, reliable and automated segmentation schemes for all organs in medical images are necessary for detecting abnormal structures and surgical planning. It has been a long-term challenge in medical imaging researches and still has been exhaustively studied over time.

Before the 1990s, various automated segmentation techniques based on data-driven approaches using simple techniques such as thresholding and voxel connectivity analysis were developed and were proven to be useful within limited clinical imaging situations.

In the 1990s, shape model-based approaches initiated by SNAKES [6] were intensively used to overcome the drawbacks of data-driven approaches. The concept uses parameter optimization to fit the model to the correct boundary of target organs. Those approaches attained some success. The key to improvement is acquiring statistics of inter-patient variations in representing shapes and image intensities of target organs.

Recent advances in medical image segmentation have mainly been based on using computational models based on a number of organ shape samples. The models are statistical representations of shapes/intensity patterns, called active shape/appearance models (ASM/AAM) [7], which are also known as statistical shape/intensity models. Recently, several mathematical tools, such as the Riemannian framework, have been successfully introduced [8], especially for statistical analysis of anatomical structures based on medical images at the population level. Those statistical approaches are effectively combined with machine learning methodologies [9] to obtain more reliable results based on a vast amount of samples.

So far, one of the most successful areas for statistical analysis of image-based anatomy is neuroimaging because of the intensive demand in this field. One of the reasons for this success is related to the challenges involved in investigating brain function. Free software for image analysis such as statistical parameter mapping (SPM) [10] has pushed research in this area forward. In brain image analysis tools, a standard brain atlas, which includes pre-segmented regions, is used. A new dataset is registered to the template and then the pre-segmented regions can be reflected in the data. Such atlas registration techniques can be regarded as another mainstream method in addition to statistical shape/intensity models.

Anatomical structures with nonpathological variations, "anomalies," can pose problems. For example, the number of the vertebrae is frequently more or less than the normal. That is, there is a discrete variation in the number of the vertebrae. In addition to continuous shape variations, these noncontinuous or discrete organ variations must also be considered in modeling anatomical structures.

Recent achievements based on analysis of a huge number of clinical image samples (so-called big data), including healthy volunteers and real patients, throw new questions at us, such as "What is the definition of abnormality?", "What is the border between normal and abnormal?", and "Is a given abnormality significant, or might its workup cause more morbidity without eliminating an important disease?". The keys to the answers are naturally in the statistical analysis.

In an attempt to answer these questions, a computational anatomy project in Japan was initiated in 2009, supported by a Grant-in-Aid for Scientific Research on Innovative Areas from the Ministry of Education, Culture, Sports, Science, and Technology, Japan [11]. As the project name "Computational Anatomy for Computer-Aided Diagnosis and Therapy: Frontiers of Medical Image Sciences ("Computational Anatomy" (CA) in short)" shows, it was aimed at establishing a mathematical framework to deal with human anatomy and diseases, primarily focused on the chest and abdomen, based on medical images with certain application-oriented aspects such as CAD and surgery.

Several related research projects are found all over the world, such as the Human Connectome Project (HCP) [12] and the Physiome Project [13]. The HCP is intended to construct a human brain map describing the complete neural connections of both structures and functions of intra- and intersubjects (over time). It is a long-term project, begun in 2011, involving the collection and sharing of multimodality images, including magnetic resonance imaging (MRI), among multiple centers. Similarly, the Physiome ("physio-" meaning "life" and the suffix "-ome" meaning "as a whole") Project is aimed at providing a quantitative description of physiological dynamics and functional behavior of the organism within anatomical structures.

Thus, computational anatomy is a developing research field involving a wide array of sciences and technologies aimed at improving human health.

1.2 Needs, Seeds, and Solutions Around Medical Imaging: History and Perspectives

1.2.1 Needs in Medical Education and Clinical Practice

1.2.1.1 From the Viewpoint of Medical Education

Sakon Noriki
Faculty of Medical Sciences, University of Fukui, Eiheiji-cho, Fukui, 910-1193, Japan
e-mail: noriki@u-fukui.ac.jp

Medical education needs to be highly structured to impart the enormous amount of information necessary just to go on to further clinical training. The basic framework is similar all over the world. It consists of basic sciences (e.g., anatomy, physiology, and biochemistry) and clinical training (e.g., internal medicine, surgery) where skills are mastered. Anatomy is considered to be basic science but is strongly related to and interwoven with clinical medicine [14, 15]. In fact, clinical instructors often interact with students taking anatomy. The educational method called "image anatomy" emphasizes the inclusion of medical imaging [16]. Image anatomy is a method for studying anatomy, in which the nondestructively visualized human body structures are used.

We describe the practice of human anatomy using cadaver imaging, which was inspired by autopsy imaging (Ai), in our facility, and then review history of anatomy education. Finally, we describe the present problems of teaching human anatomy and the future of the relationship between anatomy education and computational anatomy.

In Japan, the application of cadaver computed tomography (CT) imaging to educational anatomy training is becoming a big current, and its trials are carried out at several medical schools, where CT scanners only for the dead body are introduced.

CT images of cadavers have been used for anatomy practice by the Faculty of Medicine at Fukui University since 2010 [17–20]. Imaging data are acquired over 128 cm and reconstructed at 5-mm section thickness. The medical students can refer to the CT images as well as MRI images of cadavers under dissection in the anatomy department (Fig. 1.1). iPads connected with wireless LAN and desktop PCs connected with school LAN are used to refer to the images (Fig. 1.2).

In Japan, this is carried out at several medical schools including Fukui, Chiba, and Gunma Universities and is becoming more widespread with the growing installation of computed tomography (CT) scanners in pathology departments for postmortem examinations. An Ai conference, in which radiologists, pathologists, and forensic specialists participate, is held so that the anatomy teaching staff can understand the interpretation of the images. Then, the anatomy teaching staff pass on the CT images to the students.

Human anatomy was first described by the ancient Egyptians approximately 3500 years BCE. Minute descriptions of the cranial sutures and the brain surface are found in the Edwin Smith Papyrus, an ancient Egyptian medical text

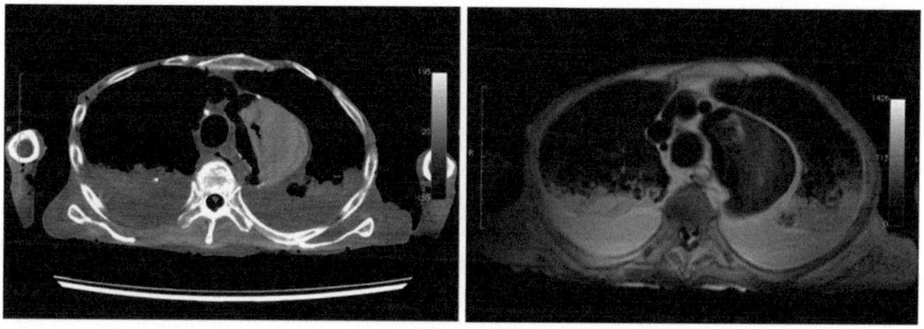

Fig. 1.1 CT image (*left*) and MR image (*right*) of a cadaver chest for anatomical study

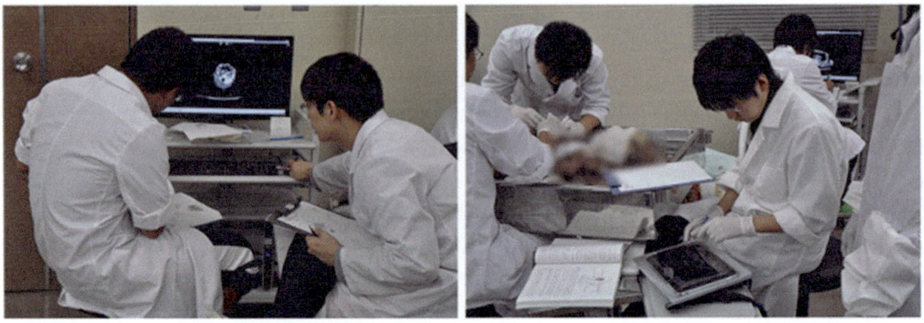

Fig. 1.2 Students can refer to CT images on the monitor, which is put adjacent to the cadaver

from approximately 1700 BCE [21]. Hippocrates, an ancient Greek physician, investigated the goat brain. Various findings about anatomy are described in the *Corpus Hippocraticum*, which the pupils of Hippocrates edited. Though it is told that Herophilos (335–280 BCE) and Erasistratus (304–250 BCE) dissected the human body in ancient Alexandria, their writings have vanished. After that, Alexandria prohibited human dissection for religious reasons, and new knowledge about anatomy did not arise for 1000 years or more. During this period, Galen (129–216 CE) dissected animals energetically [22]. He left inclusive and detailed anatomy books, mostly based on dissections of apes.

There was a revival of interest in anatomy during the Renaissance. Leonard da Vinci created human anatomical drawings, although his notebook was not published during his lifetime [23]. In the early 1500s, the study of anatomy commenced at the University of Bologna, and Andreas Vesalius (1514–1564) of the University of Padua published *De humani corporis fabrica* (*Structure of the Human Body*) based on dissections performed in 1543. This became the basis of modern anatomy.

After this, research in human anatomy advanced, and cadaver dissection became part of medical education. At the present time, physical dissection is partly replaced with image-based dissection, that is, image anatomy, such as Ai. The history of Ai is described in the other section (Chapter 4.4.1) of this book. Especially "Virtopsy" in Switzerland is actively used in the forensic field [24]. The purpose of Virtopsy is not the education of anatomy to medical students but the investigation of the cause of death. That is, there is little collaboration between experts in the Ai and educational anatomy fields in countries other than Japan. Ai and its related research areas including Virtopsy are also called "postmortem imaging (PMI)." Medical educational reform is leading to more concentration on clinical studies and less on basic science. In some areas, obtaining cadavers for dissection is difficult [25]. Human anatomy is considered a fundamental subject in medical education [26]. Adding imaging techniques, such as X-rays, CT, MRI, and new technologies (e.g., web-based learning), is becoming more common.

The "Visible Human Project," which provides cross-sectional human body imaging data including high-resolution optical, CT, and MR images [27–29], has become a resource in the study of anatomy that integrates traditional instruction with modern CT and MRI technologies realizing a three-dimensional image (http://www.nlm.nih.gov/research/visible/visible_human.html).

Moreover, there are reports on clinical anatomy training for nerve block anesthesia using a 3D anatomy system that can display the cadaver from one's preferred direction and in one's chosen depth in three dimensions [30, 31]. The advent of the 3D printer means that in addition to volume-rendered images, physical anatomical models may now be produced. Surgeons are beginning to use these to examine surgical procedures during preoperative planning.

In computational anatomy research, the comparison and registration of 3D images constructed from huge and elaborate CT/MRI image datasets (internal organs) with 3D data obtained by 3D surface scanners may be needed. A physician speculates an internal state from the body surface and performs diagnosis and treatment. A nurse also guesses the patient's condition from the body surface. Therefore, there are many points that are important to diagnose on the body surface. Some of them are called tender points, and such pain is regarded as radiating pain (referred pain). McBurney's point and Lanz point are famous in the case of appendicitis. So, if we could see through to an internal organ and an organic state from the body surface, it would be useful for medical education. Although there is a report in which transparency of the ranine (frog's) skin was achieved [32], it will likely be difficult in humans. If we could project the 3D information of the internal organs on the surface of the body by projection mapping, this would be useful for medical education. In addition, approaches using computational anatomy for not only normal cadavers but also cadavers with diseases such as malformations, tumors, and inflammation will be requested.

1.2.1.2 From the Viewpoint of Diagnostic Radiology

Shoji Kido
Department of Information Science and Engineering, Yamaguchi University, Ube-shi, Yamaguchi, 755-8611, Japan
e-mail: kido@ai.csse.yamaguchi-u.ac.jp

Before the discovery of X-rays, physicians could not look into live human bodies except during surgery, and the knowledge of anatomy could only be obtained from cadaver dissection and autopsies. X-rays enabled physicians to diagnose and treat diseases based on the findings on roentgenograms using their anatomical knowledge. In interpreting plain radiographs, which are two-dimensional projection images of three-dimensional structures, radiologists apply their knowledge of projected anatomy. For example, radiologists can interpret posterior–anterior mapped view images or lateral mapped view images in chest radiographs. They must then be able to visualize or construct the real anatomical structures from the projected images. This requires considerable training and experience.

Improvements in the technology of radiography, especially oral and intravenous contrast agents, made it possible to image previously radiolucent digestive organs and vessels. The technique of angiography was developed in 1927, 30 years after the discovery of X-rays, by the Portuguese neurologist Egas Moniz, who wished to visualize cerebral circulation [33]. Angiography, which is an imaging technique to visualize blood vessels (mainly arteries but also veins), uses contrast materials injected into vessels to opacify them on X-ray imaging. This allows visualization of arterial stenosis or aneurysm formation. It can visualize many types of arteries, including cerebral arteries and coronary arteries. It can also visualize tumor perfusion by opacifying the feeding artery or arteries. Radiologists can diagnose stenosis of coronary arteries, which causes myocardial infarctions, as well as brain and hepatic tumors. Angiography has developed a range of treatment modalities including balloons and stents to treat stenoses and special catheters to treat tumors with embolization materials.

The development of contrast agents, often combined with air or gas (double contrast) introduced by the oral or rectal routes, enables gastrointestinal radiography examinations such as the upper gastrointestinal (UGI) examination. UGI, which was developed by Shirakabe and Ichikawa, visualizes the esophagus, stomach, duodenum, and sometimes the small bowel, and the barium enema (single or double contrast) visualizes the colon [34]. UGI and barium enema visualize the surface mucosa by opacifying it (single contrast) or coating it (double contrast) with contrast. Radiologists can diagnose stomach and colon diseases with these techniques. In Japan, stomach cancer was the most common neoplasm causing death. However, the number of deaths caused by stomach cancer has decreased, in part because of screening examinations using UGI. New imaging techniques can enable radiologists to diagnose diseases in more organs than was possible with plain X-rays. The interpretation of these images requires a considerable degree of training and experience on the part of the radiologist.

Shinji Takahashi developed what was called "rotational radiography," which is regarded as an early computed tomographic apparatus in 1953 [35]. Godfrey Hounsfield furthered the concept of computed tomography (CT) in 1967 and developed the first commercially available CT scanner. The first scan using the commercially available CT scanner was performed in England for brain disease in 1971 [36]. Hounsfield won the Nobel Prize in medicine in 1979 for the development of CT. CT is a technology that can produce reconstructed images in slice (section) format of the body or brain using a computer to generate Fourier transforms or other reconstruction algorithms. In a CT scanner, a rotating X-ray tube and detector are located on the opposite sides of the target. The detector obtains multidirectional attenuated X-ray data on each rotation and the computer generates an attenuation map image. The quality of early CT images was poor, and scan times were very long. The first brain scan required 4 min to generate only two sections and required 7 min for the calculations to generate a reconstructed image. With the advent of CT, radiologists learned tomographic anatomy. CT images have many advantages over plain radiographs. Most importantly, CT eliminates the superimposition of anatomical structures. CT visualizes differences in X-ray attenuation between soft tissues that would all be radiolucent on plain radiography and constructs images that are attenuation maps. Conventional radiographs can visualize structures such as the skull and lungs in sufficient detail to allow radiologists to diagnose diseases. However, they cannot visualize soft tissues inside the brain or bronchus. Prior to the advent of CT, patients with brain or bronchial lesions had to be examined with pneumoencephalography (using air as a contrast agent) or bronchography (using barium as a contrast agent), which involved invasive procedures involving special training for the radiologist performing them. A large proportion of radiologists' training is now devoted to the various tomographic modalities.

High-resolution CT (HRCT) images can be obtained using a conventional CT scanner with a high spatial reconstruction algorithm and narrow collimation to maximize resolution. HRCT is especially important in the diagnosis of lung diseases [37]. It can reveal minute anatomical structures with high spatial resolution. It can improve the accuracy of diagnosis of diffuse lung diseases and small solitary pulmonary nodules. HRCT can also reveal microanatomic structures such as secondary lobules of lungs. Using HRCT, radiologists can distinguish among different diffuse lung diseases such as interstitial lung diseases and chronic obstructive pulmonary disease, for which microanatomical detail is important for diagnosis.

Spiral (helical) CT (SCT) and multi-detector CT (MDCT) can generate three-dimensional volume data. Spiral CT was invented by Kalender in the 1980s [38]. HRCT improves spatial resolution in the axial plane. SCT improves spatial resolution along the long axis of the body. Spiral CT, where the X-ray tube rotates continuously as the pallet moves through the gantry at an adjustable pitch, was first used with one row of detectors. The number of rows has increased steadily. MDCT can collect image data for multiple slices at once, decreasing scan time. Current MDCT such as the Aquilion One® (Toshiba Medical Systems) has 320 rows of detectors and can scan a 16-cm thickness in only 275 msec. MDCT can image minute structures with reduced motion artifact. The radiologist can obtain anatomical information equivalent to autopsies or anatomical textbooks. Radiologists can improve the accuracy of their diagnoses using a small volume of data, but CT examinations with multiple CT sections have caused many problems for radiologists, resulting in misdiagnoses. To display CT and other digital modalities, most radiologists currently use the DICOM (digital imaging and communications in medicine) standards. DICOM is a standard for handling, storing, printing, and transmitting digital information in medical imaging. Therefore, computer-aided diagnosis (CAD) can be naturally integrated to clinical practice, and many CAD algorithms have been developed. In addition, images from multiple CT sections can help produce digital atlases, which can reveal minute structures in the human body. VOXEL-MAN, which was developed at the University Medical Center Hamburg-Eppendorf, is one of the more famous digital atlases [39] (Fig. 1.3). Using such digital atlases can help radiologists learn about selected cross sections of the human body in an interactive manner. These atlases can improve the diagnostic abilities of radiologists.

MRI visualizes anatomy and pathology using strong magnetic fields and radiofrequencies to generate images. The advantage of MRI is that it does not involve

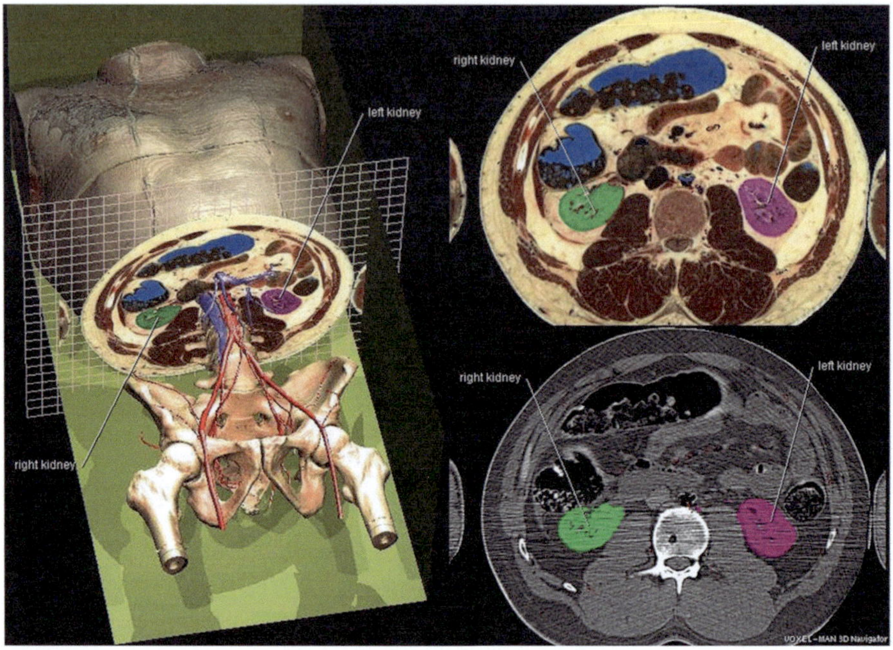

Fig. 1.3 VOXEL-MAN 3D-Navigator: Inner Organs. http://www.voxel-man.com/3d-navigator/inner_organs/

ionizing radiation. MRI is frequently used in the fields of pediatrics, obstetrics, and gynecology. Functional MRI measures signal changes in the brain and can localize brain activity. It is used in neuroscience. Diffusion-weighted MRI (DWI) measures the diffusion of water molecules in tissues and visualizes the connectivity of white matter. These applications are useful for diagnosis of neurological disorders and surgical planning. Real-time MRI can provide radiologists with four-dimensional (spatial and temporal) anatomical data. These data cannot be obtained with CT because of ionizing radiation. These four-dimensional data can provide useful information for diagnosis of conditions such as lung and heart disorders. The abovementioned VOXEL-MAN also includes applications for the creation and visualization of three-dimensional digital models of the human body using cross-sectional images from MRI/CT. MRI and CT produce tomographic images, and these are complementarily useful.

Endoscopy is a powerful diagnostic tool for the bronchus, stomach, colon, and other hollow organs. However, endoscopy preparation is often burdensome for patients and requires extensive training for endoscopists. Virtual endoscopy images can be generated using imaging data obtained from SCT or MDCT (mainly MDCT). Virtual endoscopy is different from the abovementioned imaging methods, because it uses post-processing, and synthesizes virtual images. CT colonoscopy is one of the most common virtual endoscopic examinations [40]. CT colonoscopy is less

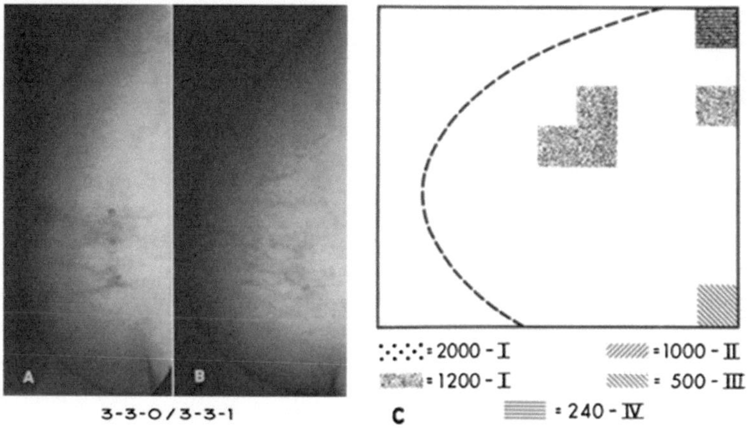

Fig. 1.4 Detection of abnormality in a breast by an early CAD algorithm (Ref. [43])

burdensome for patients, and it does not require endoscopy training for physicians (but requires training in interpretation). It is expected to be useful as a screening tool for detection of colorectal polyps. CAD algorithms exist for the detection of polyps in CT colonoscopy [41]. Virtual bronchoscopy is also useful for diagnosis of pulmonary diseases such as lung cancers and has value for surgical preoperative examination [42].

CAS is one of the most important topics in current image diagnostic diagnosis. CAD provides the quantitative diagnostic information for the detection and classification of tumors for breast cancer, lung cancer, colorectal cancer, and other cancers from medical images such as CT, MRI, and mammograms. The function of CAD is to assist radiologists and improve their diagnostic abilities; its function will not be so-called automatic diagnosis. Early studies of CAD involved mainly mammography and chest radiography [43, 44] (Fig. 1.4), and their goal was detection of abnormal lesions such as those observed in breast cancer and lung cancer. The radiographic image quality and the power of computers were not sufficient for meaningful evaluation of CAD in the early studies [43, 44]. Imaging technology and computer processing power have steadily increased. Various CAD algorithms have also been developed and improved. In 1998, an American CAD company, R2 Technology, was licensed for the first time by the US Food and Drug Administration. This was an important event for CAD development. Radiologists had recognized CAD systems as useful tools in image diagnosis. Subsequently, commercially available CAD systems for detection of colon polyps were produced.

CAD has two roles. One is computer-aided detection (CADe), and the other is computer-aided diagnosis (CADx). Quantitative imaging (QI) is a current trend in medical analysis. To improve the value and practicality of quantitative biomarkers by reducing variability across patients and time, the Quantitative Imaging Biomarkers Alliance (QIBA) was formed by the Radiological Society of North America (RSNA) in 2007. It currently has six active technical committees: MRI, functional MRI, FDG-PET, CT Volumetry, COPD-Asthma, and Ultrasound. Many QI algorithms are now installed on diagnostic workstations in departments of radiology [45].

CAD systems currently available are for single organs and single diseases, such as detection of calcification and masses on mammography, detection of polyps on CT colonoscopy, and detection and classification of pulmonary nodules on chest X-rays or thoracic CT. These algorithms are mainly based on local image features and machine learning algorithms and are not strictly based on anatomy. This diagnostic logic is different from that of radiologists. Radiologists usually interpret radiographs based on local features and local and global information of anatomical structures. In daily clinical practice, radiologists have to diagnose multiple diseases in multiple organs. For example, in the case of lung cancer, diagnoses are based on primary lesions in the lungs. However, radiologists must also diagnose other pulmonary lesions such as those of diffuse lung disease, and they also need to diagnose metastatic lesions in other organs, such as the liver or brain. In such cases, information regarding multiple anatomical structures is needed. In the Japanese research project entitled "Intelligent Assistance in Diagnosis of Multi-dimensional Medical Images (2003–2007)," researchers developed many CAD algorithms for multiple disease features and entities in multiple organs [46, 47] (Fig. 1.5). At a symposium held in Tokyo in 2007, one of these CAD systems was demonstrated using plug-ins developed by researchers on the common platform named "Pluto." Many applications of the CAD algorithms were integrated into one system on the common platform as plug-ins.

The computational anatomical models are important for development of local/global CAD systems. Anatomical models such as statistical atlases provide various kinds of information about human organs [48]. Statistical atlases are already popular in the field of neuroscience. One of them is statistical parametric mapping (SPM) [49]. Computational anatomical models are needed for the development of advanced CAD algorithms [50]. The diagnosis made by CAD algorithms

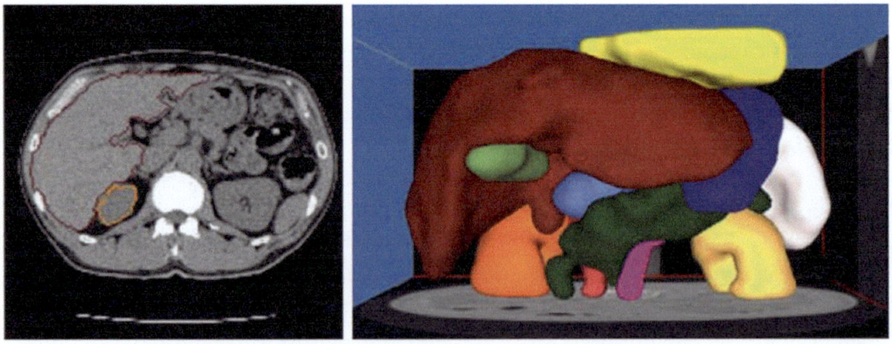

Fig. 1.5 Twelve organs detected by the simultaneous segmentation method (Refs. [46, 47]). *Left*: Typical axial abdominal CT section with segmented liver and gallbladder. *Right*: Twelve segmented organs in a 3D display

should be based on the computational anatomy as well as the interpretation of the radiologist. These CAD systems, using computational anatomical models, can deal with multiple diseases in multiple organs. Moreover, the high-resolution volume data of CT and MRI can be used to make digital atlases such as VOXEL-MAN. Radiologists can look at selected cross sections of the human body interactively, which they could not do previously. They can improve their knowledge of anatomy using such digital atlases. The computational anatomical models can provide statistical and structural information about human anatomy. Therefore, it is expected that computational anatomy can increase radiologists' knowledge of anatomy and the functionalities of CAD algorithms.

1.2.1.3 From the Viewpoint of Therapeutic Radiology

Hidetaka Arimura
Department of Health Sciences, Kyushu University, Fukuoka, 812-8582, Japan
e-mail: arimurah@med.kyushu-u.ac.jp

The biological effects of radiation were investigated within a year after Wilhelm Conrad Röntgen discovered X-rays in 1895, and then it was recognized as beneficial effect for curing malignant tumor [51], which may be considered the beginning of radiation therapy. What are required in radiation therapy from medical physics point of view are (1) high conformity of dose distributions to tumor regions and (2) accurate tumor localization and patient positioning. To achieve these requirements, radiation therapy researchers have dedicated their efforts since around 1960 to the development of novel technologies such as conformal radiotherapy [52], intensity-modulated radiation therapy (IMRT) [53], real-time tracking radiotherapy (RTRT) [54], and image-guided radiation therapy (IGRT) [55, 56].

Since radiation therapy can preserve organ function and is useful in patients, particularly elderly patients, unsuited to surgery, it has attracted greater attention. Consequently, this treatment modality is considerably important for developed countries such as Japan and the United States of America, whose populations are rapidly aging. In Japan, the percentage of older people (65 years and over) was estimated to be around 23% in 2011 [57]. Therefore, great benefits from radiation therapy can be provided for many patients, including elderly patients, whose quality of life could increase.

The primary aim of radiation therapy is to deliver as high a dose as possible to the tumor while causing as little damage as possible to normal tissues and organs at risk (OARs) [55, 56]. The OARs are normal tissues whose radiation sensitivity may significantly affect radiation treatment planning (RTP) and/or the prescribed dose [58]. To achieve these goals, high-precision radiation therapy methods have been developed, such as stereotactic body radiation therapy (SBRT), intensity-modulated radiation therapy (IMRT), adaptive radiotherapy (ART), real-time tracking radiotherapy (RTRT), and image-guided radiation therapy (IGRT). These advanced techniques have recently led to impressive progress regarding the

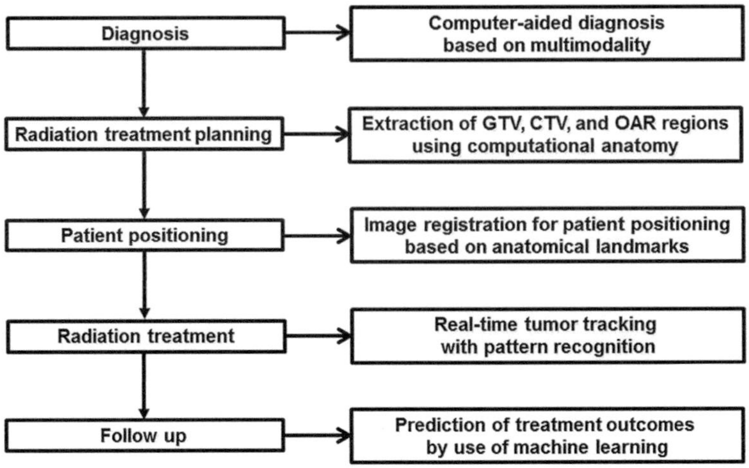

Fig. 1.6 Five steps of radiation therapy and examples of image engineering techniques, including computational anatomy, in each step

precision of radiation delivery. As a result, high-precision radiation therapy has been reported to provide outcomes comparable to surgery for some cancers [59]. In these high-precision radiation therapies, novel methods of image analysis, including computer graphics, image processing, and pattern recognition, play considerable roles in improving the accuracy of radiation therapy and assisting in treatment planning.

Radiation therapy procedure includes five steps: diagnosis, treatment planning, patient setup, radiation administration, and follow-up. Computational image engineering techniques are employed to assist the radiation oncology staff in decision-making at each step. Figure 1.6 describes the steps and provides examples of image processing techniques including computational anatomy in each step.

The first step is diagnosing the cancer. Computer-aided diagnostic technologies may be useful if the oncologist employs multiple imaging modalities. Then the radiation oncologist and clinicians determine the radiation treatment goal, i.e., curative treatment or palliative treatment.

The second step involves developing the treatment plan, in which the gross tumor volume (GTV), clinical target volume (CTV, the volume including gross tumor and areas of likely microscopic involvement), and planning target volume (PTV, the actual volume to be irradiated to ensure that the entire CTV receives a therapeutic dose, assuming some target motion and other inaccuracies) are defined. A plan is created by arranging beam paths to maximize the tumor dose and minimize the OAR dose. Figure 1.7 depicts illustrations of a radiation treatment plan: (a) a beam's eye view with a GTV region and organs at risk (light blue, bladder; pink, rectum) and (b) dose distribution images produced using CT. Image engineering techniques are used to help define these volumes and OARs. In current clinical practice, GTV regions have been manually delineated by radiation oncologists using treatment planning

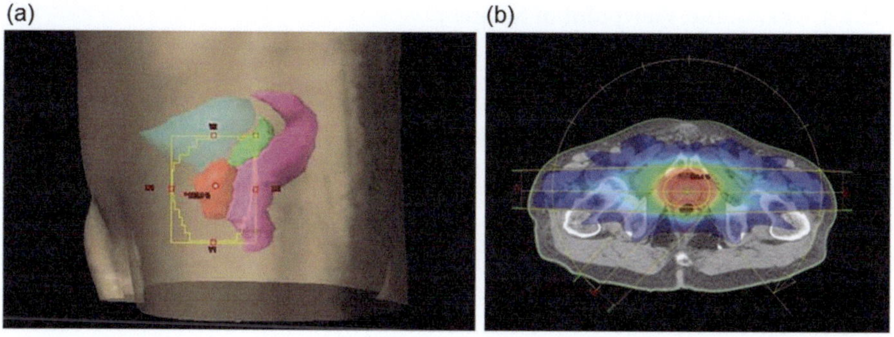

Fig. 1.7 Illustrations of a prostate radiation treatment plan: (**a**) a beam's eye view with a gross tumor volume (GTV) and OARs outlined (*light blue*, bladder; *pink*, rectum) and (**b**) dose distribution using IMRT treatment planning

computed tomography (CT). However, the subjective manual contouring of a tumor region is tedious and time-consuming, and its reproducibility is relatively low, which could cause inter- and intra-observer variability [60–62]. A number of automated contouring methods for determining the GTV have been proposed to reduce this variability and planning time and increase the segmentation accuracy of the GTV [63–65].

The third step is patient setup. In this step, the radiation technologist positions the patient manually on the treatment couch as accurately as possible, often using immobilization devices. After that, fine-tuning of target position is performed by using image registration techniques, which register a moving image with a reference image with respect to the target. In general, in the patient setup phase, digitally reconstructed radiographic (DRR) images and planning CT images are used as the reference images. An electronic portal imaging device (EPID) and cone-beam CT (CBCT) images produced using kilovoltage or megavoltage X-rays during treatment are employed as the moving images. Previous studies have revealed that these techniques are effective for reducing setup error [66, 67].

The fourth step is the actual radiation treatment. A photon or particle beam is delivered to the planning target volume (PTV) in a patient according to the treatment plan. For tumors subject to respiratory motion such as lung and some liver cancers, respiratory gating or fiducial markers are used to follow the target. For example, an RTRT system has been developed, which employs pattern recognition techniques to follow gold markers within the tumor to track it and switch the treatment beam on and off [54].

Computational anatomy-based technologies could make it possible to develop statistical models of events happening in tumors and human bodies as well as mathematical prediction models of tumor or normal tissue responses during radiation treatment. If these novel technologies were developed, they could make large impacts on automated segmentation of GTV, CTV, and OAR and optimization of radiation treatment planning.

1.2.1.4 From the Viewpoint of Surgery

Morimasa Tomikawa
Department of Surgery, Fukuoka City Hospital, Fukuoka, 812-0046, Japan
e-mail: mtomikaw@surg2.med.kyushu-u.ac.jp

Medical imaging is one of the techniques that have made a great contribution to the development of surgery. As an example, in neurosurgery it is important for the surgeon to avoid damage to the brain structures that will cause dramatic functional deficits (including limb paralysis, language function, and visual). An integrated protocol of new MRI methods may, prior to the surgery, delineate important functional areas in the brain and their vital nerve fiber connections. This is of great help to the neurosurgeon in the planning of the operation and can also be displayed for the neurosurgeon in the operating theater and updated during the surgical intervention. Image-guided techniques assist with precise instrument guidance.

Surgical navigation is in some ways the same as commonly used car navigation, attempting to localize or determine a position in the physical space in the context of its surroundings [68]. Surgical navigation is usually "image-based," meaning that imaging data such as preoperative CT or MRI images are used in the operating room [69]. Before surgery, surgical planning can entail delineating regions of interest within the images and producing datasets for use in the operating room. The preoperative image data need to be matched to the current patient position via registration, establishing a relation between the "physical" coordinate system as defined by the patient's reference array and the "virtual" coordinate system of the imaging data. Modern surgical navigation systems use a stereoscopic camera emitting infrared light which can determine the position of fiduciary reflective marker spheres, allowing for real-time tracking. During the surgery, the marker spheres are attached to the patient and to surgical instruments to enable an exact localization in the space. The computer can calculate the position and orientation of each instrument. Correct localization and virtual display of the instrument on the computer screen is ensured by firmly attaching a reference array to the patient. The surgeon then virtually "sees" both the current situation and the imaging datasets side by side by volume rendering or by directly superimposing them, that is, augmented reality visualization.

Neurosurgery was the first surgical discipline to adopt navigation and integrate it successfully into surgical routine, followed by maxillofacial surgery, dental surgery, and orthopedic surgery. The neurosurgical procedures supported by surgical navigation range from intracranial tumor resections and frameless biopsies to pedicle screw placement and stabilizations in the spine. In the field of thoracoabdominal surgery (general surgery), navigation was less often employed, because open thoracotomy or laparotomy makes target organs visible enough. However, surgical navigation has been gradually adopted since the advent of minimally invasive surgery (MIS), thoracoscopic or laparoscopic surgery, in the late 1980s. The benefits of MIS to the patient have been shown in terms of reduced postoperative pain, lower risk of wound infection, shorter hospital stay, and quicker return to normal physical activities. Conversely, a limited field of view, dealing with essentially two-dimensional images, challenges to eye–hand coordination, and inability to achieve tactile feedback are limitations. Therefore, in general surgery, new imaging techniques for real-time enhanced laparoscopic surgical guidance are currently the subject of research worldwide. The implementation of intraoperative imaging methods can be of great assistance to surgeons in training, and these methods can help to improve the safety and efficacy of MIS.

Compared with other fields of surgery, general surgery has several factors of lower affinity with conventional surgical navigation. Deformation of soft tissues, such as the skin and abdominal wall, can affect the accuracy of registration. The peritoneal cavity is usually inflated with CO_2 to enlarge the field, and the leakage of gas from the port site easily changes the intraperitoneal pressure and contour of the abdominal wall. Peristalsis can also present problems. Respiratory motion always needs to be addressed. Because of these factors, preoperative images can be less helpful. Although navigation accuracy is reduced for exact localization of targets, it remains valuable for intraoperative orientation. To address these problems, intraoperative imaging was employed over the last decade to provide the navigation system with real-time images. Intraoperative imaging solutions can range from live ultrasound images to intraoperative MRI (iMRI) or intraoperative CT (iCT). iMRI offers the best soft tissue contrast among them. Figure 1.8a shows an operating room equipped with an open-configuration MRI scanner, which is particularly suitable for iMRI [70–72]. Figure 1.8b shows a laparoscopic image in which an anatomical structure reconstructed from iMR images is overlaid [72] upon the patient's image.

While iMRI is a powerful tool to achieve optimal operation control in combination with the navigation, it remains the most expensive imaging option and

(a) (b)

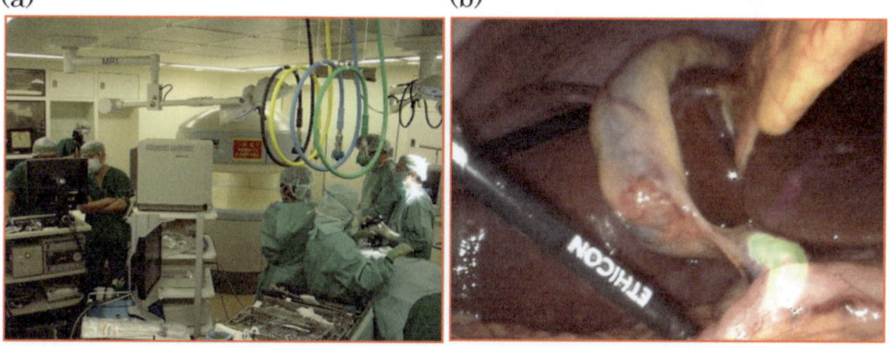

Fig. 1.8 Laparoscopic augmented reality visualization combined with open-configuration MRI system [72]. (**a**) Laparoscopic cholecystectomy in the open MRI therapeutic room. (**b**) Augmented reality laparoscopic image during a laparoscopic cholecystectomy. The laparoscopic monitor clearly shows the augmented reality model of the common bile duct (green semitransparent area), allowing the surgeon to avoid it during the procedure

requires significant building costs. As a trade-off among soft tissue image quality, versatility, and affordability, iCT has emerged. iCT allows for minimal interruption of the surgical workflow since its scan time is significantly shorter than MRI and patient positioning is less limited. Newer generations of portable iCT scanners are designed specifically for intraoperative use and enable the surgeon to verify the surgical progress and automatically update the navigation.

The current level of accuracy is acceptable for laparoscopic surgery despite inferior accuracy compared with other fields such as neurosurgery. Although at present it looks difficult to improve the accuracy, novel methodologies for nonrigid registration that can account for soft tissue shifts and deformations as well as estimation for soft tissue deformation may be realized in the future to improve the accuracy of the surgical navigation system. Statistical and physiological modeling of deformation patterns specific to organs will be required to improve the navigation accuracy and to widen applicability in the future. Computational anatomy, which may include statistical modeling of organ shapes and deformations, is expected to be a powerful tool for this purpose.

1.2.2 Seeds and Solutions in Science, Technology, and Engineering

Hidekata Hontani
Department of Computer Science, Nagoya Institute of Technology, Aichi, 466-8555, Japan
e-mail: hontani@nitech.ac.jp

As described in the previous subsection, medical imaging has played an increasingly prominent role in diagnosis and therapy since the 1960s. Exponentially increasing computer power and digital imaging technology have allowed the design of computer-aided systems, which efficiently extract information useful for diagnosis and surgery from given medical images.

Many methods for medical image analysis have been proposed since 1960, but research on CAD systems started in 1980 [73]. Before 1980, one of the goals of medical image analysis was to develop automated diagnosis systems, which were different from CAD. The automated diagnosis systems aimed to simulate the decision-making processes of physicians, while CAD systems provide a kind of "second opinion." Pattern recognition techniques can be used for the simulation of decision-making. Given input data, pattern recognition methods output symbols that denote the decision. The automated diagnosis systems are not clinically useful until their performances are as accurate as those of physicians.

CAD systems, on the other hand, can help physicians reading medical images by extracting only some of the image features needed for the decision-making and by displaying the extracted features [73–75]. It is important for CAD systems to

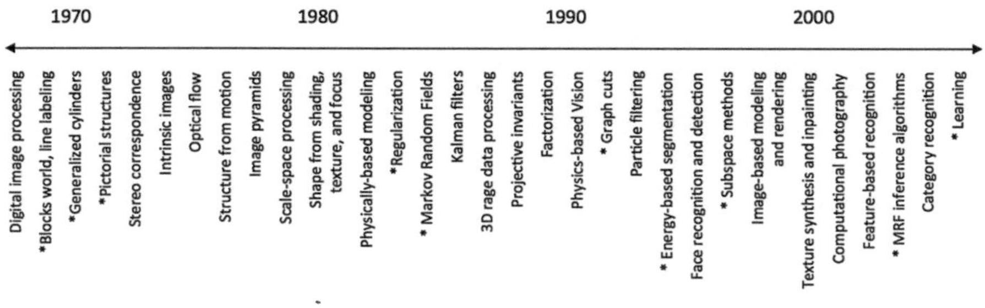

Fig. 1.9 Timeline of some of the most active research topics in computer vision appeared in Ref. [3]. Topics with *asterisks* are mentioned in this subsection

decrease the amount of data to be read by physicians without affecting diagnostic accuracy. For example, a method for computing temporal subtraction images from pairs of successive whole-body bone scans, developed by Shiraishi et al., enhances the interval changes between a previous image and a corresponding current one and can help physicians to distinguish new cold and hot lesions [76]. Only by enhancing some important image features and displaying them appropriately, CAD systems largely decrease the number of pixels the physician must carefully analyze.

The techniques for extracting useful information from given images have been developed in computer vision since about 1960. Figure 1.9, which appeared in [3], shows the changes in popular research topics in the field of computer vision.

Research in medical image analysis started in 1960. The main objective of the research was to extract primitives, such as edges and skeletons, that are useful for recovering physical properties of targets, e.g., the surfaces of the organs, from given images. In the 1960s and 1970s, studies aiming at realizing general-purpose vision systems were more popular than today, and many of these investigations employed bottom-up approaches. That is, the primitives extracted from given images were independent from targets to be described and were common to any vision system using this approach, which roughly correspond to Marr's 2–1/2D sketch [77, 78], which is a rich description of object surfaces projected in given images. These processes in early stages of visual recognition are called "early vision."

In the 1970s, blocks world [79, 80] and generalized cylinders [81, 82] handled models for representing the three-dimensional structure of objects. Using these models, the primitives extracted from given images were assembled into the descriptions of the objects. It should be noted that, different from many models used for medical image analysis today, these models were not directly applied to given images (i.e., matrixes of pixel values) but were applied to the general-purpose primitives (i.e., symbolic descriptions) extracted from the images (see Fig. 1.10). The approaches in which models of targets are directly applied to given images became more widespread in the late 1970s.

In medical image analysis, it is important to describe the organ structures, and it is necessary to determine the boundaries of the organs in given images. Detection of edges, at which the pixel values rapidly change around the organ boundaries,

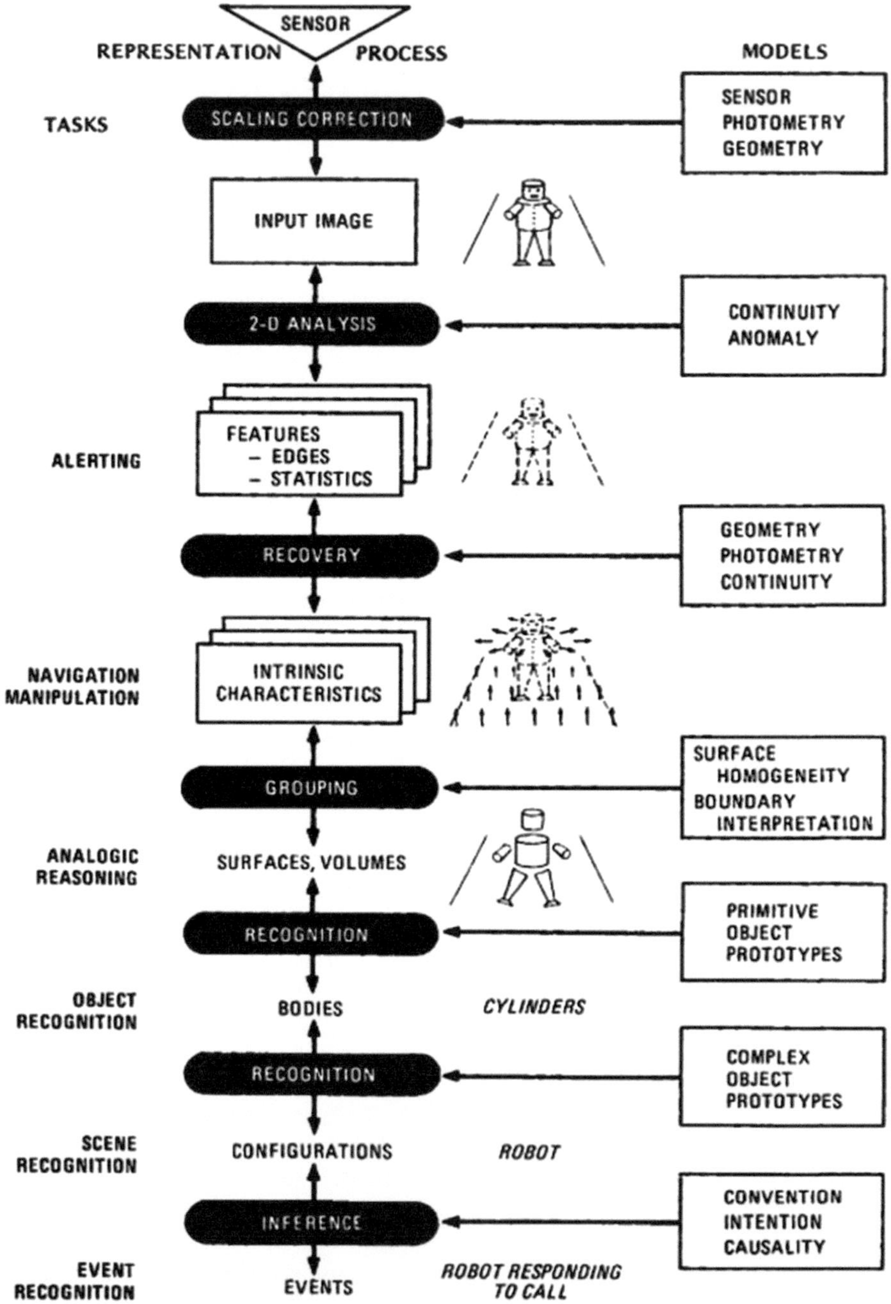

Fig. 1.10 Computational architecture for a general-purpose vision system appeared in Ref. [109]

has been an important research topic. It was supposed that an edge was one of the primitives for general-purpose vision systems. Poggio et al. pointed out in 1985 that many problems in early vision, including edge detection, were ill-posed. Some prior knowledge of the targets is needed to constrain the admissible solutions of the problems and to obtain unique solutions [83, 78]. In other words, it is necessary to apply models of the targets for the extraction of the primitives and hence it is difficult to realize general-purpose vision systems with the bottom-up approach.

Pictorial structures [84, 85] are also used as models of targets and more directly applied to given images to detect the targets in the images. A pictorial structure model represents a target by a combination of components. In Fischler and Elschlager's work [84], the distances between pairs of the components are represented by linking them with a spring, and the model is registered to given images by minimizing a cost function using a dynamic programming technique. The pictorial structures constrain the admissible locations of the components in given images and help the stable determination of their locations.

In the 1980s, a framework of regularization was introduced for solving the ill-posed early vision problems [78, 86]. In this framework, the early vision problems are solved by minimizing cost functions with regularization terms, which constrained the admissible solutions. Here, from a perspective of medical image analysis, active contour models (ACMs) [6] and active shape models (ASMs) [87] have played an important role. The boundaries of targets in given images can be determined by registering these models to the images. In the registration, cost functions with shape regularization terms are minimized. The regularization terms in ACMs represent the smoothness and/or the curvedness of the boundaries. The relationships between ACMs and some popular edge detectors such as those proposed by Marr and Hildreth [88] and by Canny [89] are described in Kimmel and Bruckstein's work [90]. In ASMs, the shapes of the boundaries are regularized by using statistical shape models (SSMs) constructed from sets of training data of target shapes. The model shapes are limited by constraining them on the subspace constructed from the training data. The ASM is one of the most fundamental methods for segmenting organ regions in medical images [91–93]. Figure 1.11 shows examples of such training data for an SSM of the liver [94]. The approaches that use the SSMs for the region segmentation do not aim to construct general-purpose vision systems but to construct specific-purpose vision systems, e.g., CAD systems. Not only the generalization ability but also the specification is employed for evaluating the performance of the SSMs [95]. The models used for determining the boundaries are required to represent only those of the specific targets. Analogous to the regularization approaches, Bayesian ones can constrain the admissible solutions and have been more widely employed for solving the early vision problems [83]. For example, the SSMs used for region segmentation represent the prior probability distributions of the shapes of the target boundaries, and the specification evaluates the accuracy of the prior distributions constructed from the training data; an SSM with better specification corresponds to an accurate prior probability distribution, where the contours that have specific shapes peculiar to the target appear with higher probabilities.

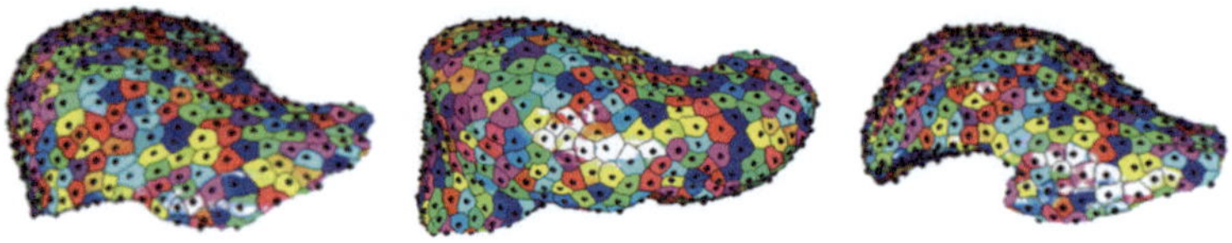

Fig. 1.11 Examples of training data for constructing an SSM for the liver (Ref. [94]). A set of corresponding points are indicated on each of the surfaces

Markov random fields (MRFs) were employed for the representation in CV in the 1980s [96–98]. For example, Geman and Geman proposed an image segmentation method that used an MRF model for representing the distribution of pixel labels and that segmented given images into regions by inferring the label for each pixel on the MRF model [97]. A Gibbs sampling technique was used in [97] for the inference, but graph-cut techniques [99] were later widely employed for inference on MRF models; if the MRF models satisfy some conditions, the optimal solution is obtained by using the graph-cuts or other MRF inference algorithms such as belief propagation [100, 101]. Many methods of energy-based segmentation in the 1990s [102, 103] segmented images by minimizing energy functions which can be derived from the MRF models. Pairs of an MRF model and a graph-cut technique are also one of the most fundamental tools for image region segmentation today. ASMs can be applied for determining the boundaries of the organs, followed by graph-cut techniques for improving the precision of the segmentation. Specifying targets can largely improve the performance of vision systems. It looks seriously difficult to construct the general-purpose vision systems.

The goal of vision systems, including those for medical applications, is to generate compact descriptions of targets from given image data by using models: Given image data $\boldsymbol{x}$, you describe the state of the world, $\boldsymbol{w}$, by inferring $\boldsymbol{w}$ by using models that represent the target world and the relationships between $\boldsymbol{x}$ and $\boldsymbol{w}$. The image data, $\boldsymbol{x}$, consists of many pixel values, and the compact descriptions, $\boldsymbol{w}$, consist of a smaller number of numerical values or symbols. The inference is called "recognition" when the descriptions, $\boldsymbol{w}$, are discrete and is called "regression" when they are continuous [104].

One difficulty that is specific to computer vision comes from the fact that, in many cases, a large portion of each given image is not useful for the inference. An image consists of many voxel values and not all these values contribute to the inference. For example, to recognize characters in an image, one should first detect the locations of the characters in the image, and only the local images around the detected locations should be processed for the inference of the character codes. Only the local images contribute to inferring the character codes, and other large portions of the given image, the backgrounds, do not contribute to the inference. One has to select appropriate portions of pixels in given image data before the final inference is processed. Such pixel selection is not easy, especially when the targets to be inferred and the backgrounds to be excluded can have varieties of appearances in images, because it is not easy to find features of appearances that can distinguish between the targets and the backgrounds.

The variety of images of a target entity also makes the inference in computer vision difficult. The appearance of a single 3D object, of a human face, for example, largely varies depending on its relative pose and distance with respect to the camera and on the lighting conditions. Occlusions also cause large varieties of appearances. The variety among multiple target entities that should be described by an identical description, $\boldsymbol{w}$, also makes the problems difficult. For example, a system of human face detection should be able to detect the face of any person. These large varieties require high degrees of freedom (DoF) of the models to represent the appearances. The processes of the inference are more complex and the accurate construction of the models is more difficult when the DoF is higher.

For solving the abovementioned problems, a computer vision system, in general, processes given images successively; at each stage, information useful for the next process is extracted from the descriptions output by the previous process, and the newly extracted information is described and is input to the next process. Roughly speaking, there exist two approaches for designing such systems, the bottom-up approach and the top-down approach. Today, top-down approaches are often employed for realizing the practical applicability of vision systems. In the top-down approaches, a model that specifically represents the global aspects of each target is introduced and the processes at earlier stages are also specifically designed for each target. The data, which are extracted at consecutive multiple stages in a bottom-up system, are often extracted by one process at a single stage. For example, local primitives are simultaneously extracted, while a global shape model of a target object is registered to a given image. Most medical image processing systems are also designed using top-down approaches.

Medical image processing is different from other image processing in the following aspects:

1. Input images are, in many situations, three-dimensional.
2. Various kinds of targets are included in a given image, the targets are located close together, their shapes are extremely varied, and they interact with each other.
3. It is vital to determine the location and the boundary of each anatomical target accurately.

The first aspect, the difference in the dimensions, means that medical images contain a more number of voxels than pixels in conventional two-dimensional images, and the second aspect makes the identification of the boundary of each target (e.g., a target organ) difficult. The third aspect, the crucial importance of the boundary determination, characterizes the research field of medical image processing. Many computer vision systems that have been realized today do not need to accurately determine the boundaries of targets with such a high degree of accuracy. Character recognition systems do not detect the contours of character strokes, face recognition systems do not detect exact face boundaries, and image searching systems do not detect the boundaries of targets in images. Some of the systems, such as character recognition and face detection systems, determine a bounding box of each target, and the appearance in the bounding box is directly input to the final process for obtaining the final description of the targets. Others, such as image searching systems, compute histograms of some image features, e.g., the "bag of features," and do not explicitly use any information of the boundaries of targets for the final inference. The need for highly accurate determination of object boundaries is one of the bottlenecks in the realization of the medical computer vision systems' practical applicability.

One can classify the top-down methods of image segmentation into two categories: In one category, the values of the parameters of the target global models are inferred by regression. In the other, a label of each pixel that denotes a target ID number is inferred by recognition. One of the main methods included in the former category is the registration of the models to given images, which include the ASMs. As described above, the ASMs are statistical models that represent the global shapes of targets and are generated from training data of the targets. The variety of the global shapes is represented by a set of shape parameters, $\boldsymbol{w}$, and the values of $\boldsymbol{w}$ are inferred so that the resultant curves (or surfaces) are located along the boundaries of the targets. One of the main methods in the latter, the recognition category, is voxel label inference with the minimization of cost functions. A set of labels that denote the target IDs is defined and a label is inferred for each voxel in the given images. The cost functions quantitatively evaluate the appropriateness of the labels of voxels based on the voxel values in the given images and on the label combinations between neighboring voxels.

Both recognition and regression have been well studied in the research field of machine learning, which is one of the most popular research topics for computer vision today, as shown in Fig. 1.9. The progress of learning algorithms makes it possible efficiently to construct models with higher DoF, which can represent statistical properties of targets more accurately, from a set of larger number of training data. For example, learning algorithms for deep neural networks have been proposed [105, 106], and the resultant networks have demonstrated state-of-the-art performance of pattern recognition and of regression in many applications. As the number of medical images available for learning is monotonically increasing, the role of learning in medical image processing is becoming more important. It should be noted, though, that the statistical properties derived from training data are not the only knowledge that can be used for medical image processing; such knowledge can be provided from many existing fields, e.g., anatomy, pathology, and physics [107, 108].

Computational anatomy is a system of automatically generating medical descriptions of patients from their medical images: The system consists of the algorithms for generating the descriptions and the models used in the algorithms. In the computer vision research field, algorithms for generating the descriptions of given images and models used in the algorithms are studied, and their progress improves the algorithms for medical image analysis and accelerate the development of computational anatomy. Among the large variety of research topics of computer vision, image segmentation is one of the most fundamental ones and its improvement is vital for CAD systems. It largely helps physicians to accurately and automatically describe each anatomical structure in a given medical image. As described above, image segmentation needs models of targets, and hence the segmentation of anatomical structures needs computational anatomical models. This is why the algorithms for the registration of the models to given images are more intensely studied in medical image analysis research. Though the computational anatomical models are useful only for segmenting medical images, the medical image segmentation is not isolated from other research fields. The progress of machine learning and statistical inference, for example, can directly improve the construction of statistical anatomical models and the performance of model registration, and the progress of computational anatomy can contribute to the development of computer vision, especially through the development of new algorithms and models for accurately and stably describing target regions.

1.3 Whole-Body Computational Anatomy

1.3.1 Impact of Whole-Body Imaging

Yoshinobu Sato
Graduate School of Information Science, Nara Institute of Science and Technology, Ikoma-shi, Nara, 630-0192, Japan
e-mail: yoshi@is.naist.jp

This section discusses how whole-body imaging has influenced radiological diagnosis, medical image analysis, medical education, and related research fields.

As a result of rapid improvement in scanning speeds and resolution of MDCT, whole-body (or full-body) imaging is now widely available in clinical practice. CT imaging provides high-resolution image data which precisely delineate the anatomical structures of the whole body, and it can be regarded as the best imaging modality for extracting and understanding macroscopic anatomical information. Since these data may contain a huge and increasing amount of valuable information, it is becoming difficult for radiologists and surgeons to fully utilize the information even with currently available computer assistance. In addition, these data will be potentially useful not only for diagnosis and therapy of the patient to be imaged in the data but also for a wide range of basic and clinical medicine as well as other sciences related to the human body if they are accumulated in the database and processed as population data of human anatomy. Visible Human (VH) data [110] influenced both the medicine and computer science research fields. VOXEL-MAN [5, 111] is a digital (or electronic) anatomical atlas generated from VH data and one of the most successful applications of VH data. Figure 1.12 shows VH data processing (Fig. 1.12a) and typical visualizations of VOXEL-MAN (Fig. 1.12b, c). The resolution of current CT data is high enough to regard anatomical identification

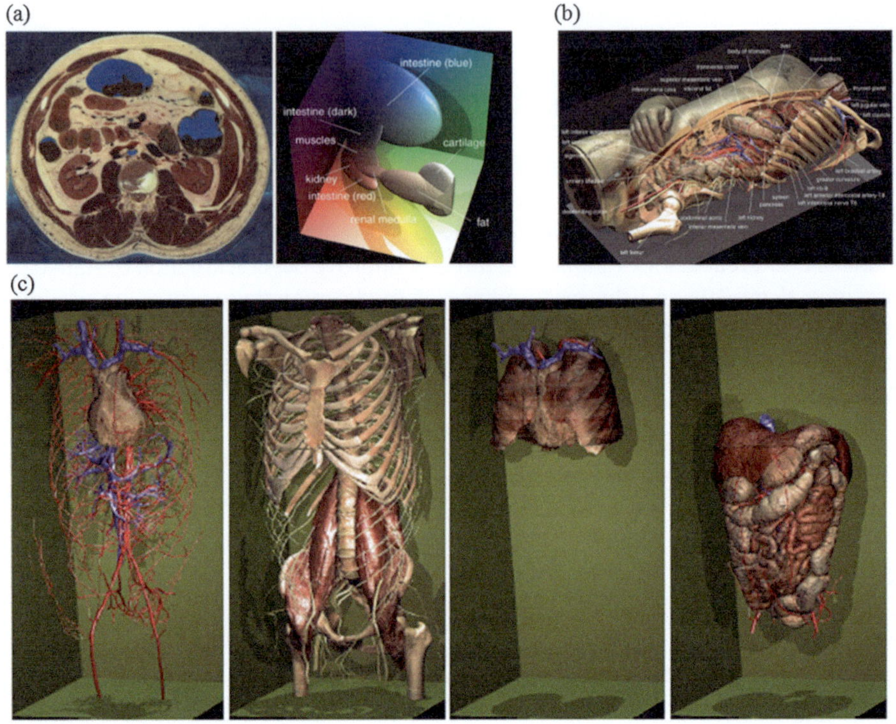

Fig. 1.12 VOXEL-MAN 3D digital atlas reconstructed from Visible Human (VH) data. (**a**) VH data analysis (Ref. [111]). (**b**) Digital (or electronic) 3D anatomical atlas. (**c**) 3D reconstruction of the vasculature, bones, chest, abdomen, and pelvis

in CT data as "virtual" dissection of the human body, similar to VOXEL-MAN generated from VH data. Therefore, development of methodologies for automated identification of the whole-body CT data is a key issue for fully utilizing potential information inherent in the whole-body CT data.

Whole-body imaging has influenced radiological diagnosis. In early studies, trade-offs between advantages and disadvantages of whole-body CT screening of healthy individuals have been discussed, and some criticisms have been made because of the radiation dose and limited real benefits [112]. However, in populations with illnesses and a higher pretest likelihood of positive findings, whole-body CT combined with PET is often used for tumor staging [113]. Whole-body PET/CT is now recognized as a useful modality to find unexpected cancers for patients in clinical practice. Apart from cancer diagnosis, whole-body CT has been suggested for serious trauma patients [114]. Whole-body diffusion MRI, which does not involve any radiation, has also been shown to be potentially useful for tumor staging [115] and nerve imaging [116]. More recently, from the technical point of view, dose reduction is becoming possible through innovations in CT reconstruction algorithms [117] and rapid reduction of computational cost in recent years. Therefore, in the

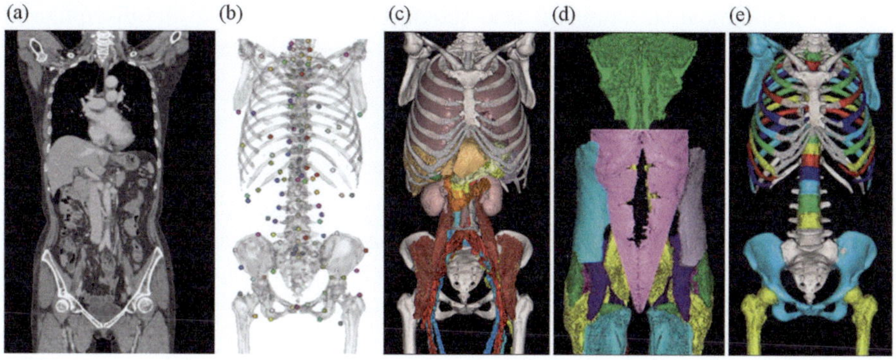

Fig. 1.13 Fully automated segmentation and anatomical identification of whole-body non-contrast CT data. (**a**) Original CT image. (**b**) Landmarks. (**c**) Chest, abdomen, and pelvis. (**d**) Muscles. (**e**) Skeletal system

long run, whole-body imaging will become more advantageous, while disadvantages will decrease.

Whole-body imaging is becoming more widespread and opens new research opportunities for the medical image processing field. The concept of multi-organ multi-disease CAD was tested by a Japanese nationwide project [47], which also provided a strong motivation for whole-body image analysis. The importance to computational anatomy is that this involves whole-body 3D data of a large population, rather than of a single subject, such as VOXEL-MAN. Computational anatomy models, which represent intersubject variability of anatomical structures, are systematically constructed from population data, as shown in Fig. 1.13.

The computer vision research field has already developed key technologies for modeling and recognizing anatomical structures from images. In contrast to medical image analysis, which mainly analyzes 3D images inside the body, computer vision deals with outer appearances of persons in addition to other real-world objects. Typical anatomical structures (or parts of the human body) addressed by computer vision include faces, hands, and body motion. Eigenface [118], ASMs [87], and AAMs were typical approaches developed in computer vision. They initially addressed 2D images and then were extended to 3D domains. The morphable face model [119] is one of these extensions and represents intersubject variability of 3D shape of the human face. Similarly, 3D shapes and image intensity distributions of organs and anatomical structures, such as the lung, liver, and pelvis, are modeled by 3D versions of ASMs, SSMs, and AAMs [93]. These technologies will also be essential for whole-body modeling. Although previous approaches assumed single-structure modeling, the body consists of various types of structures, such as the parenchymal organs, vessels, lymphatic system, musculoskeletal system, and nerves, which are interrelated with each other. These structures may be well-described by using different shape modeling schemes. The challenges of whole-body computational anatomy models include the problem of multi-structure modeling of the different types of anatomical structures. It also should be congruent with the discipline of anatomy which has been established for several hundred years. In this book, these efforts will be described.

Education of medical students is one of the important applications of whole-body imaging. As described before, digital or online atlases of the whole body, such as VOXEL-MAN, are now widely available. Furthermore, whole-body autopsy imaging realizes combined virtual and real cadaver dissection. Several papers have evaluated the usefulness of pre-dissection autopsy imaging in the cadaver dissection course in anatomy education [120, 121]. Automated segmentation and anatomical identification of cadaver CT data will enhance the usefulness of pre-dissection autopsy imaging. While conventional digital atlases represent one subject, computational anatomy models typically represent intersubject variability in addition to average shapes. Another potential benefit for education will be to use whole-body parametric computational anatomy modeling, which will generate an individual anatomy having arbitrary height, weight, and variable organ shapes while maintaining consistent relations among anatomical structures. These models will be effective for learning not only typical anatomy but also its variability.

The Physiome is an emerging project for comprehensive modeling of the human body from the aspect of physiology [122]. One of the goals of the Physiome will be patient-specific multi-scale simulations of the whole body to predict patient function after treatment to optimize treatment planning. Whole-body imaging will provide patient-specific anatomy required for multi-scale simulation as aimed at in the Physiome project. In the musculoskeletal system, whole-body analysis is effective in assessing human movement even if the patient has a problem in a specific joint. Patient-specific biomechanical simulations of the whole musculoskeletal system will be meaningful for preoperative prediction of postoperative patient function. Figure 1.14 shows anatomy modeling in OpenSim [123], one of several well-developed platforms for musculoskeletal analysis. As another example, simulation should be performed for the entire cardiovascular and circulatory system [124]. These simulations currently require huge computational power. However, considering rapid reduction of computing cost, the problem will not be permanent. To

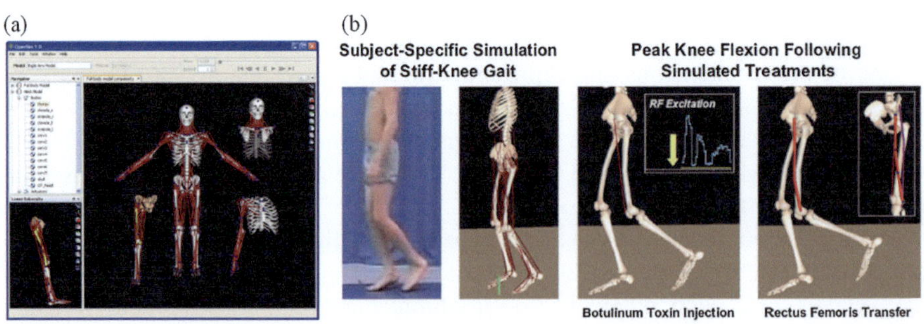

Fig. 1.14 OpenSim platform for subject-specific musculoskeletal simulation (Ref. [123]). (**a**) Anatomy modeling. (**b**) Simulation

realize these scenarios in clinical practice, patient-specific anatomical models need to be reconstructed with sufficient accuracy and a minimum amount of burdens for the medical staff. Therefore, patient-specific modeling from whole-body imaging and other sensory data is regarded as an important issue in the Physiome project.

Whole-body imaging plays an important role in human anatomy modeling for simulations of radiation dosimetry and radio-frequency wave propagation. Several voxel and surface models were developed for this purpose [125–127]. More recently, a whole-body statistical shape model was used to perform simulations of types of the human body [128]. Further improvement of whole-body computational anatomy models will contribute to more accurate, systematic, and comprehensive dosimetry simulations.

Impacts of whole-body imaging on several domains were described, including radiological diagnosis, medical image analysis, simulations for predictive medicine, and dosimetry simulations. Initially, cost and radiation dose were considered to be disadvantageous for clinical practice. However, cost and radiation dose were able to be reduced as a result of technological developments and improvements. Therefore, whole-body imaging will play an increasing role on simulations related to the human body as well as radiological diagnosis. Developments in whole-body computational anatomy and its application to subject-specific anatomy modeling will be a key issue to fully utilize whole-body image data.

1.3.2 Toward Complete Medical Image Understanding

Yoshitaka Masutani
Faculty of Information Science, Hiroshima City University, Hiroshima, 731-3194, Japan
e-mail: masutani@hiroshima-cu.ac.jp

Before the new discipline of "computational anatomy" emerged, medical image understanding is central to medical imaging research. Most research has been aimed at recognition of anatomical (and/or pathological) objects in medical images, that is, image segmentation. As introduced later in this book, an array of techniques has been developed. Medical image segmentation research is generally targeted to specific organs such as the brain, lung, and liver, while several attempts at simultaneous segmentation of multiple organs have been performed. Also in some approaches, segmentation results of other organs are utilized to yield more reliable extraction of main target. However, the rest of the image, outside of the organ(s) of interest, is neglected.

By contrast, "complete medical image understanding" involves the extraction of all the data from organs and tissues within given medical images. The motivation for complete understanding is based on maximization of information extraction from each clinical dataset. As radiologists are always required to read the entire image (for instance, detection of bone metastasis in the spine is expected while reading a

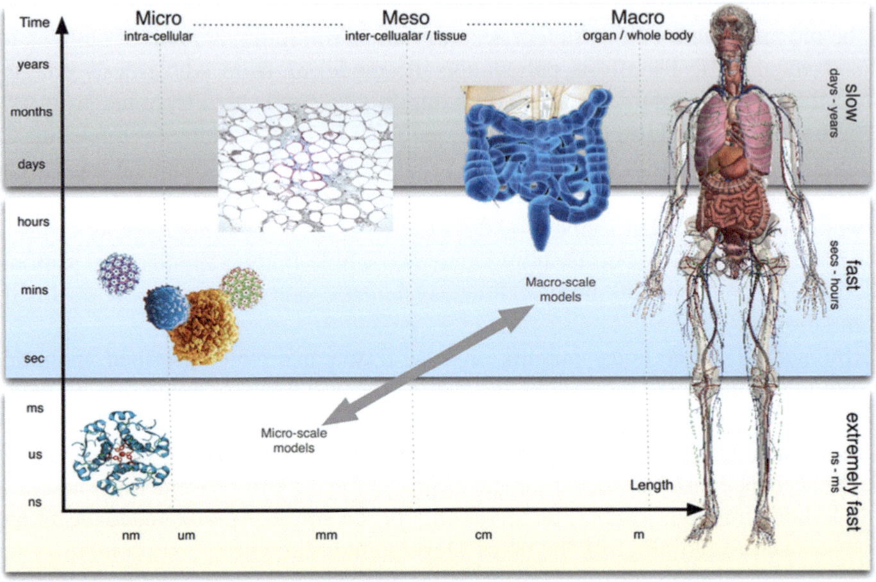

Fig. 1.15 Anatomical classification: organ level (macro) to cell level (micro) (Ref. [131])

chest X-ray), computational understanding also needs to cover the entire area while, at the same time, not devoting attention to harmless findings or "incidentalomas."

1. Classification Completeness

Anatomical objects have been classified into vast classes of objects from organ level to cell level (Fig. 1.15). However, when we introduce the notion of classes to medical image understanding, the feasibility of classification strongly depends on the spatial resolution of a given image. To achieve true completeness in classification by covering all classes, almost infinite image resolution might be required in each image acquisition. This is almost impossible in clinical situations, and therefore, we need a practical solution, which is discussed later.

2. Spatial Completeness

This is quite simple. In terms of the minimum unit of medical images, voxel, there must be no voxel unclassified. In other words, all the voxels in a given image must be anatomically labeled (exhaustive labeling) according to the anatomical classification including surrounding air and even artificial objects such as cannulas or catheters. Considering the partial volume effect, a single voxel can be a mixture of several anatomical structures, and therefore, multiple labels may need to be assigned to a single voxel.

3. Completeness as Correctness

As well as conventional image understanding including partial understanding of medical images, the analysis results from complete understanding of the scheme must be correct. That is, the anatomical label(s) must be assigned to every voxel without errors. However, the correctness is strongly dependent on the classifications used for anatomical labeling.

An important characteristic in classification of anatomical objects is that there are hierarchical structures of the classes. For instance, the liver can be recognized on several levels such as the entire liver, segments defined by Couinaud [129, 130], tissues, and cells. Such levels of detail required in medical image understanding may depend on the purposes, which are also restricted by spatial resolution of images.

Basically, the stance of complete medical image understanding is to make the maximum effort to classify in as much detail as possible. However, considering the computing time cost in clinical situations, a practical level as far as matching the needs can be chosen. As a matter of fact, it should be achieved by the balance of maximum detail of the classification and the correctness of the results. For example, if we pursue only the correctness, the classification can be the two classes of "human body" and "air (rest of the human body)." In most cases, this classification is nonsense for any diagnostic or therapeutic purposes. Instead, cell-level classification in CT images with a spatial resolution of submillimeter level is not feasible. In other words, the classification completeness should be defined adaptively depending on several factors, such as imaging modality, image resolution, purpose, and computational cost permissible to the purpose. This is an important discussion for practical and application-oriented aspects of computational anatomy.

The remainder of this book details complete medical image understanding, including how to achieve the results.

1.4 Book Organization

Yoshitaka Masutani
Faculty of Information Science, Hiroshima City University, Hiroshima, 731-3194, Japan
e-mail: masutani@hiroshima-cu.ac.jp

Computational anatomy is related to a wide variety of established disciplines from applied mathematics to computer science, to medical sciences. To learn the new discipline systematically in association with the related and established disciplines is not a simple task. In this book, the authors tried to refer to established disciplines so that readers are not required to prepare with such backgrounds.

Following this chapter describing the introduction of the new discipline of computational anatomy, the rest of this book is organized as follows:

Chapter 2: Fundamental Theories and Techniques
Chapter 3: Computational Anatomy Model-Based Medical Image Understanding
Chapter 4: Applied Technologies and Systems
Chapter 5: Perspectives

First, the fundamentals of theories and techniques related to computational anatomy are systematically presented in Chap. 2. The theoretical fundamentals cover basic and applied mathematics and statistics, while fundamental techniques include pattern recognition, computer vision, and computer vision applied to medical image understanding. This chapter can be a catalogue for the state-of-the-art techniques for medical image analysis. The essence of the fundamentals is described in minimal detail with references for further details. In addition, we tried to describe anatomical structures from a mathematical viewpoint, including anomaly as the discrete diversity of the structure.

In Chap. 3, based on the fundamental techniques introduced in Chap. 2, concrete technical examples of pattern recognition and image segmentation for each structure of regional anatomy are described in detail, from the brain to the digestive system and from the skeleton to the abdominal organs. In this chapter, readers may appreciate that many techniques are proposed for each anatomical structure. Next, in addition to the theories and techniques, instances of applied systems useful to clinical application are also presented in the chapter, which covers anatomy-guided systems of diagnostic support and surgical navigation in Chap. 4. Finally, perspectives on the new field of computational anatomy are discussed in Chap. 5, covering future directions of the new discipline.

References

1. Thompson DAW (1917) On Shape and Growth. Cambridge University Press, Cambridge
2. Fukunaga K (1972) Introduction to statistical pattern recognition. Academic press, New York
3. Szeliski R (2010) Computer vision: algorithms and applications. Springer, New York
4. Russell S, Norvig P (2009) Artificial intelligence: a modern approach, 3rd edn. Prentice-Hall
5. Höhne KH et al (1995) A new representation of knowledge concerning human anatomy and function. Nat Med 1(6):506–511
6. Kass M, Witkin A, Terzopoulos D (1988) Snakes: active contour models. Int J Comput Vis 1(4):321–331
7. Cootes TF, Edwards GJ, Taylor GJ (1999) Comparing active shape models with active appearance models. In BMVC
8. Pennec X (2009) Statistical computing on manifolds: from Riemannian geometry to computational anatomy. In: Emerging trends in visual computing. Springer, Berlin, pp 347–386
9. Suzuki K (2012) Machine learning in computer-aided diagnosis: medical imaging intelligence. Medical Information Science Reference, Hershey
10. SPM. Available from: http://www.fil.ion.ucl.ac.uk/spm/software/

11. Computational anatomy for computer-aided diagnosis and therapy: frontiers of medical image sciences. Available from: http://www.comp-anatomy.org/wiki/index.php?Computational%20Anatomy
12. The Human Connectome Project. Available from: http://www.humanconnectomeproject.org/
13. Physiome Project. Available from: http://physiomeproject.org
14. Winkelmann A (2007) Anatomical dissection as a teaching method in medical school: a review of the evidence. Med Educ 41(1):15–22
15. Turney B (2007) Anatomy in a modern medical curriculum. Ann R Coll Surg Engl 89(2):104
16. Purkayastha S, Paraskevas P, Darzi A (2007) Anatomy crisis: make surgeons more active in teaching anatomy at all levels. Br Med J 334(7585):110
17. Tanaka M et al (2013) IT system and network required for operation of Ai. INNERVISION 28(1):69–71
18. Noriki S et al (2012) Positioning of Ai as a tool of the lifelong education which used IT as the base. INNERVISION 27(1):30–33
19. Ai center of University of Fukui. Available from: http://ai.labos.ac/
20. Iino S (2014) Present condition of anatomy practice and image education in the university of Fukui. INNERVISION 29(1):46–48
21. Porter R (1999) The greatest benefit to mankind: a medical history of humanity (The Norton history of science). WW Norton & Company, New York
22. Duckworth W (1962) Galen: on anatomical procedures: the later books. Cambridge University Press, Cambridge
23. Mason SF (1962) A history of the sciences. Collier Books, New York
24. Thali MJ et al (2003) Virtopsy, a new imaging horizon in forensic pathology: virtual autopsy by postmortem multislice computed tomography (MSCT) and magnetic resonance imaging (MRI)–a feasibility study. J Forensic Sci 48(2):386–403
25. Korf H-W et al (2008) The dissection course–necessary and indispensable for teaching anatomy to medical students. Ann Anat Anat Anz 190(1):16–22
26. McLachlan JC (2004) New path for teaching anatomy: living anatomy and medical imaging vs. dissection. Anat Rec B New Anat 281(1):4–5
27. Slavin K (1997) The visible human project. Surg Neurol 48(6):638
28. Ackerman M et al (1994) The visible human data set: an image resource for anatomical visualization. Medinfo MEDINFO 8:1195–1198
29. Ackerman MJ (1991) The visible human project. J Biocommun 18(2):14
30. Nobuoka D et al (2014) Surgical education using a multi-viewpoint and multi-layer three-dimensional atlas of surgical anatomy (with video). J Hepatobiliary Pancreat Sci 21(8):556–561
31. Matsuo T, Takeda Y, Ohtsuka A (2013) Stereoscopic three-dimensional images of an anatomical dissection of the eyeball and orbit for educational purposes. Acta Med Okayama 67:87–91
32. Tainaka K et al (2014) Whole-body imaging with single-cell resolution by tissue decolorization. Cell 159(4):911–924
33. Moniz E, de Carvalho L, Lima A (1931) Angiopneumographie. Presse Med 39:996–999
34. Ichikawa H (1967) X-Ray diagnosis of early gastric cancer. J Gastroenterol 2(4):277–281
35. Takahashi S (1949) A new device to get a radiological section figure of body. Tohoku J Exp Med 51(1–2):70–70
36. Hounsfield GN (1973) Computerized transverse axial scanning (tomography): Part 1. Description of system. Br J Radiol 46(552):1016–1022
37. Todo G et al (1982) [High resolution CT (HR-CT) for the evaluation of pulmonary peripheral disorders]. Rinsho hoshasen. Clin Radiogr 27(12):1319–1326
38. Kalender W et al (1990) Spiral CT: an innovative method of volumetric recording. Pt. 1. Roentgenpraxis 43 : 323 – 330
39. Hohne K, Riemer M, Tiede U (1988) Volume rendering of 3D–tomographic imagery. In: Information processing in medical imaging. Springer, New York

40. Sosna J et al (2003) CT colonography of colorectal polyps: a metaanalysis. Am J Roentgenol 181(6):1593–1598
41. Yoshida H, Nappi J (2001) Three-dimensional computer-aided diagnosis scheme for detection of colonic polyps. IEEE Trans Med Imaging 20(12):1261–1274
42. Vinning D, Shitrin R, Haponik E (1994) Virtual bronchoscopy. Radiology 193(P):261
43. Meyers PH et al (1964) Automated computer analysis of radiographic images 1. Radiology 83(6):1029–1034
44. Winsberg F et al (1967) Detection of radiographic abnormalities in mammograms by means of optical scanning and computer analysis 1. Radiology 89(2):211–215
45. The QIBA website. Available from: https://www.rsna.org/QIBA.aspx
46. Shimizu A et al (2007) Segmentation of multiple organs in non-contrast 3D abdominal CT images. Int J Comput Assist Radiol Surg 2(3–4):135–142
47. Kobatake H (2007) Future cad in multi-dimensional medical images:–project on multi-organ, multi-disease cad system. Comput Med Imaging Graph 31(4):258–266
48. Shimizu A, Sato Y (2006) Construction of statistical atlas of abdominal organs and its application to multi-organ segmentation. Med Imaging Technol 24(3):153–160
49. The SPM website. Available from: http://www.fil.ion.ucl.ac.uk/spm/
50. Kobatake H (2011) Aims of the research project "computational anatomy". Med Imaging Technol 29(3):99–103
51. MacKee GM (1921) X-rays and radium in the treatment of diseases of the skin. Lea & Febiger, Philadelphia
52. Takahashi S (1965) Conformation radiotherapy. Rotation techniques as applied to radiography and radiotherapy of cancer. Acta Radiol Diagn Suppl 242: 1+−1+
53. Webb S (2001) Intensity-modulated radiation therapy. CRC Press, Chapman Hall
54. Shirato H et al (2000) Four-dimensional treatment planning and fluoroscopic real-time tumor tracking radiotherapy for moving tumor. Int J Rad Oncol Biol Phys 48(2):435–442
55. Dawson LA, Sharpe MB (2006) Image-guided radiotherapy: rationale, benefits, and limitations. Lancet Oncol 7(10):848–858
56. Evans PM (2008) Anatomical imaging for radiotherapy. Phys Med Biol 53(12):R151
57. Ministry of Internal Affairs and Communications (MIC). 2011; Available from: http://www.stat.go.jp/data/jinsui/pdf/201102.pdf
58. Wambersie A, Landber T (1999) ICRU report 62: prescribing, recording and reporting photon beam therapy. Supplement to ICRU report, 50
59. Onishi H et al (2011) Stereotactic body radiotherapy (SBRT) for operable stage I non–small-cell lung cancer: can SBRT be comparable to surgery? Int J Rad Oncol Biol Phys 81(5):1352–1358
60. Van de Steene J et al (2002) Definition of gross tumor volume in lung cancer: inter-observer variability. Radiother Oncol 62(1):37–49
61. El Naqa I et al (2007) Concurrent multimodality image segmentation by active contours for radiotherapy treatment planninga. Med Phys 34(12):4738–4749
62. Nakamura K et al (2008) Variation of clinical target volume definition among Japanese radiation oncologists in external beam radiotherapy for prostate cancer. Jpn J Clin Oncol 38(4):275–280
63. Geets X et al (2007) A gradient-based method for segmenting FDG-PET images: methodology and validation. Eur J Nucl Med Mol Imaging 34(9):1427–1438
64. Day E et al (2009) A region growing method for tumor volume segmentation on PET images for rectal and anal cancer patients. Med Phys 36(10):4349–4358
65. Hatt M et al (2011) PET functional volume delineation: a robustness and repeatability study. Eur J Nucl Med Mol Imaging 38(4):663–672
66. Ploquin N, Rangel A, Dunscombe P (2008) Phantom evaluation of a commercially available three modality image guided radiation therapy system. Med Phys 35(12):5303–5311
67. Wang Z et al (2009) Refinement of treatment setup and target localization accuracy using three-dimensional cone-beam computed tomography for stereotactic body radiotherapy. Int J Rad Oncol Biol Phys 73(2):571–577

68. Hong J, Hashizume M (2010) An effective point-based registration tool for surgical navigation. Surg Endosc 24(4):944–948
69. Souzaki R et al (2013) An augmented reality navigation system for pediatric oncologic surgery based on preoperative CT and MRI images. J Pediatr Surg 48(12):2479–2483
70. Maeda T et al (2009) Tumor ablation therapy of liver cancers with an open magnetic resonance imaging-based navigation system. Surg Endosc 23(5):1048–1053
71. Tomikawa M et al (2010) Real-time 3-dimensional virtual reality navigation system with open MRI for breast-conserving surgery. J Am Coll Surg 210(6):927–933
72. Tsutsumi N et al (2013) Image-guided laparoscopic surgery in an open MRI operating theater. Surg Endosc 27(6):2178–2184
73. Doi K (2007) Computer-aided diagnosis in medical imaging: historical review, current status and future potential. Comput Med Imaging Graph 31(4):198–211
74. Ayache N (1995) Medical computer vision, virtual reality and robotics. Image Vis Comput 13(4):295–313
75. Rangayyan RM, Ayres FJ, Desautels JL (2007) A review of computer-aided diagnosis of breast cancer: toward the detection of subtle signs. J Frankl Inst 344(3):312–348
76. Shiraishi J et al (2007) Development of a computer-aided diagnostic scheme for detection of interval changes in successive whole-body bone scans. Med Phys 34(1):25–36
77. Marr D (1976) Early processing of visual information. Philos Trans R Soc Lond B Biol Sci 275(942):483–519
78. Poggio T, Torre V, Koch C (1989) Computational vision and regularization theory. Image Underst 3(1–18):111
79. Roberts LG (1963) Machine perception of three-dimensional soups. Massachusetts Institute of Technology
80. Slaney J, Thiébaux S (2001) Blocks world revisited. Artif Intell 125(1):119–153
81. Binford TO (1971) Visual perception by computer. In: IEEE conference on systems and control
82. Ponce J, Chelberg D, Mann WB (1989) Invariant properties of straight homogeneous generalized cylinders and their contours. IEEE Trans Pattern Anal Mach Intell 11(9):951–966
83. Marroquin J, Mitter S, Poggio T (1987) Probabilistic solution of ill-posed problems in computational vision. J Am Stat Assoc 82(397):76–89
84. Fischler MA, Elschlager RA (1973) The representation and matching of pictorial structures. IEEE Trans Comput 22(1):67–92
85. Felzenszwalb PF, Huttenlocher DP (2005) Pictorial structures for object recognition. Int J Comput Vis 61(1):55–79
86. Terzopoulos D (1986) Regularization of inverse visual problems involving discontinuities. IEEE Trans Pattern Anal Mach Intell PAMI-8(4):413–424
87. Cootes TF et al (1995) Active shape models-their training and application. Comput Vis Image Underst 61(1):38–59
88. Marr D, Hildreth E (1980) Theory of edge detection. Proc R Soc Lond Ser B 207(1167):187–217
89. Canny J (1986) A computational approach to edge detection. IEEE Trans Pattern Anal Mach Intell 8(6):679–698
90. Kimmel R, Bruckstein AM (2003) Regularized Laplacian zero crossings as optimal edge integrators. Int J Comput Vis 53(3):225–243
91. Cootes TF et al (1993) The use of active shape models for locating structures in medical images. In: Information processing in medical imaging. Springer, Berlin
92. Audette MA, Ferrie FP, Peters TM (2000) An algorithmic overview of surface registration techniques for medical imaging. Med Image Anal 4(3):201–217
93. Heimann T, Meinzer H-P (2009) Statistical shape models for 3D medical image segmentation: a review. Med Image Anal 13(4):543–563

94. Hontani H, Tsunekawa Y, Sawada Y (2013) Accurate and robust registration of nonrigid surface using hierarchical statistical shape model. In: Computer Vision and Pattern Recognition (CVPR), 2013 IEEE Conference on. 2013. IEEE
95. Davies RH et al (2010) Building 3-d statistical shape models by direct optimization. IEEE Trans Med Imaging 29(4):961–981
96. Cross GR, Jain AK (1983) Markov random field texture models. IEEE Trans Pattern Anal Mach Intell 5(1):25–39
97. Geman S, Geman D (1984) Stochastic relaxation, Gibbs distributions, and the Bayesian restoration of images. IEEE Trans Pattern Anal Mach Intell 6(6):721–741
98. Derin H, Elliott H (1987) Modeling and segmentation of noisy and textured images using Gibbs random fields. IEEE Trans Pattern Anal Mach Intell 9(1):39–55
99. Adelson EH, Bergen JR (1991) The plenoptic function and the elements of early vision. Comput Model Vis Process **1**(2):3–20
100. Felzenszwalb PF, Huttenlocher DP (2006) Efficient belief propagation for early vision. Int J Comput Vis 70(1):41–54
101. Szeliski R et al (2008) A comparative study of energy minimization methods for markov random fields with smoothness-based priors. IEEE Trans Pattern Anal Mach Intell 30(6):1068–1080
102. Blake A, Zisserman A (1987) Visual reconstruction, vol 2. MIT press, Cambridge
103. Mumford D, Shah J (1989) Optimal approximations by piecewise smooth functions and associated variational problems. Commun Pure Appl Math 42(5):577–685
104. Prince SJ (2012) Computer vision: models, learning, and inference. Cambridge University Press, New York
105. Hinton G, Osindero S, Teh Y-W (2006) A fast learning algorithm for deep belief nets. Neural Comput 18(7):1527–1554
106. Bengio Y (2009) Learning deep architectures for AI. Found Trends® Mach Learn 2(1):1–127
107. Müller H et al (2004) A review of content-based image retrieval systems in medical applications—clinical benefits and future directions. Int J Med Inform 73(1):1–23
108. Sagerer G, Niemann H (2013) Semantic networks for understanding scenes. Springer, New York
109. Tenenbaum JM, Witkin A (1983) On the role of structure in vision. In: Human and machine vision. Academic, New York, pp 481–543
110. Spitzer V et al (1996) The visible human male: a technical report. J Am Med Inform Assoc 3(2):118–130
111. Pommert A et al (2001) Creating a high-resolution spatial/symbolic model of the inner organs based on the Visible Human. Med Image Anal 5(3):221–228
112. Beinfeld MT, Wittenberg E, Gazelle GS (2005) Cost-effectiveness of Whole-Body CT Screening 1. Radiology 234(2):415–422
113. Antoch G et al (2003) Whole-body dual-modality PET/CT and whole-body MRI for tumor staging in oncology. JAMA 290(24):3199–3206
114. Huber-Wagner S et al (2009) Effect of whole-body CT during trauma resuscitation on survival: a retrospective, multicentre study. Lancet 373(9673):1455–1461
115. Takahara T et al (2003) Diffusion weighted whole body imaging with background body signal suppression (DWIBS): technical improvement using free breathing, STIR and high resolution 3D display. Radiat Med 22(4):275–282
116. Yamashita T, Kwee TC, Takahara T (2009) Whole-body magnetic resonance neurography. N Engl J Med 361(5):538–539
117. Silva AC et al (2010) Innovations in CT dose reduction strategy: application of the adaptive statistical iterative reconstruction algorithm. Am J Roentgenol 194(1):191–199
118. Turk M, Pentland A (1991) Eigenfaces for recognition. J Cogn Neurosci 3(1):71–86
119. Blanz V,Vetter T (1999) A morphable model for the synthesis of 3D faces. In: Proceedings of the 26th annual conference on computer graphics and interactive techniques. ACM Press/Addison-Wesley Publishing Co., New York

120. Hisley KC et al (2008) Coupled physical and digital cadaver dissection followed by a visual test protocol provides insights into the nature of anatomical knowledge and its evaluation. Anat Sci Educ 1(1):27–40
121. Jacobson S et al (2009) Creation of virtual patients from CT images of cadavers to enhance integration of clinical and basic science student learning in anatomy. Med Teach 31(8):749–751
122. Hunter PJ, Borg TK (2003) Integration from proteins to organs: the Physiome Project. Nat Rev Mol Cell Biol 4(3):237–243
123. Delp SL et al (2007) OpenSim: open-source software to create and analyze dynamic simulations of movement. IEEE Trans Biomed Eng 54(11):1940–1950
124. Xiao N, Humphrey JD, Figueroa CA (2013) Multi-scale computational model of three-dimensional hemodynamics within a deformable full-body arterial network. J Comput Phys 244:22–40
125. Nagaoka T et al (2004) Development of realistic high-resolution whole-body voxel models of Japanese adult males and females of average height and weight, and application of models to radio-frequency electromagnetic-field dosimetry. Phys Med Biol 49(1):1
126. Caon M (2004) Voxel-based computational models of real human anatomy: a review. Radiat Environ Biophys 42(4):229–235
127. Christ A et al (2010) The virtual family—development of surface-based anatomical models of two adults and two children for dosimetric simulations. Phys Med Biol 55(2):N23
128. Lee SL et al (2011) A whole body statistical shape model for radio frequency simulation. In: Engineering in Medicine and Biology Society, EMBC, 2011 Annual International Conference of the IEEE. 2011. IEEE
129. Couinaud C (1957) Le foie: études anatomiques et chirurgicales. Masson & Cie, Paris
130. Bismuth H (1982) Surgical anatomy and anatomical surgery of the liver. World J Surg 6(1): 3–9
131. Castiglione F et al (2013) The onset of type 2 diabetes: proposal for a multi-scale model. JMIR Res Protoc 2(2):e44

Chapter 2
Fundamental Theories and Techniques

Hidekata Hontani, Yasushi Hirano, Xiao Dong, Akinobu Shimizu, and Shohei Hanaoka

Abstract In this section, fundamental theories and techniques for understanding computational anatomy are described. First, the mathematical foundations of a signal processing and of statistics are discussed. Signal processing is the basis of the image processing required for extracting local image features that are useful for the identification of the organ regions in medical images, which is one of the most important tasks in CA. A knowledge of statistics is needed for understanding the statistical shape models (SSMs) of the organs and the registration of the models to given medical images, which is one of the most basic techniques used for the organ region identification. Second, model representations of the organs, e.g., point distribution models (PDMs), medial representations (m-reps), and nonuniform rational basis splines (NURBS), are described. Different models, e.g., a point distribution model (PDM), a medial representation (m-rep), or nonuniform rational basis splines (NURBS), can be employed for representing a target organ, and a region of a target organ in a given image can be identified by registering the

H. Hontani (✉)
Department of Computer Science, Nagoya Institute of Technology, Gokiso, Showa-ku, Aichi 466-8555, Nagoya, Japan
e-mail: hontani@nitech.ac.jp

Y. Hirano
Graduate School of Medicine, Yamaguchi University, 1677-1, Yoshida, Yamaguchi-shi, 735-8511, Yamaguchi, Japan
e-mail: yhirano@yamaguchi-u.ac.jp

X. Dong
Faculty of Computer Science and Engineering, Southeast University, 2 Sipailou, Jiangsu, 210018, Nanjing, China
e-mail: 101011465@seu.edu.cn

A. Shimizu
Institute of Engineering, Tokyo University of Agriculture and Technology, Naka-cho 2-24-16, Koanei, 184-0012, Tokyo, Japan
e-mail: simiz@cc.tuat.ac.jp

S. Hanaoka
Department of Radiology, The University of Tokyo Hospital, 7-3-1, Hongo, Bunkyo-ku, 113-8655, Tokyo, Japan
e-mail: hanaoka-tky@umin.ac.jp

H. Kobatake, Y. Masutani (eds.), *Computational Anatomy Based on Whole Body Imaging*, DOI 10.1007/978-4-431-55976-4_2

employed model to the image: Several techniques for the model registration are also discussed in this chapter. The performance of the organ region identification can change depending on the employed representation and on the employed registration technique. Finally, the difficulties posed by multiple organ registration and the handling of anatomical anomalies are considered.

Keywords Image segmentation • Model registration • Image registration • Statistical shape model • Point distribution model

2.1 From Anatomy to Computational Anatomy

Hidekata Hontani and Yasushi Hirano

2.1.1 Introduction

Today, the dissection of human bodies rarely brings new knowledge of human anatomy but is required for observing the anatomical structures of each patient. One can observe the structures without dissecting the patient body by observing detailed medical images. Imaging plays a very important role in medicine because it enables the observation of the form and structure of the organs specific to each living patient. For accurate assessment, one needs to identify the boundaries of the organs and doctors, e.g., radiologists, identify the organ boundaries in images based on the knowledge of the anatomy of human body, and imaginarily reconstruct the 3D boundaries of the organs in their heads. The imaginary reconstruction of the organs, though, is not useful for obtaining geometric information of the organs and for computer-aided diagnosis (CAD) systems or other clinical applications. One needs to explicitly describe the 3D boundaries of the organs, but it is prohibitively time consuming to describe the boundaries of the organs by manually labeling organ regions in given images.

Computational anatomy (CA) is the study of computational methodologies for medical image analysis, and one of the main purposes of the analysis is to accurately and automatically segment all of the organs included in the images: The goal is to label every voxel in the images with the name of the organ to which the corresponding voxel belongs. For every modality with known spatial resolution, one can generate a set of labels of the organs; one should identify their regions in the images. Let the label be denoted by l_i ($i = 1, 2, \cdots, M$), and let the set be denoted by $\boldsymbol{L} = \{l_i | i = 1, 2, \cdots, M\}$ where M denotes the total number of the labels. Given a medical image and the set of the labels, $\boldsymbol{L}$, it is required to label voxel as the name of the corresponding organ, l_i. Labeling the organs in this way and identifying the boundaries of the organs require prior knowledge of human anatomy and of image patterns that are needed for accurate segmentation. Computational models of the organs supply the *prior knowledge* for the segmentation. In the reminder of this section, some important topics fundamental in the medical image segmentation are described.

2.1.2 *Simple Examples*

In this subsection, the outline of one of the most fundamental methods for the segmentation of the organs in medical images, active shape model (ASM) [1], is described along with other techniques.

2.1.2.1 Outline of ASM

The ASM segments a target organ region in a given image by registering a statistical shape model (SSM) of the boundary of the target organ to the given image. A SSM parametrically represents the statistical variety of the shapes of the boundary of the target organ, and the model is registered to the given image by estimating the values of the parameters of the SSM. One can divide the ASM algorithm into two steps: (1) construction of a SSM and (2) registration of the SSM. The basics of outlining the ASM are similar to those of many other methods for medical image segmentation. Figure 2.1 shows the process of (1) construction of a SSM and of (2) of registration.

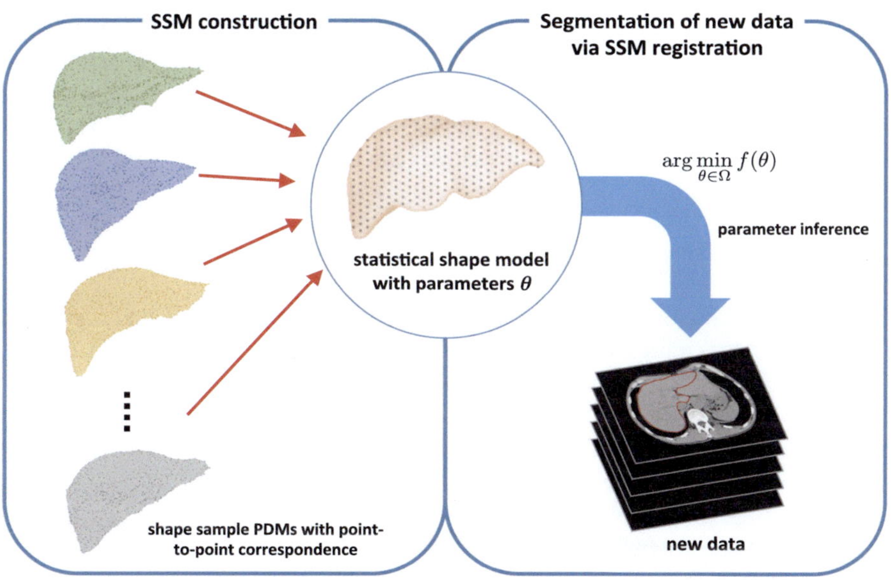

Fig. 2.1 Flow of the construction of a SSM and of its registration to given images

1. Cootes and Taylor [2] first employed a point distribution model (PDM) to represent the boundary of a target. The boundary of the target organ in each set of training images is manually labeled, and a set of corresponding points is distributed on each of the labeled boundaries. The SSM of the PDM represents the probability distribution of the coordinates of the points. The statistical model has shape parameters that are the statistical random variables, and one can vary the shape of the model by changing the values of these parameters.
2. Registration of SSM
   The objective of ASM is to determine the boundary of a target region in a new given image by registering the constructed SSM to the image. The SSM is registered by estimating appropriate values of the parameters of the SSM. When the SSM is registered accurately, each point on the surface represented by the SSM is located on the voxels around which the local image pattern has some features specific to the boundary of the target organ.

2.1.2.2 Required Techniques

SSM construction and registration require details of the following issues:

Shape representation

The method of representation of a target to be extracted from given images needs to be determined. In addition to a point distribution model (PDM) used in [2], other methods include medial representations (m-reps) [3], spherical harmonics (SPHARM) boundary description [4], nonuniform rational basis splines (NURBS) [5, 6], or a set of characterizing shape descriptors [7]. For example, the m-reps represent a target with its medial lines and the diameters on each point of the medial lines. Representation of objects is described in Sect. 2.3.2.

Construction of Statistical Shape Model

For the construction of SSMs using training images, each of the labeled regions or their boundaries is represented with one of the above methods. For example, each of the boundaries of the labeled regions can be represented using a PDM. It is necessary to determine (1) a mathematical model to describe the space and variability of anatomical structures, (2) a computable distance metric to measure the difference between shapes, and (3) statistical analysis tools for the shapes. Representation of objects is described further in Sect. 2.3.2. A diffeomorphism-based framework provides one of the most mathematically fundamental procedures for handling these issues, as will be described in Sect. 2.3.4: Using this framework allows generation of an invertible function that maps one boundary surface of an organ in one training image to another surface in a different training image. Such correspondence is vital

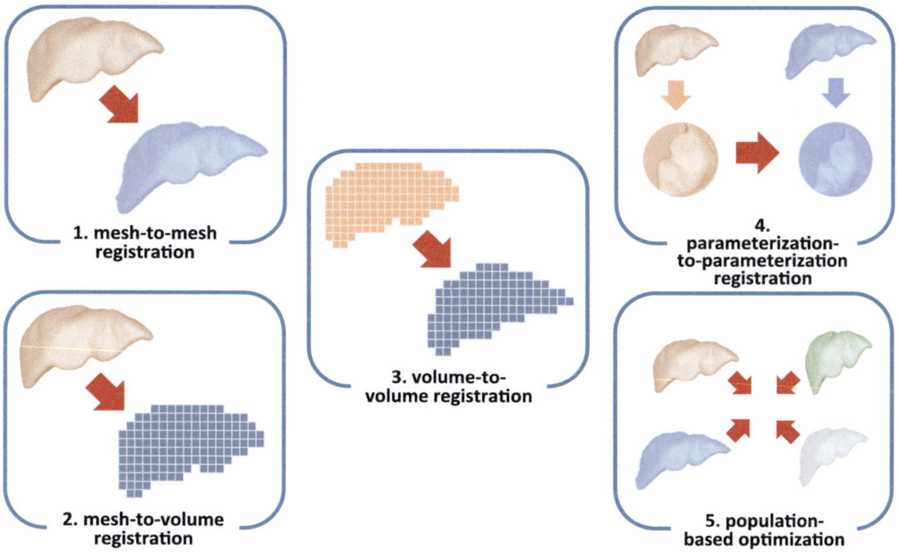

Fig. 2.2 Different approaches for the registration between two shapes

to SSM construction, and many strategies other than the diffeomorphism-based framework have been proposed for making this one-to-one correspondence. These strategies can be classified into five categories (see also Fig. 2.2).

1. Mesh-to-Mesh Registration
   A straightforward method to establish correspondence between two surfaces is to correspond points distributed on the surfaces directory. In mesh-to-mesh strategy, each of the boundaries is first represented using a mesh (or a set of points), and then the meshes are registered together for the correspondence generation. For example, a standard rigid matching algorithm such as the iterative closest point (ICP) algorithm [8] or the softassign Procrustes [9] would be applicable. Given two surfaces, one of the surfaces is transformed to match the other. Once the two surfaces are matched, it is easy to generate a set of pairs of corresponding points. Both methods accept different numbers of initial points distributed on the surfaces, and the optimal similarity transformation from one surface to the other surface is defined. Nonrigid registration of shapes can also be utilized (see, e.g., [10, 11]).
2. Mesh-to-Volume registration
   In this strategy, one surface of an organ in one training image is represented by a mesh, and it is registered not to a surface in other training image but to a labeled organ region in other training image [12–15], where the voxel values are equal to one (1) in the target regions and are equal to zero outside of the regions. The key issue in this strategy is robustness of the deformable template algorithm. Techniques to ensure this robustness include the multi-resolution

approach [16], gradient vector flow [13], and regularization of internal energies [12, 14, 15]. Dam et al. proposed a bootstrap approach to segment learning data and to iteratively refine correspondences [17]. For all these approaches to mesh-to-volume registration, homeomorphic mapping between the input shapes is guaranteed unless the template does not fold itself in the adaptation process.

3. Volume-to-Volume Registration
   Correspondence between the boundaries can be achieved by nonrigidly registering the images. In this strategy, a volumetric atlas or an average of the training images is first computed. Here, the average is defined as an image that has the smallest bias toward any specific training image. The computed average is then utilized to generate the correspondence. Frangi et al. describe a method to warp the atlas to binary training volumes using multi-resolution B-spline deformation [18]. Rueckert et al. describe a method that constructs a statistical model of the deformation fields [19].
4. Volume-to-Volume Registration
   Correspondence between the boundaries can be achieved by nonrigidly registering the images. In this strategy, a volumetric atlas or an average of the training images is first computed. Here, the average is defined as an image that has the smallest bias toward any specific training image. The computed average is then utilized to generate the correspondence. Frangi et al. describe a method to warp the atlas to binary training volumes using multi-resolution B-spline deformation [18]. Rueckert et al. describe a method that constructs a statistical model of the deformation fields [19].
5. Parameterization-to-Parameterization Registration
   In this strategy, the boundary surfaces are represented by meshes, and they are not registered together but are registered to one base domain. Most of these methods select a sphere as the base domain. A method proposed by Kelemen et al. uses a spherical harmonics mapping (SPHARM) for the registration to the sphere [20]. In a method proposed by Brett and Taylor [21], all shapes are mapped to 2D disks that are then aligned to generate the correspondence. These methods guarantee a diffeomorphism among all datasets, but it is difficult for users to control the resultant correspondences. Other methods have been proposed that control the parameterization through a small set of known or assumed correspondences [22–26].
6. Population-Based Optimization
   As mentioned above, it is not easy to define *good* correspondence. Kotcheff and Taylor [27] proposed a method that generates the correspondences so that the resultant SSM has *good* properties: The authors defined the compactness of the resultant model as an index of *goodness*. The compactness is evaluated through the determinant of the covariance matrix. Although the method improves the performance, the objective function minimized when the corresponding points generated have no theoretical foundation. Using the same strategy, Davies et al. proposed an objective function defined with the minimum description length [28]. This objective function has a theoretical foundation, but the calculation cost is high. Thodberg describes a simplified version of the cost function [29].

Local Image Feature Detection

In ASM [30], edge points, at which the voxel values rapidly change, are detected, and the model surface is deformed so that points on the surface are located at the edges. Edges, ridges, or other local image features specific to the points on the target boundaries must be detected in given images to enable accurate model-to-image registration. Some details of fundamental operations for the image feature detection will be described in Sect. 2.2.1.

Statistical Inference

Given a new image, one can register the SSM to the image by inferring the values of the parameters of the SSM: One can deform the model by changing the values of the parameters, and one can register the model by finding appropriate values of the parameters that describe the shape of the target organ in the given image. One of the most fundamental approaches for the inference of the parameter values is to construct an objective function of the parameters in a statistical framework and to compute the optimal parameter values that minimize (or maximize) the objective function.

2.2 Mathematical Foundation

Hidekata Hontani and Yasushi Hirano

2.2.1 *Signal Processing*

Knowledge of signal processing provides a mathematical foundation for describing the relationships between physical objects and (image) data obtained by measuring the objects by (imaging) sensors.

2.2.1.1 Digital Images

In many cases, a medical image is defined over a bounded three-dimensional rectangular lattice. Let a three-vector $\boldsymbol{X} = (X_1, X_2, X_3)^T$ denote a location in a real three-dimensional bounded space, $\Omega \in \mathbb{R}^3$, and let a physical quantity distribution in Ω measured by an imaging system be represented by a function, $f(\boldsymbol{X})$: $\mathbb{R}^3 \mapsto \mathbb{R}$. For example, one uses a computed tomography (CT) scanner to measure physical quantities, $f(\boldsymbol{X})$, which, in the case of CT, are the degree of attenuation of the X-ray beam at each location, $\boldsymbol{X}$. In general, an image is a set of measurements obtained at rectangular lattice points in Ω. Let the coordinates of each of the lattice points be denoted by $\hat{X} = [\Delta_1 x_1, \Delta_2 x_2, \Delta_3 x_3]^T$, where Δ_s $(s = 1, 2, 3)$ is the regular interval between two neighboring lattice points along the s-th axis line and where x_s $(s = 1, 2, 3)$ is an integer. Let $\boldsymbol{x} = [x_1, x_2, x_3]^T$. Then an (ideal) imaging system captures an image, $I(\boldsymbol{x})$, such that

$$I(\boldsymbol{x}) \propto f(\hat{X}). \tag{2.1}$$

Each point represented by the tuple, $\boldsymbol{x}$, is called a *voxel*. Let us assume that x_s is bounded as $1 \leq x_s \leq W_s$ and that the image size is $W_1 \times W_2 \times W_3$. Concatenating all values, $I(\boldsymbol{x})$, of all voxels into one column, one obtains a D-vector, $\boldsymbol{I}$, which is widely used for describing an image, where $D = W_1 \times W_2 \times W_3$.

2.2.1.2 Linear Operation

Let $\mathscr{O}[f]$ denote an operator applied to a signal $f(\boldsymbol{u})$ of a d-dimensional variable, $\boldsymbol{u} = [u_1, u_2, \ldots, u_d]^T$. It is said to be a linear operator if it satisfies the following conditions for any pair of signals, $f(\boldsymbol{u})$ and $g(\boldsymbol{u})$:

$$\mathscr{O}[f(\boldsymbol{u}) + g(\boldsymbol{u})] = \mathscr{O}[f(\boldsymbol{u})] + \mathscr{O}[g(\boldsymbol{u})], \tag{2.2}$$

and

$$\mathscr{O}[af(\boldsymbol{u})] = a\mathscr{O}[f(\boldsymbol{u})], \tag{2.3}$$

where a is a scalar [30].

2.2.1.3 Convolution

Convolution between a target function and a filter function represents a linear filtering operation. Let $f(\boldsymbol{u})$ denote a target (input) function and let $g(\boldsymbol{u})$ be functions defined over a three-dimensional space, where $\boldsymbol{u} \in \mathbb{R}^d$, and $f(\cdot)$ and $g(\cdot)$ are real functions: $\mathbb{R}^d \mapsto \mathbb{R}$, the convolution of $f(\boldsymbol{u})$ and $g(\boldsymbol{u})$ is defined as follows:

$$h(\boldsymbol{x}) = f(\boldsymbol{x}) * g(\boldsymbol{x}) = \int_{-\infty}^{\infty} f(x_1 - s_1, x_2 - s_2, \ldots, x_d - s_d) g(s_1, s_2, \ldots, s_d) ds_1 ds_2 \ldots ds_D. \tag{2.4}$$

The convolution defined above is linear because it satisfies Eqs. (2.2) and (2.3).

Let an image be denoted by $I(x_1, x_2, x_3)$ $(1 \leq x_i \leq W_i,\ i = 1, 2, 3)$ and let a filter be denoted by $G(x_1, x_2, x_3)$, where $G()$ is a real function: $\mathbb{Z}^3 \mapsto \mathbb{R}$. Assuming that the value of the filter $G(x_1, x_2, x_3)$ is equal to zero when $\boldsymbol{x} = (x_1, x_2, x_3)$ is outside of a bounded region, V=$\{(x_1, x_2, x_3)^T | -V_s \leq x_s \leq V_s, s = 1, 2, 3\}$. Then, following

Eq. (2.4), the convolution of the image and the filter is defined as

$$H(x_1, x_2, x_3) = \sum_{i_1=-V_1}^{V_1} \sum_{i_2=-V_2}^{V_2} \sum_{i_3=-V_3}^{V_3} I(x_1 - i_1, x_2 - i_2, x_3 - i_3) G(i_1, i_2, i_3). \quad (2.5)$$

For computing $H(\boldsymbol{x})$ in (2.4) at all voxels in the image, one needs values of voxels outside of the image regions. Let $R = \{(x_1, x_2, x_3)^T | 1 \leq x_s \leq W_s, s = 1, 2, 3\}$ denote the domain of the image. The outside of R can be filled with zeros or $I(\boldsymbol{x}_{\text{out}}) = I(\boldsymbol{x}^*_{\text{in}})$ can be set, where $\boldsymbol{x}_{\text{out}} \notin R$ and $\boldsymbol{x}^*_{\text{in}}$ denotes the inside voxel closest to $\boldsymbol{x}_{\text{out}}$.

2.2.1.4 Cross Correlation

Let us assume that $\boldsymbol{x} = (x_1, x_2, \ldots, x_d)^T \in \mathbb{R}^d$, and let us assume that $f(\boldsymbol{x})$ and $g(\boldsymbol{x})$ are real functions, $\mathbb{R}^d \mapsto \mathbb{R}$. Cross correlation between the two functions, $f(\boldsymbol{x})$ and $g(\boldsymbol{x})$, is used for measuring the similarity of their waveforms and is defined as

$$c(\boldsymbol{x}) = f(\boldsymbol{x}) \star g(\boldsymbol{x}) = \int_{-\infty}^{+\infty} f(x_1 + s_1, x_2 + s_2, \ldots, x_d + s_d) g(s_1, s_2, \ldots, s_d) ds_1 ds_2 \ldots ds_d. \quad (2.6)$$

The cross correlation is linear because it satisfies Eqs. (2.2) and (2.3). Comparing with Eq. (2.4), it should be noted that the sign of s_i in the arguments of $f()$ on the right side is different, obtaining $f(\boldsymbol{x}) * g(\boldsymbol{x}) = f(\boldsymbol{x}) \star g'(\boldsymbol{x})$ if $g'(\boldsymbol{x}) = g(-\boldsymbol{x})$.

Let an image be denoted by $I(x_1, x_2, x_3)$ $(1 \leq x_i \leq W_i, i = 1, 2, 3)$ and let a filter be denoted by $G(x_1, x_2, x_3)$, where $G()$ is a real function: $\mathbb{Z}^3 \mapsto \mathbb{R}$. Assuming that the value of the filter $G(x_1, x_2, x_3)$ is equal to zero when $\boldsymbol{x} = (x_1, x_2, x_3)$ is outside of a bounded region, V=$\{(x_1, x_2, x_3)^T | -V_s \leq x_s \leq V_s, s = 1, 2, 3\}$. Then, the cross correlation of $I()$ and $G()$ is defined as follows:

$$C(\boldsymbol{x}) = I(\boldsymbol{x}) \star G(\boldsymbol{x}) = \sum_{i_1=-V_1}^{V_1} \sum_{i_2=-V_2}^{V_2} \sum_{i_3=-V_3}^{V_3} I(x_1 + i_1, x_2 + i_2, x_3 + i_3) G(i_1, i_2, i_3). \quad (2.7)$$

Let R^* denote a local region around $\boldsymbol{x}^* = (x_1^*, x_2^*, x_3^*)^T$ in the image domain where $x_s^* - V_s \leq x_s \leq x_s + V_s$ $(s = 1, 2, 3)$, and let the part of the image in R^* be denoted by a $V_1 \times V_2 \times V_3$-vector, $\boldsymbol{I}^*$, which is obtained by concatenating the voxel values, $I(\boldsymbol{x})$, where $\boldsymbol{x} \in R^*$. Let $G(\boldsymbol{x})$ be denoted by a $V_1 \times V_2 \times V_3$-vector, $\boldsymbol{G}$, which are obtained by concatenating the voxel values, $G(\boldsymbol{x})$, where $-V_s \leq x_s \leq V_s$ $(s = 1, 2, 3)$. Then the value of the cross correlation between $I(\boldsymbol{x})$ and $G(\boldsymbol{x})$ is nothing but the inner product between $\boldsymbol{I}^*$ and $\boldsymbol{G}$:

$$C(\boldsymbol{x}^*) = \boldsymbol{I}^* \cdot \boldsymbol{G} = (\boldsymbol{I}^*)^T \boldsymbol{G} = \sum_{i_1=-V_1}^{V_1} \sum_{i_2=-V_2}^{V_2} \sum_{i_3=-V_3}^{V_3} I(x_1^*+i_1, x_2^*+i_2, x_3^*+i_3) G(i_1, i_2, i_3). \tag{2.8}$$

A normalized cross correlation, $C_{\text{norm}}(\boldsymbol{x}^*)$, between $\boldsymbol{I}^*$ and $\boldsymbol{G}$ is widely employed for matching a template $\boldsymbol{G}$ to the given image I and is typically defined as

$$C_{\text{norm}}(\boldsymbol{x}^*) = \frac{\boldsymbol{I}^* - \bar{\boldsymbol{I}}^*}{\sigma_I} \cdot \frac{\boldsymbol{G} - \bar{\boldsymbol{G}}}{\sigma_G}, \tag{2.9}$$

where $\bar{\boldsymbol{I}}^*$ and $\bar{\boldsymbol{G}}$ are the averages of $\boldsymbol{I}^*$ and $\boldsymbol{G}$, respectively, and σ_I and σ_G are the standard deviations of them, respectively. When the template and the local appearance are more similar, the larger value of the normalized cross correlation is obtained. It should be noted that the value of C_{norm} is invariant against a linear change of the brightness of the image: $I(\boldsymbol{x}) \leftarrow aI(\boldsymbol{x})$, where $a > 0$.

2.2.1.5 Fourier Series Expansion

The Fourier series expansion is used for analyzing the frequencies of given signals. Let $\boldsymbol{x}$ denote a D-vector where $\boldsymbol{x} \in \mathbb{R}^D$ and let $f(\boldsymbol{x})$ ($\mathbb{R}^D \mapsto \mathbb{R}$) denote an absolutely integrable real function that satisfies

$$\int |f(\boldsymbol{x})|^2 d\boldsymbol{x} < \infty. \tag{2.10}$$

then, the Fourier transform of the function, $\mathscr{F}[f]$, is defined as

$$\mathscr{F}[f] = F(\boldsymbol{\omega}) = \int f(\boldsymbol{x}) \exp(-j\boldsymbol{\omega}^T \boldsymbol{x}) d\boldsymbol{x}, \tag{2.11}$$

where $\boldsymbol{\omega} = (\omega_1, \omega_2, \ldots, \omega_D)^T$ denotes the frequencies along the axes. The original signal, $f(\boldsymbol{x})$, can be recovered from $F(\boldsymbol{\omega})$ by the inverse Fourier transformation that is defined as

$$f(\boldsymbol{x}) = \mathscr{F}^{-1}[F(\boldsymbol{\omega})] = \frac{1}{(2\pi)^D} \int F(\boldsymbol{\omega}) \exp(j\boldsymbol{\omega}^T \boldsymbol{x}) d\boldsymbol{\omega}. \tag{2.12}$$

For simplicity, let us assume here that $D = 1$. Then, the Fourier transformation of $f(x)$ is denoted by $\mathscr{F}[f] = F(\omega)$, and the transformation is linear because it satisfies (2.2) and (2.3). The Fourier transformation has the following properties.

1. The Fourier transformation of the derivative of a function, $f(x)$, is given as follows:

$$\mathscr{F}\left[\frac{df}{dx}\right] = (-j\omega)\mathscr{F}[f]. \tag{2.13}$$

   More generally,

$$\mathscr{F}\left[\frac{d^n f}{dx^n}\right] = (-j\omega)^n \mathscr{F}[f]. \tag{2.14}$$

   As shown in Eq. (2.13), by differentiating a function, the Fourier coefficient, $\mathscr{F}[f]$, is multiplied by $j\omega$, and the components of higher frequencies are more enhanced. A function, $f(x)$, can be differentiated by computing the inverse Fourier transformation of $-j\omega F(\omega)$:

$$\frac{df}{dx} = \mathscr{F}^{-1}[-j\omega F(\omega)]. \tag{2.15}$$

2. The Fourier transformation of $f(x) * g(x)$, where $f(x)$ and $g(x)$ are real functions, is given by the product of their Fourier transformations:

$$\mathscr{F}[f * g] = \mathscr{F}[f] \times \mathscr{F}[g] = F(\omega)G(\omega). \tag{2.16}$$

   Analogously, the Fourier transformation of the multiplication of two functions, $f(x)g(x)$, is given by the convolution between the corresponding two Fourier transformations:

$$\mathscr{F}[fg] = F(\omega) * G(\omega), \tag{2.17}$$

   where $F(\omega)$ and $G(\omega)$ are the Fourier transformations of $f(x)$ and $g(x)$, respectively.
3. The Fourier transformation of a shifted function is given by rotating the phase of the Fourier transformation of the original function:

$$\mathscr{F}[f(x - x_0)] = e^{-j\omega x_0}F(\omega). \tag{2.18}$$

4. The real part of the Fourier transformation corresponds to the symmetric component of an input function, and the imaginary one corresponds to the antisymmetric component of the function. A real function is called symmetric if $f(x) = f(-x)$ and is called antisymmetric if $f(x) = -f(-x)$.

Let the real part and the imaginary part of the Fourier transformation, $F(\omega)$, be denoted by $\mathrm{Re}[F(\omega)]$ and $\mathrm{Im}[F(\omega)]$, respectively, where

$$\begin{aligned}\mathrm{Re}[F(\omega)] &= \int f(x)\cos(\omega x)dx,\\ \mathrm{Im}[F(\omega)] &= -j\int f(x)\sin(\omega x)dx,\\ F(\omega) &= \mathrm{Re}[F(\omega)] + \mathrm{Im}[F(\omega)].\end{aligned} \tag{2.19}$$

It should be noted that $\cos(\omega x)$ is symmetric and $\sin(\omega x)$ is antisymmetric and that the inverse Fourier transformations of $\mathrm{Re}[F(\omega)]$ and of $\mathrm{Im}[F(\omega)]$ generate a symmetric real function and an antisymmetric real one, respectively. Let

$$\begin{aligned}f_{\mathrm{symm}}(x) &= \mathscr{F}^{-1}[\mathrm{Re}[F(\omega)]],\\ f_{\mathrm{anti}}(x) &= \mathscr{F}^{-1}[\mathrm{Im}[F(\omega)]].\end{aligned} \tag{2.20}$$

Then $f_{\mathrm{symm}}(x)$ is the symmetric real function and $f_{\mathrm{anti}}(x)$ is the antisymmetric real function. Following Eq. (2.19) allows a unique decomposition of a target real function, $f(x)$, into its symmetric and antisymmetric components as follows:

$$f(x) = f_{\mathrm{symm}}(x) + f_{\mathrm{anti}}(x). \tag{2.21}$$

The discrete Fourier transformation (DFT) is used when a given target function is discrete and its domain is finite. Let $\boldsymbol{u} = (u_1, u_2, \ldots, u_D)^T$ denote a D-vector where all components are integers and are bounded as $0 \le u_s \le W-1$ $(s = 1, 2, \ldots, D)$, and let a target real function defined over $\boldsymbol{u}$ be denoted by $f(\boldsymbol{u})$, where $f(\boldsymbol{u})$: $\mathbb{Z}^D \mapsto \mathbb{R}$. Then, the DFT of $f(\boldsymbol{u})$ is defined as follows:

$$\mathscr{F}[f] = F(\boldsymbol{n}) = \sum_{u_1=0}^{W-1}\sum_{u_2=0}^{W-1}\cdots\sum_{u_D=0}^{W-1} f(\boldsymbol{u})\exp\left(-j\frac{2\pi\boldsymbol{u}^T\boldsymbol{n}}{W}\right). \tag{2.22}$$

Here, $\boldsymbol{n} = (n_1, n_2, \ldots, n_D)^T$ is a D-vector of which all components are integer, where $n_s = 0, 1, \ldots, W-1$, and the frequency along the s-th axis, ω_s, is proportional to n_s and is given as $\omega_s = n_s/W$ voxel^{-1}. The inverse DFT (IDFT) can reconstruct the input signal from $F(\boldsymbol{n})$ as follows:

$$\mathscr{F}^{-1}[F(\boldsymbol{n})] = f(\boldsymbol{u}) = \frac{1}{W^D}\sum_{n_1=0}^{W-1}\sum_{n_2=0}^{W-1}\cdots\sum_{n_D=0}^{W-1} F(\boldsymbol{n})\exp\left(j\frac{2\pi\boldsymbol{u}^T\boldsymbol{n}}{W}\right). \tag{2.23}$$

As shown in Eq. (2.22), the DFT is an inner product between a given discrete function and a discretized complex sinusoidal function. For simplicity, assume that

$D = 1$. Then the DFT is written as

$$\mathscr{F}[f] = F(n) = \sum_{u=0}^{W-1} f(u) \exp\left(-j\frac{2\pi un}{W}\right), \tag{2.24}$$

where $u = 0, 1, \ldots, W - 1$ and $n = 0, 1, \ldots, W - 1$. Let a W-vector, $\boldsymbol{f}$, denote the input function, where $\boldsymbol{f} = [f(0), f(1), \ldots, f(W-1)]^T$, and let a W-vector, $\boldsymbol{c}$, denote the discretized sinusoidal function, where $\boldsymbol{c} = [e^{-2\pi jn0/W}, e^{-2\pi jn1/W}, \ldots, e^{-2\pi jn(W-1)/W}]^T$. Then $\mathscr{F}[f] = F(n) = \boldsymbol{f} \cdot \boldsymbol{c} = \boldsymbol{f}^T\boldsymbol{c}$.

2.2.1.6 Differentiation of Discrete Signals

A finite difference operator is applied to a given discrete function to compute approximately the differential coefficients of a continuous function represented by the discrete given function. Let a continuous and smooth function be denoted by $f(x)$ where $x \in \mathbb{R}$, and let $f_n = f(n\Delta)$ $(n \in \mathbb{Z})$ denote a discrete function, where $\Delta > 0$ is a sampling interval.

The Taylor expansion of $f(x)$ is given as follows:

$$f(\boldsymbol{x} + \epsilon) = f(\boldsymbol{x}) + \frac{\partial f}{\partial x}\epsilon + O(\epsilon^2), \tag{2.25}$$

where the positive real number, ϵ, denotes some small perturbation.

$$\frac{\partial f}{\partial x} = \frac{f(x + \epsilon) - f(\boldsymbol{x})}{\epsilon} + O(\epsilon). \tag{2.26}$$

Assume that $x = n\Delta$. When $\epsilon = \Delta$, the following difference operation is obtained:

$$\frac{\partial f}{\partial x}(x) \simeq \frac{f(x + \Delta) - f(x)}{\Delta} = f_{n+1} - f_n. \tag{2.27}$$

Let a discrete filter be denoted by g_n such that

$$g_n = \begin{cases} +1, \; n = -1, \\ -1, \; n = 0, \\ \;\;\, 0, \text{ otherwise.} \end{cases} \tag{2.28}$$

Then, Eq. (2.27) is a convolution between f_n and g_n:

$$\frac{\partial f(x)}{\partial x} \simeq f_n * g_n. \tag{2.29}$$

A different operation for the differentiation can be obtained as follows:

$$f(\boldsymbol{x}+\epsilon) = f(\boldsymbol{x}) + \frac{\partial f}{\partial x}\epsilon + \frac{1}{2}\frac{\partial^2 f}{\partial x^2}\epsilon^2 + O(\epsilon^3), \tag{2.30}$$

$$f(\boldsymbol{x}-\epsilon) = f(\boldsymbol{x}) - \frac{\partial f}{\partial x}\epsilon + \frac{1}{2}\frac{\partial^2 f}{\partial x^2}\epsilon^2 + O(\epsilon^3). \tag{2.31}$$

Computing Eq. (2.30) minus Eq. (2.31) obtains

$$\frac{\partial f}{\partial x} = \frac{f(\boldsymbol{x}+\epsilon) - f(\boldsymbol{x}-\epsilon)}{2\epsilon} + O(\epsilon^2). \tag{2.32}$$

Substituting $\epsilon = \Delta$ obtains

$$\frac{\partial f}{\partial x}(X) \simeq \frac{f_{n+1} - f_{n-1}}{2}. \tag{2.33}$$

The above equation can be rewritten as

$$\frac{\partial f(x)}{\partial x} \simeq f_n * g'_n, \tag{2.34}$$

where

$$g'_n = \begin{cases} +1/2, \; n = -1, \\ -1/2, \; n = +1, \\ \quad 0, \; \text{otherwise}. \end{cases} \tag{2.35}$$

Adding Eqs. (2.30)–(2.31) obtains an operator for approximating the second derivative:

$$\frac{d^2 f}{dx^2}(X) \simeq f_{n-1} - f_0 + f_{n+1}. \tag{2.36}$$

As shown in (2.1), an image, $I(\boldsymbol{x})$, is a discrete signal obtained by measuring a spatial distribution of some physical quantity, $f(X)$, at the lattice points. The spatial differential coefficients of $f(X)$ can be computed approximately by convolving the given image, $I(x)$, with a filter, g, for computing the finite differentiation. The finite differentiation of an image is used for approximately computing differentials, e.g., the gradient at each location, and is one of the most important operations in medical image analysis. The accuracy of the approximation of the finite differentiation varies depending on the coefficients of the filter. It is known, for example, that one can approximate the direction of the computed gradient vector more accurately by applying a *consistent filter* [31].

2.2.2 *Fundamental Transformations*

2.2.2.1 Coordinate Transformation

Let $\boldsymbol{x} = [x_1, x_2, \ldots, x_D]^T$ and $\boldsymbol{y} = [y_1, y_2, \ldots, y_D]^T$ denote two D-vectors. Their inner product is defined as

$$\boldsymbol{x} \cdot \boldsymbol{y} = \boldsymbol{x}^T \boldsymbol{y} = \sum_{i=1}^{D} x_i y_i. \tag{2.37}$$

Let a norm of $\boldsymbol{x}$ be defined as

$$\|\boldsymbol{x}\| = \sqrt{\sum_{i=1}^{D} x_i^2}, \tag{2.38}$$

and the angle between the directions of $\boldsymbol{x}$ and $\boldsymbol{y}$ be denoted by θ. Then, $\boldsymbol{x}^T \boldsymbol{y} = \|\boldsymbol{x}\| \|\boldsymbol{y}\| \cos\theta$. Two vectors, $\boldsymbol{x}$ and $\boldsymbol{y}$, are said to be orthogonal if $\theta = \pi/2$. The inner product of $\boldsymbol{x}^T \boldsymbol{y}$ is equal to zero if $\boldsymbol{x}$ and $\boldsymbol{y}$ are orthogonal. If $\|\boldsymbol{x}\| = 1$, then $\boldsymbol{x}$ is said to be a unit vector.

A basis of a D-dimensional space is a set of D vectors of which linear combination can represent any D-vector in the space. Let a basis of D-dimensional space be denoted by $\{\boldsymbol{u}_1, \boldsymbol{u}_2, \ldots, \boldsymbol{u}_D\}$, and let the origin of the space and a point other than the origin be denoted by O and P, respectively. Let a D-vector, $\boldsymbol{p}$, denote the location of the point, P, where $\boldsymbol{p} = \boldsymbol{OP}$, and assume that the D-vector, $\boldsymbol{p}$, is represented by a linear combination such that

$$\boldsymbol{p} = x_1 \boldsymbol{u}_1 + x_2 \boldsymbol{u}_2 + \ldots x_D \boldsymbol{u}_D. \tag{2.39}$$

Then, $[x_1, x_2, \ldots, x_D]^T$ is said to be the coordinates of $\boldsymbol{x}$ under the employed basis. Let a $D \times D$ vector that is constructed from the D basis vectors be denoted by U, where

$$U = \left[\begin{array}{c|c|c|c|c} \boldsymbol{u}_1 & \boldsymbol{u}_2 & \boldsymbol{u}_3 & \ldots & \boldsymbol{u}_D \end{array} \right], \tag{2.40}$$

and let $\boldsymbol{x}$ denote a D-vector such that $\boldsymbol{x} = [x_1, x_2, \ldots, x_D]^T$. Then, (2.39) can be represented as follows:

$$\boldsymbol{p} = U\boldsymbol{x}. \tag{2.41}$$

An identical point in the space can be represented by different coordinates if the basis changes. Let a changed basis be denoted by $\{\boldsymbol{v}_1, \boldsymbol{v}_2, \ldots, \boldsymbol{v}_D\}$ and let $\boldsymbol{p} = y_1\boldsymbol{v}_1 + y_2\boldsymbol{v}_2 + \ldots y_D\boldsymbol{v}_D$. Then,

$$\boldsymbol{p} = V\boldsymbol{y}, \tag{2.42}$$

where

$$V = \begin{bmatrix} \left.\boldsymbol{v}_1\middle|\boldsymbol{v}_2\middle|\boldsymbol{v}_3\middle|\ldots\middle|\boldsymbol{v}_D\right. \end{bmatrix}, \tag{2.43}$$

and $\boldsymbol{y} = [y_1, y_2, \ldots, y_D]^T$. The change of the coordinates with respect to the change of the basis can be computed from the equation, $\boldsymbol{p} = U\boldsymbol{x} = V\boldsymbol{y}$, as follows:

$$\boldsymbol{y} = V^{-1}U\boldsymbol{x} = V^{-1}\boldsymbol{p}. \tag{2.44}$$

It is said to be an orthonormal basis if every basis vector has a unit length and any two of the basis vectors are orthogonal. Assume that the basis, $\{\boldsymbol{v}_1, \boldsymbol{v}_2, \ldots, \boldsymbol{v}_D\}$, is orthonormal. Then the corresponding matrix, V, is orthonormal and $V^TV = VV^T = I$, where I is a unit matrix, and $\boldsymbol{y} = V^TU\boldsymbol{x} = V^T\boldsymbol{p}$. This means the y-th component of the coordinates, y_i, is equal to the inner product of the i-th basis vector and the target vector.

2.2.2.2 Linear Subspace

Let S denote a set of D-vectors, $S = \{\boldsymbol{u}_1, \boldsymbol{u}_2, \ldots, \boldsymbol{u}_M\}$, where $\boldsymbol{u}_i$ $(i = 1, 2, \ldots, M)$ are D-vectors and M is the number of the vectors in S. A span of S is a set of points that is represented by a linear combination of the vectors included in S as follows:

$$\text{span}(S) = \{a_1\boldsymbol{u}_1 + a_2\boldsymbol{u}_2 + \ldots a_M\boldsymbol{u}_M | a_i \in R, \boldsymbol{u}_i \in S\}. \tag{2.45}$$

A span of a basis is the whole space. Let a subset of a basis be denoted by S_{sub}. Then $\text{span}(S_{\text{sub}})$ is a linear subspace, and its dimension is determined by the number of vectors included in the subset, S_{sub}.

2.2.2.3 Affine Transformation

A D-dimensional affine transformation is a combination of a linear mapping and a translation and transfers a target D-vector, $\boldsymbol{x}$, to a D-vector, $\boldsymbol{y}$ as follows:

$$\boldsymbol{y} = A\boldsymbol{x} + \boldsymbol{t}, \tag{2.46}$$

where A is a $D \times D$ matrix and $\boldsymbol{t}$ is a D-vector. The former matrix, A, represents the linear mapping of a target and the latter vector, t, represents the translation.

A linear mapping, $\boldsymbol{y} = \mathcal{O}[\boldsymbol{x}]$, satisfies the following conditions:

$$\mathcal{O}[\boldsymbol{x}_1 + \boldsymbol{x}_2] = \mathcal{O}[\boldsymbol{x}_1] + \mathcal{O}[\boldsymbol{x}_2], \tag{2.47}$$

$$\mathcal{O}[a\boldsymbol{x}] = a\mathcal{O}[\boldsymbol{x}], \tag{2.48}$$

where $\boldsymbol{x}_1$ and $\boldsymbol{x}_2$ are D-vectors and a is a scalar. When the above conditions are satisfied, then the following condition is also satisfied:

$$\mathcal{O}[a_1\boldsymbol{u}_1 + a_2\boldsymbol{u}_2 + \cdots + a_M\boldsymbol{u}_M] = a_1\mathcal{O}[\boldsymbol{u}_1] + a_2\mathcal{O}[\boldsymbol{u}_2] + \ldots a_M\mathcal{O}[\boldsymbol{u}_M]. \tag{2.49}$$

Let $S = \{\boldsymbol{u}_1, \boldsymbol{u}_2, \ldots, \boldsymbol{u}_M\}$ and let $S_{\mathcal{O}} = \{\mathcal{O}[\boldsymbol{u}_1], \mathcal{O}[\boldsymbol{u}_2], \ldots, \mathcal{O}[\boldsymbol{u}_M]\}$. Then, linear mapping maps the linear subspace spanned, span(S), to the other linear subspace, span($S_{\mathcal{O}}$). A mapping represented as $A\boldsymbol{x}$ is a linear one: $\mathcal{O}[\boldsymbol{x}] = A\boldsymbol{x}$.

Let $\boldsymbol{e}_i = [0, \ldots, 0, 1, 0, \ldots, 0]^T$, where only the i-th component is one and the other ones are equal to zero, and let $\boldsymbol{u}_i = A\boldsymbol{e}_i$. Then,

$$U = AI, \tag{2.50}$$

where

$$I = \left[\begin{array}{c|c|c|c|c} \boldsymbol{e}_1 & \boldsymbol{e}_2 & \boldsymbol{e}_3 & \ldots & \boldsymbol{e}_D \end{array}\right], \text{ and } U = \left[\begin{array}{c|c|c|c|c} \boldsymbol{u}_1 & \boldsymbol{u}_2 & \boldsymbol{u}_3 & \ldots & \boldsymbol{u}_D \end{array}\right]. \tag{2.51}$$

Equation (2.50) shows that $A = U$ and that the i-th column of the matrix A is identical with $\boldsymbol{u}_i = A\boldsymbol{e}_i$. Linear mappings $A\boldsymbol{x}$ preserve the shapes and sizes of targets if and only if the mapped basis, $\{A\boldsymbol{e}_1, A\boldsymbol{e}_2, \ldots, A\boldsymbol{e}_D\}$, is also an orthonormal one and such that the mapping consists of rotation mapping and reflection mapping. The mapping reflects targets when det(A) < 0 and rotates when det(A) > 0. The features of linear mapping can be described more clearly using a singular value decomposition of A, which will be explained below.

2.2.2.4 Singular Value Decomposition

A singular value decomposition (SVD) is powerful for analyzing matrices and for analyzing linear mappings. Applying the SVD transforms an $M \times N$ real matrix $\boldsymbol{A}$ into the following form:

$$A = U\Sigma V^T, \tag{2.52}$$

where U is a $M \times N$ matrix of which column vectors are unit vectors and are orthogonal to each other and V is an $N \times N$ orthonormal matrix such that

$$U = \begin{bmatrix} \\ \boldsymbol{u}_1\ \boldsymbol{u}_2\ \boldsymbol{u}_3\ \dots\ \boldsymbol{u}_N \\ \\ \end{bmatrix}, \text{ and } V = \begin{bmatrix} \\ \boldsymbol{v}_1\ \boldsymbol{v}_2\ \boldsymbol{v}_3\ \dots\ \boldsymbol{v}_N \\ \\ \end{bmatrix}. \tag{2.53}$$

Here, $\boldsymbol{u}_i$ $(i = 1, 2, \dots, M)$ is a unit M-vector and $\boldsymbol{u}_i^T \boldsymbol{u}_j = 0$ if $i \neq j$, and $\boldsymbol{v}_i$ $(i = 1, 2, \dots, M)$ is a unit N-vector and $\boldsymbol{v}_i^T \boldsymbol{v}_j = 0$ if $i \neq j$. It should be noted that $\{\boldsymbol{v}_i | i = 1, 2, \dots, N\}$ is an orthonormal basis of the N-dimensional space. Σ in (2.52) is an $N \times N$ diagonal matrix,

$$\Sigma = \begin{bmatrix} \sigma_1 & & & \\ & \sigma_2 & & \\ & & \ddots & \\ & & & \sigma_N \end{bmatrix}, \tag{2.54}$$

where the scalars, σ_i $(i = 1, 2, \dots, N)$, are called *singular values*. In the following equations, it is assumed that the singular values are in decreasing order: $\sigma_1 \geq \sigma_2 \geq \dots \geq \sigma_N$.

Let $\boldsymbol{y}$ denote an M-vector generated from an N-vector by a linear mapping such that

$$\boldsymbol{y} = A\boldsymbol{x}, \tag{2.55}$$

where A is a $M \times N$ matrix. Applying the SVD to A, the range of the linear mapping and of its zero-space can be derived, as will be described below. Substituting $A = U\Sigma V^T$ produces the following equation:

$$\boldsymbol{y} = U\Sigma V^T \boldsymbol{x}. \tag{2.56}$$

The last two factors, $V^T\boldsymbol{x}$, are transforming the coordinates of $\boldsymbol{x}$ using the orthonormal basis $\{\boldsymbol{v}_i | i = 1, 2, \dots, N\}$, as described in the Sect. 2.2.2.1. Let the new coordinates be denoted by a N-vector, $\boldsymbol{z} = [z_1, z_2, \dots, z_N]^T = V^T\boldsymbol{x}$. Now, the linear mapping shown in (2.56) is represented as

$$\boldsymbol{y} = U\Sigma\boldsymbol{z}. \tag{2.57}$$

Substituting $\Sigma = \mathrm{diag}(\sigma_1, \sigma_2, \dots, \sigma_N)$ and $U = [\boldsymbol{u}_1 | \boldsymbol{u}_2 | \dots | \boldsymbol{u}_N]$, (2.57) can be rewritten as follows:

$$\boldsymbol{y} = (\sigma_1 z_1)\boldsymbol{u}_1 + (\sigma_2 z_2)\boldsymbol{u}_2 + \dots + (\sigma_N z_N)\boldsymbol{u}_N. \tag{2.58}$$

Let a set of indexes that indicates the singular values of zeros be denoted by $\mathscr{Z} = \{j | j \in \{1, 2, \ldots, N\}, \sigma_j = 0\}$, and let a set of indexes for nonzero singular values be denoted by $\mathscr{N} = \{j | j \in \{1, 2, \ldots, N\}, \sigma_j \neq 0\}$, where $\mathscr{Z} \cup \mathscr{N} = \{1, 2, \ldots, N\}$ and $\mathscr{Z} \cap \mathscr{N} = \phi$. Then, the M-vectors, $\boldsymbol{u}_j$, that correspond to the nonzero singular values, $j \in \mathscr{N}$, determine the domain of the linear map shown in (2.58): for any input vector, $\boldsymbol{x}$, the mapped vector, $\boldsymbol{y}$, that is in the subspace spanned by $\{\boldsymbol{u}_j | j \in \mathscr{N}\}$. The N-vectors, $\boldsymbol{v}_k$, that correspond to the zero-singular values, $k \in \mathscr{Z}$, determine the zero-space of A: the mapped vector, $\boldsymbol{y}$, is always zero if the input vector, $\boldsymbol{x}$, is in a subspace that is spanned by the N-vectors that correspond to the nonzero singular values, that is, $\boldsymbol{x} \in \mathrm{span}(V_{\mathscr{Z}})$ where $V_{\mathscr{Z}} = \{\boldsymbol{v}_s | s \in \mathscr{Z}\}$. This is because, in (2.58), $\sigma_i = 0$ for $i \in \mathscr{Z}$ and $z_i = \boldsymbol{v}_i^T \boldsymbol{x} = 0$ for $i \in \mathscr{N}$.

2.2.2.5 Principal Component Analysis

Principal component analysis (PCA) is widely used for constructing a subspace that approximates a distribution of training data and has a strong relationship with the SVD mentioned above. Let a set of training data be denoted by $\mathscr{D} = \{\boldsymbol{x}_1, \boldsymbol{x}_2, \ldots, \boldsymbol{x}_M\}$ where $\boldsymbol{x}_i$ $(i = 1, 2, \ldots, M)$ are D-vectors. Let $\boldsymbol{x}$ denote the mean of $\boldsymbol{x}_i$ $(i = 1, 2, \ldots, M)$ and let an $D \times D$ covariance matrix be denoted by Σ_{emp}, where

$$\Sigma_{\mathrm{cov}} = \frac{1}{M} \sum_{i=1}^{M} (\boldsymbol{x}_i - \hat{\boldsymbol{x}})(\boldsymbol{x} - \hat{\boldsymbol{x}})^T. \tag{2.59}$$

It is clear that Σ_{cov} is a symmetric real matrix. In many cases, the intrinsic dimension of the distribution of the training data is lower than the dimension of the datum, D, and the PCA is used to obtain a low-dimensional subset that approximates the distribution.

In the PCA, a set of D pairs of an eigenvalues and an eigenvector is computed from the covariance matrix of the training data, Σ_{cov}. Let $\lambda_i \in R$ $(i = 1, 2, \ldots, D)$ denote the eigenvalues and let $\boldsymbol{w}_i$ $(i = 1, 2, \ldots, D)$ denote the eigenvectors. It is known that, given a symmetric real matrix, Σ_{cov}, there exist D pairs of $(\lambda_i, \boldsymbol{w}_i)$ that satisfy the following equation:

$$\Sigma_{\mathrm{cov}} \boldsymbol{w}_i = \lambda_i \boldsymbol{w}_i, \tag{2.60}$$

and that a set of the eigenvectors, $\{\boldsymbol{w}_i | i = 1, 2, \ldots, D\}$, is an orthonormal basis of a D dimensional space: $\|\boldsymbol{w}_i\| = 1$ and $\boldsymbol{w}_i^T \boldsymbol{w}_j = 0$ if $i \neq j$. Assume that the eigenvalues are in decreasing order, $\lambda_1 \geq \lambda_2 \geq \cdots \geq \lambda_D$, and let a set of the K largest $(K \leq D)$ eigenvalues be denoted by $\mathscr{W}_K = \{\boldsymbol{w}_i | 1 \leq i \leq K\}$.

When approximating the training data using a subspace, Π, every training datum, $\boldsymbol{x}_i$, is projected to the subspace, and the original data are approximated the projected data. Let $\tilde{\boldsymbol{x}}_i$ denote the projected data, and let the approximation error, $E(\Pi)$, be defined as follows:

$$E(\Pi) = \sum_{i=1}^{M} \|\boldsymbol{x}_i - \tilde{\boldsymbol{x}}_i\|^2. \tag{2.61}$$

Then, under a condition that the dimension of a subspace is fixed to K, the subspace that minimizes $E(\Pi)$ is span($\mathscr{W}_K$), and the unit vectors, $\boldsymbol{w}_i$ ($i = 1, 2, \ldots, K$) that span the subspace are called the *principal components* of the distribution of the training data. The projected data, $\tilde{\boldsymbol{x}}_i$, can be represented as

$$\tilde{\boldsymbol{x}}_i = z_1\boldsymbol{w}_1 + z_2\boldsymbol{w}_w + \cdots + z_K\boldsymbol{w}_K, \tag{2.62}$$

where $\boldsymbol{z}_i = [z_1, z_2, \ldots, z_K]^T$ is the coordinate of $\tilde{\boldsymbol{x}}_i$ described based on the basis, $S_K = \{\boldsymbol{w}_1, \boldsymbol{w}_2, \ldots, \boldsymbol{w}_K\}$, and can be computed as follows:

$$\boldsymbol{z}_i = W_K^T\boldsymbol{x}_i, \tag{2.63}$$

where W_K is $D \times K$ matrix such that

$$W_K = \begin{bmatrix} \Bigg| & \Bigg| & & \Bigg| \\ \boldsymbol{w}_1 & \boldsymbol{w}_2 & \ldots & \boldsymbol{w}_K \\ \Bigg| & \Bigg| & & \Bigg| \end{bmatrix}. \tag{2.64}$$

Substituting (2.63) to (2.62) results in

$$\tilde{\boldsymbol{x}}_i = W_K\boldsymbol{z}_i = W_K W_K^T\boldsymbol{x}_i. \tag{2.65}$$

When the training data follow a Gaussian distribution, then each eigenvalue indicates the variance of the distribution along the corresponding eigenvector. Let a D-dimensional Gaussian distribution be denoted by $\mathscr{N}(\boldsymbol{x}|\bar{\boldsymbol{x}}, \Sigma_{\text{cov}})$, where the D-vector, $\bar{\boldsymbol{x}}$, denotes the mean and the $D \times D$ matrix, Σ_{cov}, denotes the covariance.

$$\mathscr{N}(\boldsymbol{x}|\bar{\boldsymbol{x}}, \Sigma_{\text{cov}}) = \frac{1}{(2\pi)^{D/2}(\det \Sigma_{\text{cov}})^{1/2}} \exp\left\{-\frac{1}{2}(\boldsymbol{x}-\bar{\boldsymbol{x}})^T \Sigma_{\text{cov}}^{-1}(\boldsymbol{x}-\bar{\boldsymbol{x}})\right\} \tag{2.66}$$

From (2.60), the following equation about the covariance matrix holds:

$$\Sigma_{\text{cov}} W = W\Lambda, \tag{2.67}$$

where

$$\Lambda = \begin{bmatrix} \lambda_1 & & & \\ & \lambda_2 & & \\ & & \ddots & \\ & & & \lambda_D \end{bmatrix}, \tag{2.68}$$

and

$$W = W_D = \left[\boldsymbol{w}_1 \middle| \boldsymbol{w}_2 \middle| \cdots \middle| \boldsymbol{w}_D \right]. \tag{2.69}$$

Multiplying both sides by $W^T = W^{-1}$ from the right, one obtains

$$\Sigma_{\text{cov}} = W \Lambda W^T, \tag{2.70}$$

and

$$\Sigma_{\text{cov}}^{-1} = W^T \Lambda^{-1} W, \tag{2.71}$$

where

$$\Lambda^{-1} = \text{diag}(1/\lambda_1, 1/\lambda_2, \ldots, 1/\lambda_D). \tag{2.72}$$

Substituting (2.71) to (2.66) results in

$$\mathcal{N}(\boldsymbol{x}|\bar{\boldsymbol{x}}, \Sigma_{\text{cov}}) = \frac{1}{(2\pi)^{D/2}(\det \Sigma_{\text{cov}})^{1/2}} \exp\left\{ -\frac{1}{2}(\boldsymbol{x} - \bar{\boldsymbol{x}})^T W^T \Lambda^{-1} W (\boldsymbol{x} - \bar{\boldsymbol{x}}) \right\}. \tag{2.73}$$

Here, let $\boldsymbol{y} = [y_1, y_2, \ldots, y_D]^T = W\boldsymbol{x}$. Because $\{\boldsymbol{w}_i | i = 1, 2, \ldots, D\}$ is an orthonormal basis, W is a rotation matrix or a reflection matrix. Substituting $\boldsymbol{y} = W\boldsymbol{x}$ into Eq. (2.73) results in

$$\mathcal{N}(\boldsymbol{y}|\bar{\boldsymbol{y}}, \Lambda) = \frac{1}{(2\pi)^{D/2}(\prod_i \lambda_i)^{1/2}} \exp\left\{ -\frac{1}{2}(\boldsymbol{y} - \bar{\boldsymbol{y}})^T \Lambda^{-1} (\boldsymbol{y} - \bar{\boldsymbol{y}}) \right\}, \tag{2.74}$$

where $\bar{\boldsymbol{y}} = W\bar{\boldsymbol{x}}$. Because Λ^{-1} is a diagonal matrix, the result is

$$\mathcal{N}(\boldsymbol{y}|\bar{\boldsymbol{y}}, \Lambda) = \prod_{i=1}^{D} \frac{1}{(2\pi\lambda_i)^{1/2}} \exp\left\{ -\frac{1}{2}\frac{(y_i - \bar{y}_i)^2}{\lambda_i} \right\} = \prod_{i=1}^{D} \mathcal{N}(y_i|\bar{y}_i, \lambda_i). \tag{2.75}$$

The D-dimensional Gaussian distribution is a product of D one-dimensional Gaussian distributions of y_i, of which the mean is $\bar{y}_i$ and the variance is λ_i. Here, it should be remembered that $\boldsymbol{y} = W\boldsymbol{x}$ and that $\boldsymbol{y}$ is the coordinates of $\boldsymbol{x}$ represented using the basis, $\{\boldsymbol{w}_i | i = 1, 2, \ldots, D\}$. The one-dimensional Gaussian distribution, $\mathcal{N}(y_i|\bar{y}_i, \lambda_i)$, represents the distribution along the eigenvector, $\boldsymbol{w}_i$, and the corresponding eigenvalue, λ_i, represents the variance along the direction. The principal components that span the subspace for the data approximation correspond to the eigenvectors along which the distribution has larger variances.

Comparing the SVD of a matrix shown in (2.56) with (2.70), it can be seen that U and V in (2.56) are identical with W in (2.70) and the singular values are identical with the eigenvalues, $\sigma_i = \lambda_i$.

2.2.3 *Probability and Statistics: Foundations of CA*

Probability and statistics are essential to CA, in which targets are represented and, in given images, are described by their statistical models. A framework of probability theory is employed for constructing the statistical models and for describing the targets in given images. In this subsection, some basics of the probability theory are described.

2.2.3.1 Sum Rule and Product Rule of Probability

First, discrete probability is described. In this case, the random variable X takes discrete values such as $x_1, x_2, \ldots, x_n$. If the frequency in the case that X takes x_i is c_i, the probability that X takes x_i is represented as

$$p(X = x_i) = \frac{c_i}{N}, \tag{2.76}$$

where $N = \sum_{i=1}^{n} c_i$. When there is another random variable Y that represents another aspect of the event mentioned above, Y takes values of $y_1, y_2, \ldots, y_m$. The frequency in the case that Y takes y_j is d_j , and the frequency in the case that X takes xi and Y takes yj simultaneously is r_{ij}. The joint probability is described as

$$p(X = x_i, Y = y_j) = \frac{r_{ij}}{N} \tag{2.77}$$

and

$$c_i = \sum_{j=1}^{m} r_{ij}. \tag{2.78}$$

Then, the following equation is derived from Eqs. (2.76), (2.77), and (2.78):

$$p(X = x_i) = \sum_{j=1}^{m} p(X = x_i, Y = y_j). \tag{2.79}$$

This equation is called the *sum rule of probability*. Because the left side of Eq. (2.79) means marginalization in terms of the random variable Y, it is called a *marginalprobability*. Assuming that X is fixed on x_i, the ratio of frequency of the case of $Y = y_j$ to the cases of all Y is described as $p(Y = y_j|X = x_i)$. This is called a *conditional probability*, because this is the probability of $Y = y_j$ under the condition of $X = x_i$. This conditional probability is calculated as

$$p(Y = y_j|X = x_j) = \frac{r_{ij}}{c_i}. \tag{2.80}$$

By substituting Eqs. (2.76) and (2.80) into Eq. (2.77), the equation is transformed as

$$p(X = x_i, Y = y_j) = \frac{r_{ij}}{N} = \frac{r_{ij}}{c_l}\frac{c_i}{N} = p(Y = y_j|X = x_i)p(X = x_i). \tag{2.81}$$

This equation is called the *product rule of probability*. The probabilistic distribution of the random variable X is denoted as $p(X)$, and the probability in the case that X takes a specific value is denoted as $p(x_i)$. By using these notations, the sum rule and the product rule of probability are written thus: The sum rule of probability is represented as

$$p(X) = \sum_{Y} p(X, Y), \tag{2.82}$$

and the product rule of probability is represented as

$$p(X, Y) = p(Y|X)p(X). \tag{2.83}$$

Because $p(X, Y)$ is symmetrical with respect to the random variables X and Y, $p(X, Y) = p(Y, X)$ or $p(Y|X)p(X) = p(X|Y)p(Y)$. By transforming this equation, the relationship between two conditional probabilities is derived:

$$p(Y|X) = \frac{p(X|Y)p(Y)}{p(X)}. \tag{2.84}$$

This relationship is referred to as *Bayes' theorem*. By substituting Eq. (2.83) into Eq. (2.82) after swapping X with Y on the right side of Eq. (2.83), the denominator of the right side of Eq. (2.84) is written as

$$p(X) = \sum_{Y} p(X|Y)p(Y). \tag{2.85}$$

Assuming the necessity of finding the probabilities of the random variable Y in Eq. (2.84), $p(Y)$ and $p(Y|X)$ are referred to as a *prior probability distribution* (or simply, a "prior") and a *posterior probability*. These probabilities are so named because the former is a known probability before the actual value of the random variable X is known, and the latter is known after the actual value of the random variable X is known. When the joint probability $p(X, Y)$ is equal to a product of $p(X)$ and $p(Y)$, i.e., $p(X, Y) = p(X)p(Y)$ holds, the random variables X and Y are independent. In this case Eq. (2.84) (the product rule of probability) is transformed thus:

$$p(X)p(Y) = p(Y|X)p(X), \tag{2.86}$$

and hence $p(Y) = p(Y|X)$. This means that the probability of Y is unaffected by X, and vice versa. When the random variables have continuous value, Bayes' theorem is described thus:

$$p(\theta|x) = \frac{f(x|\theta)q(\theta)}{\int f(x|\theta)q(\theta)d\theta}, \tag{2.87}$$

where x and θ are random variables, p is a posterior probability density function, q is a prior probability density function of θ, and f is a likelihood function. The denominator of the right-hand side of Eq. (2.87) is a marginal probability density function.

2.2.3.2 Expectation and Variance

One of the most basic statistics of the distribution is an expectation. When x is a random variable, $f(x)$ is a function of x, and $p(x)$ is a probabilistic discrete distribution of x, the expectation of the value of $f(x)$ is defined as

$$E[f] = \sum p(x)f(x). \tag{2.88}$$

When $p(x)$ is continuous, the expectation is defined as

$$E[f] = \int p(x)f(x)dx. \tag{2.89}$$

In this case, $p(x)$ is a probability density function. If a limited number of samples are used, the expectation is approximated as

$$E[f] \simeq \frac{1}{N}\sum_{n=1}^{N} f(x_n). \tag{2.90}$$

In the case that the probabilistic distribution is conditional, the expectation also becomes conditional, and it is referred to as a *conditional expectation*:

$$E_x[f|y] = \sum_x p(x|y)f(x) \quad \text{(discrete case)}, \tag{2.91}$$

$$E_x[f|y] = \int p(x|y)f(x)dx \quad \text{(continuous case)}. \tag{2.92}$$

The suffix means that the expectation or the summation are calculated with respect to x.

Another of the most basic statistics of the distribution is *variance*. It is defined by the following equation:

$$\text{var}[f] = E[(f(x) - E[f(x)])^2] = E[(f(x))^2] - (E[f(x)])^2 \tag{2.93}$$

The square root of variance is called the *standard deviation*. Assuming that there are two random variables x and y, the covariance of x and y is calculated as

$$\text{cov}[x, y] = E_{x,y}[(x - E[x])(y - E[y])] = E_{x,y}[xy] - E[x]E[y]. \tag{2.94}$$

The value, $\text{cov}[x, y]$, evaluates how x and y are statistically dependent together. For two random variable vectors $\boldsymbol{x}$ and $\boldsymbol{y}$ instead of random variables, a covariance matrix is defined by

$$\text{cov}[\boldsymbol{x}, \boldsymbol{y}] = E_{\boldsymbol{x},\boldsymbol{y}}[(\boldsymbol{x} - E[\boldsymbol{x}])(\boldsymbol{y}^T - E[\boldsymbol{y}^T])] = E_{\boldsymbol{x},\boldsymbol{y}}[\boldsymbol{x}\boldsymbol{y}^T] - E[\boldsymbol{x}]E[\boldsymbol{y}^T] \tag{2.95}$$

The covariance between the components in $\boldsymbol{x}$ is calculated as $\text{cov}[\boldsymbol{x}] \equiv \text{cov}[\boldsymbol{x}, \boldsymbol{x}]$. The diagonal components of this matrix are variance, and the non-diagonal components are covariance.

2.2.3.3 Gaussian Distribution

The Gaussian distribution or normal distribution is important, because it can be used as a model to simplify naturally complex phenomena. In the continuous case, the Gaussian distribution for a single random variable x is defined by

$$\mathcal{N}(x|\mu, \sigma^2) = \frac{1}{\sqrt{2\pi}\sigma} \exp\left(-\frac{1}{2}\left(\frac{x-\mu}{\sigma}\right)^2\right). \tag{2.96}$$

where μ and σ^2 are the mean and the variance of the distribution, respectively. The inverse of the variance, $\lambda = 1/\sigma^2$, is called the *precision of the distribution*.

The Gaussian distribution for the random D variable, $\boldsymbol{x}$, is given as

$$\mathcal{N}(\boldsymbol{x}|\boldsymbol{\mu}, \Sigma) = \frac{1}{(2\pi)^{D/2}|\Sigma|^{1/2}} \exp\left(-\frac{1}{2}(\boldsymbol{x}-\boldsymbol{\mu})^T \Sigma^{-1}(\boldsymbol{x}-\boldsymbol{\mu})\right). \tag{2.97}$$

The quadratic form

$$\Delta^2 = (\boldsymbol{x}-\boldsymbol{\mu})^T \Sigma^{-1}(\boldsymbol{x}-\boldsymbol{\mu}) \tag{2.98}$$

is called the Mahalanobis distance between $\boldsymbol{\mu}$ and $\boldsymbol{x}$. If Σ is an identity matrix, the Mahalanobis distance, Δ, is identical to the Euclidean distance between $\boldsymbol{\mu}$ and $\boldsymbol{x}$. Σ is a symmetric and positive semidefinite matrix. Let $\boldsymbol{u}_i$ and λ_i $(i = 1, 2, \ldots, D)$ denote the eigenvectors and their corresponding eigenvalues where $\lambda_1 \geq \lambda_2 \geq \ldots \geq \lambda_D \geq 0$. Then, the following equation is satisfied:

$$\Sigma \boldsymbol{u}_i = \lambda_i \boldsymbol{u}_i \ \ (i = 1, 2, \ldots, D). \tag{2.99}$$

A set of the eigenvectors, $\{\boldsymbol{u}_i | i = 1, 2, \ldots, D\}$, is the orthonormal basis of the space of $\boldsymbol{x}$ because Σ is symmetric and positive definite:

$$\boldsymbol{u}_i^T \boldsymbol{u}_j = \begin{cases} 1, & \text{if} i = j, \\ 0, & \text{if} i \neq j. \end{cases} \tag{2.100}$$

The covariance matrix can be constructed using these eigenvectors:

$$\Sigma = \sum_{i=1}^{D} \lambda_i \boldsymbol{u}_i \boldsymbol{u}_i^T, \tag{2.101}$$

and its inverse matrix is obtained thus:

$$\Sigma^{-1} = \sum_{i=1}^{D} \frac{1}{\lambda_i} \boldsymbol{u}_i \boldsymbol{u}_i^T. \tag{2.102}$$

Substituting Eq. (2.102) into Eq. (2.98) results in

$$\Delta^2 = \sum_{i=1}^{D} \frac{y_i^2}{\lambda_i}, \tag{2.103}$$

where

$$y_i = \boldsymbol{u}_i(\boldsymbol{x} = \boldsymbol{\mu}). \tag{2.104}$$

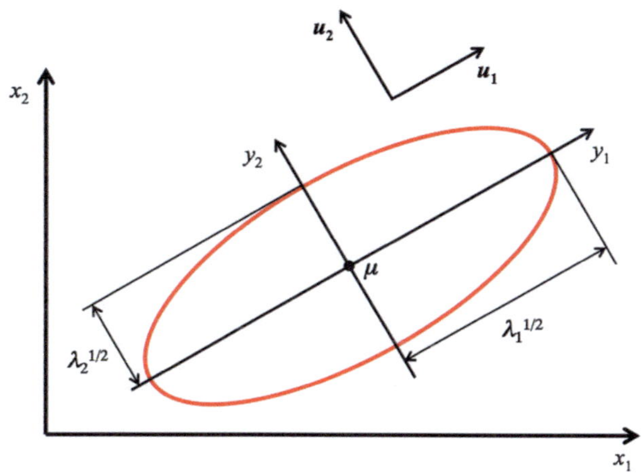

Fig. 2.3 A level set of a Gaussian distribution

Let $\boldsymbol{y} = [y_1, y_2, \ldots, y_D]^T$ and $U = [\boldsymbol{u}_1, \boldsymbol{u}_2, \ldots, \boldsymbol{u}_D]^T$; then,

$$\boldsymbol{y} = U(\boldsymbol{x} - \boldsymbol{\mu}). \tag{2.105}$$

If all eigenvalues are positive, in other words, if the covariance matrix is positive definite, a set of points that locate at the same Mahalanobis distance from $\boldsymbol{\mu}$ forms an elliptic quadratic surface with its center located at μ in a space of $\boldsymbol{x}$. This is a level-set surface of the Gaussian distribution. As shown in Fig. 2.3, the directions of the axes of the surface are parallel to the eigenvectors, and their lengths are proportional to the corresponding eigenvalues.

Although the Gaussian distribution is widely used for density models, there are some limitations. Generally, a symmetric covariant matrix has $D(D + 3)/2$ independent parameters, and when D is large, it is not easy to estimate accurately a covariance matrix from the data of a limited number of training samples and to compute the precision matrix, i.e., the inverse of the covariance matrix. To avoid over-fitting to the training data, a graphical lasso, for example, can be employed [32] that uses a regularization technique for accurately estimating precision matrices. The estimation accuracy can also be improved by approximating a covariance matrix with smaller numbers of parameters. For instance, a diagonal matrix, $\Sigma_{\text{diag}} = \text{diag}(\sigma_i^2)$, has only D parameters and can be used for approximating a covariance matrix that has in general nonzero off-diagonal components. Representing a covariance matrix with a diagonal matrix can avoid over-fitting, and its inverse can be computed more easily. This approximation, though, ignores the mutual correlations between different variables. Figure 2.4 shows examples of isocontours of the Gaussian distributions with covariance matrices approximated by a diagonal matrix (a) and by an isotropic matrix (b), $\Sigma_{\text{iso}} = \sigma^2 I$, where I denotes a unit matrix.

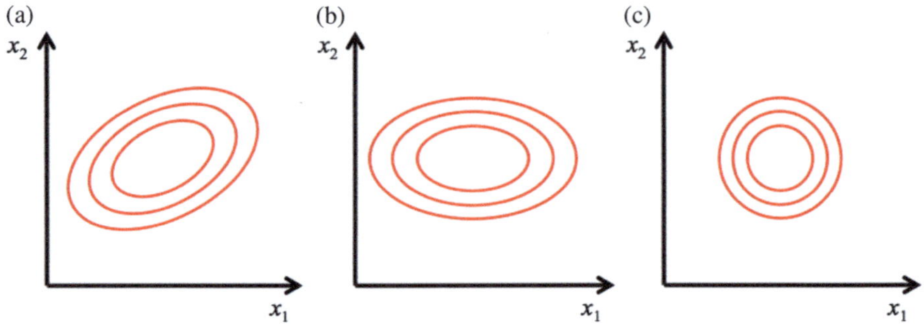

Fig. 2.4 Examples of isocontours of Gaussian distributions. (**a**) The covariance matrix is approximated by a diagonal matrix. (**b**) The covariance matrix is approximated by an isotropic matrix [33]

2.2.4 Foundations of Pattern Recognition

Computational anatomy (CA) is a system of automatically generating medical descriptions of patients from their medical images and the medical description can be generated by classifying each pixel in the images into one of some pre-defined classes. This classification problem can be solved by solving a pattern recognition problem in which, when representing data as points in a high-dimensional space, the space must be divided into regions, each of which represents each of the classes. A classifier then judges in which region a given data point is included and outputs the class that corresponds to the region.

2.2.4.1 Bayes Decision Theory

The best classifier minimizes the classification error under the assumption that the probability distributions of the data points in the space are known. Assuming only two categories, w_1 and w_2, let x denote a given datum. Then, the probability of misclassification of the data, $p(\text{error}|x)$, is given as [34]

$$p(\text{error}|x) = \begin{cases} P(w_1|x), & \text{if we decide } w_2, \\ \\ P(w_2|x), & \text{if we decide } w_1. \end{cases} \tag{2.106}$$

The probability of the classification error can be minimized by selecting the category that maximizes $P(w_j|x)$. This selection maximizes the average probability of misclassification for all data such that

$$P(\text{error}) = \int P(\text{error}|x)P(x)dx = \int P(\text{error}, x)dx. \tag{2.107}$$

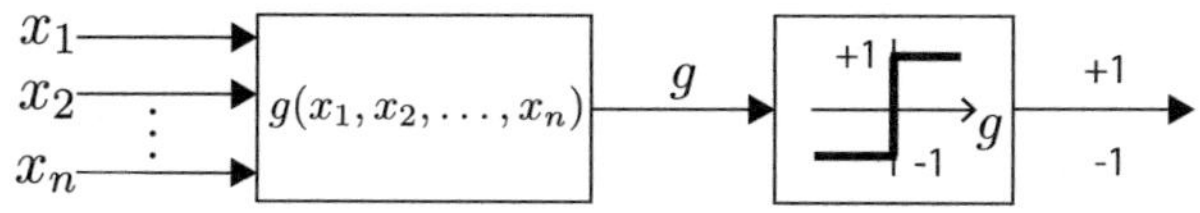

Fig. 2.5 Block diagram of a classifier in [35]

It is called a *Bayes decision theory* that the average probability of the can be minimized by following this rule:

Rule: Decide w_j if $P(w_j|x)$ is larger than $P(w_k|x)$ for any $k \neq j$.

2.2.4.2 Classifier Design

A *classifier* decides the class of a given input datum. Assume that there are only two classes, w_1 and w_2. Then, a block diagram of such a classifier has the form shown in Fig. 2.5. Such a classifier first computes some feature of the input data, x, and assigns the datum to the class w_1 if the sign of $g(x)$ is positive. According to [33], the following three different approaches work for the construction of classifiers:

1. Construction of a Generative Model
   In this approach, a posterior probability distribution of each class is computed as

$$P(w_j|x) = \frac{P(w_j)P(x|w_j)}{P(x)}, \tag{2.108}$$

   where all of the terms on the right side are estimated from a set of training data. Once the posterior probability distributions of all classes are computed, a classifier can be constructed by following the Bayes' decision theorem. This approach is called generative because input data can be artificially generated based on the estimated distributions.
2. Construction of a Discriminative Model
   In a discriminative model approach, the posterior probability distribution of each class is directly estimated from a set of training data, and a classifier is constructed based on Bayes' decision theorem.
3. Direct Construction of Classifier
   In this approach, resultant classification functions output a class of an input datum not necessarily based on the input data's posterior probability distributions. Such classification functions can be constructed by using training data with their desired outputs.

For example, the function $g(x)$ in Fig. 2.5 can be designed as follows, employing the first or second approach:

$$g(x) - \frac{P(w_1|x)}{P(w_2|x)} - 1. \tag{2.109}$$

Classifiers are constructed from sets of training data and the training process is called *learning*. A classifier has parameters whose values are estimated from a set of training data in the learning process. In supervised learning, for example, a label of a desired class is assigned to each set of the training data, and the parameter values are estimated so that the outputs of the classifier are consistent with the labels. In the first two approaches above, the parameters of the classifier describe the probability distributions of data and of classes. In the last approach, the classification function is parametrically represented, and the parameter values are estimated so that the probability of the misclassification is minimized.

A flowchart of the process of classifier design is shown in Fig. 2.6, which appeared in [35]. In the flowchart, gathered data are first normalized and/or registered together. This important step is not easy. When a classifier of character images is constructed, for example, images of characters should be gathered and the

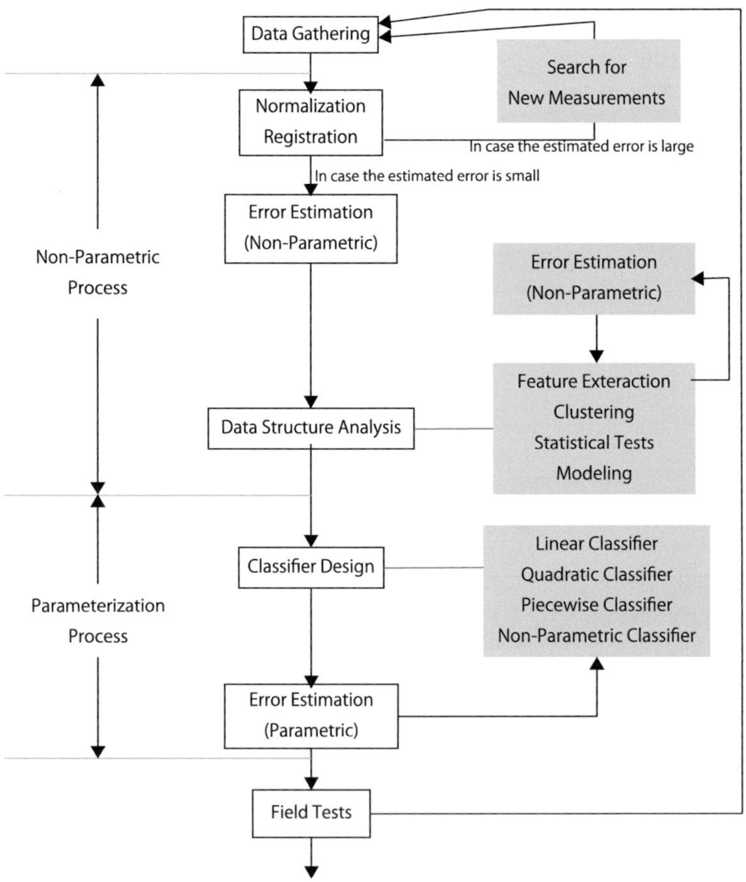

Fig. 2.6 A flowchart of the process of classifier design appeared in [35]

character images should be normalized by bounding each character with a rectangle followed by resizing all of the rectangles to identical dimensions.

This normalization cancels the variations in character images with respect to the locations, sizes, and patient position, and a classifier can be constructed to classify each given character image based only on the difference of the shapes/patterns intrinsic to each character. This normalization/registration of gathered data is required not only for the construction of classifiers but also for the construction of regressors. The training data of organs need to be normalized/registered when constructing computational anatomical models that are used for medical image analysis. To cancel the differences of the locations, sizes, and positions of patients' bodies, the gathered training images must be non rigidly registered: Each voxel in a training image is matched to a voxel in each of the other training images from an anatomical point of view, or each point on the surface of a training organ region is matched to a point on the surface of each of the other training organ regions. For the former nonrigid registration, anatomical landmarks are essential and will be described in Sect. 2.3.3.1. For the computation of the matching of surfaces, diffeomorphism supplies the mathematical foundation and will be described in Sect. 2.3.4. After normalization/registration is successfully applied to all of the data, the next step is the process of data structure analysis followed by classification design.

One of the main purposes of data structure analysis is to determine the representation of targets. In [36], a representation is defined as "a formal system for making explicit certain entities or types of information, together with a specification of how the system does this." The result of using a representation to describe a given entity is called a description of the entity in that representation. Feature extraction from measured data, for example, is a representation of the data. Applying different representations to given measured data results in different descriptions. For constructing recognition systems, representations that improve the separability of the descriptions of data into different classes need to be employed. Recognition problems or regression problems should be solved when segmenting organs in given medical images. The performance of the segmentation also varies depending on the representations of targets. A classifier/regressor processes the descriptions and hence should be designed based on the representation of measured data as is shown in Fig. 2.6.

Performance Evaluation

Figure 2.6 displays the methodology by which classification error of a designed classifier is evaluated. The results of the evaluation are used to improve the classifier design. Techniques appropriate for the performance evaluation are vital because the amount of data available for the classifier learning and for the performance evaluation is limited. If the performance of a classifier is evaluated by testing it with a dataset that is also used for the construction of the classifier, then the performance evaluation will be biased; in many cases, the classification error will be underestimated. For achieving unbiased performance evaluation, a n-fold cross validation or a bootstrap method is employed.

n-Fold Cross Validation

First, a set of samples is divided into n subsets. Then $n - 1$ subsets are used for the learning, and the remaining one subset is used for the performance evaluation. In subsequent steps, the subsets are shuffled and combinations of learning and the evaluation are repeated n times. Assuming that the error probability in each test step is e_i, the generalization error of the classifier is estimated as $(\sum_i e_i)/n$. When the number of samples is equal to the number of subsets (i.e., each subset contains only a single sample), the method is called the *leave-one-out method* or the *jackknife method*. The leave-one-out method often results in a more satisfactory evaluation [34], though it is more computationally complex especially when the amount of data is large.

Bootstrap Method

A bootstrap dataset is created by randomly selecting m data from a set of the data with replacement. Selecting the bootstrap data set B times independently enables evaluation of the classification error as $\sum_i e'_i/B$, where e'_i denotes the classification error evaluated by using the i-th bootstrap data set.

Examples of Classifiers

Many methods for constructing classifiers have been proposed. In the followings, some of the most popular and powerful methods for the classification are briefly described.

Decision Tree

Decision trees are represented by tree-structured directed graphs (see Fig. 2.7) [37]. A decision tree enables classification of input data through a sequence of questions along the tree. In the tree, each internal node represents a test, and each directed branch links a parent node to a child node and represents a result of the test at the parent node, with each leaf node representing a class. Starting from a root node, the data are classified by applying a test represented by the current node and by moving toward the leaf node by pursuing each branch representing the result of the test at each node. Several algorithms may be employed, such as classification and regression trees (CART) [38], ID3 [37], and C4.5 [38], for constructing decision trees from a set of training data. Regression trees can also be constructed [38] from sets of training data for solving regression problems.

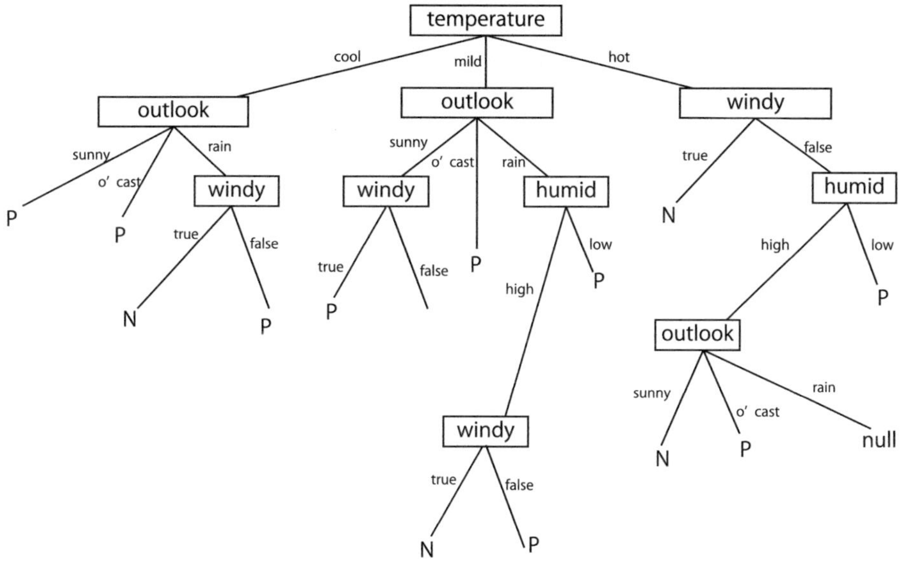

Fig. 2.7 An example of a decision tree of *Saturday morning attributes* appeared in [37]. *P* denotes a positive class of the mornings suitable for some activity and *N* denotes a negative class

Support Vector Machine (SVM)

An SVM is a method for solving two-class problems and is included in discriminative models [33, 39]. Linear classifiers classify data, x, into two categories based on the sign of the following linear classification function:

$$f(x) = w^T \phi(x) + b, \tag{2.110}$$

where $\phi(x)$ denotes a vector of which components are features extracted from the data x, and w and b denote the coefficients of the linear function. The features to be extracted are determined in advance at the step of the data structure analysis shown in Fig. 2.6, and in the designing step of classifiers, the values of these coefficients are estimated by using a set of given training data. One of the strong points of SVMs is the ability to estimate the coefficient values by solving a convex optimization problem: If the training data are linearly separable, then the globally optimal values of the coefficients can be attained. Assuming that the data are linearly separable, the convex optimization problem is derived from a criterion of the goodness of the decision boundary, $f(x) = 0$, which is a hyperplane in a feature space represented by $\phi(x)$. Let the term *margin* denote the distance between the decision boundary, $f(x) = 0$, and the training data closest to the decision boundary. An SVM estimates the values of the coefficients, which generate a decision boundary that maximizes the margin (see Fig. 2.8). The generalization error of the resultant linear classifier can be minimized using the strategy of the maximization of the margin. The

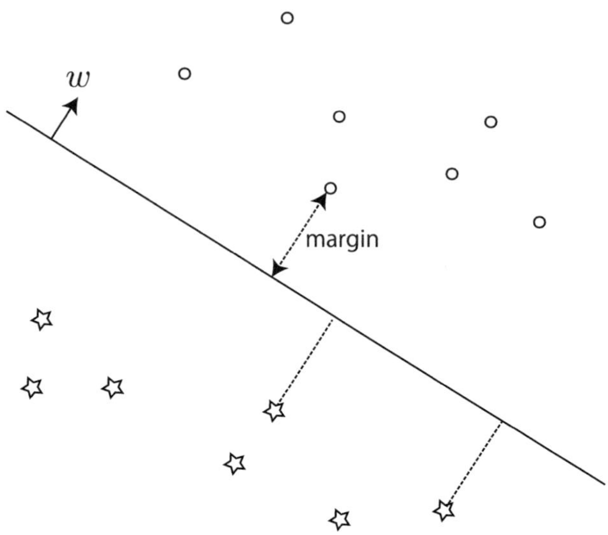

Fig. 2.8 An example of a decision boundary computed by an SVM

assumption of the linear separability can be relaxed by using soft- margin techniques [39], and nonlinear classification functions can be constructed by using a kernel trick [40].

Random Forest

Random forest was proposed by Leo Breiman [38], and it is used for pattern classification and regression. A random forest is an ensemble learning algorithm consisting of decision trees as weak classifiers, hence the name of the random forest. The procedure of the learning steps for random forest is:

1. Select *m* subsets from learning samples (Bootstrap sample).
2. Generate *m* decision trees using each subset.
3. Generate nodes until the number of the nodes reaches a specified number in the following way:
   a. Select *k* explanatory variables from learning samples randomly.
   b. Decide split function of the node using explanatory variable which has the best accuracy to classify learning samples and the threshold.

This procedure generates a set of decision trees that have low correlations with one another.

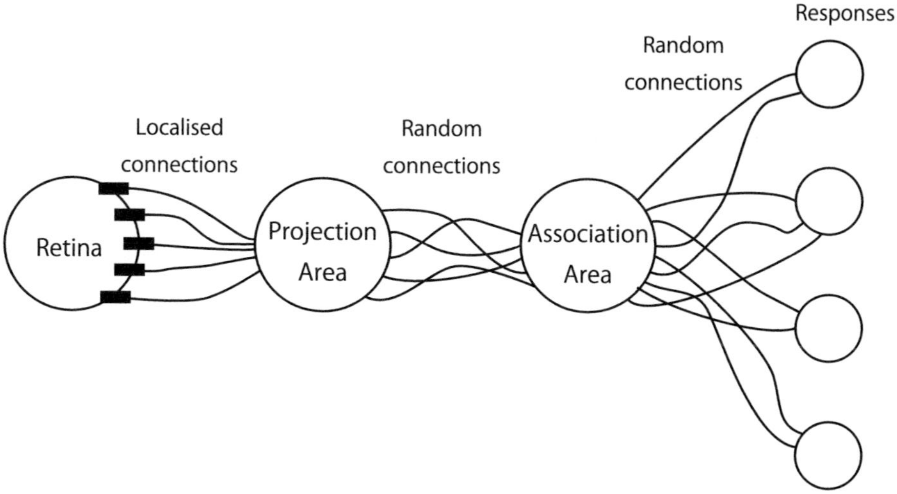

Fig. 2.9 Organization of perceptron appeared in [41]

Artificial Neural Network (ANN)

ANN is a generic name for models with the ability to solve problems by varying the coupling strength between artificial synapses in networks. The ANN often is capable of relatively high performance in multidimensional and nonlinearly separable problems. One of the origins of the research in ANNs was that of a single-layered *perceptron* algorithm (see Fig. 2.9) proposed by Rosenblatt in 1958 [41], which has the capability of learning classifiers only for linearly separable data [42]. The number of layers and of the synapses of ANNs increases to achieve the capability of learning nonlinear classifiers, as shown in the multilayered neural network known as the *neo-cognitron* [43] and in a convolution neural network (CNN) [44]. One can design all steps of image processing from image feature extraction to the classification simultaneously by employing a multiple-layered ANN. A serious difficulty was found, however, in training of the coupling strengths of the synapses. A framework of back propagation was invented for training in the 1980s [45, 46], but it did not have enough capability for appropriately varying the coupling strengths of the many synapses. Later, several important techniques for the training, for example, sparse coding [47] and layerwise pretraining [48], were invented, making it much easier to construct strong nonlinear classifiers with multilayered ANNs (Fig. 2.10).

Boosting

Boosting algorithms are used for improving the performance of any given learning algorithm: Assuming that a given learning algorithm can construct a weak classifier that can classify data slightly more accurately than a classifier that randomly

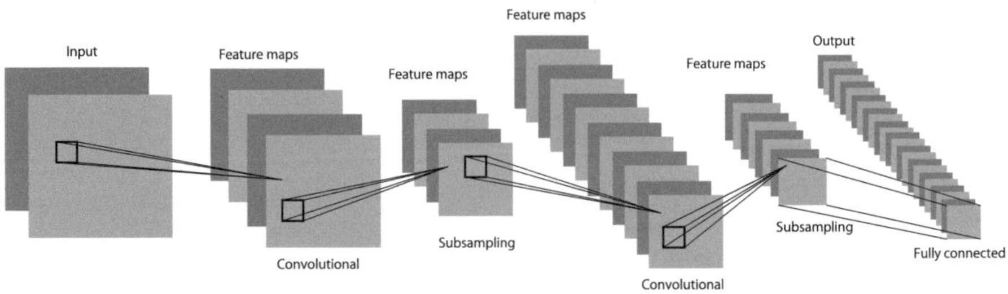

Fig. 2.10 The typical architecture of a convolution neural network (CNN) appeared in [49]

determines the class of input data independently from the data, boosting algorithms construct a strong classifier that classifies much more accurately than the weak classifiers by combining them [34].

Assuming that a set of n training data is given and that the data can be classified into two classes, a simple boosting algorithm constructs a strong classifier from three weak ones as follows: Let a subset of the training data be denoted by D_1 and a classifier constructed by the given algorithm be denoted by C_1. Then, applying C_1 to all of the training data, a new subset of the training data D_2 is constructed such that half of D_2 are misclassified by C_1, with the other half classified correctly. Let a classifier constructed based on the new dataset D_2 be denoted by C_2. Next, applying C_1 and C_2 to all of the training data, the third data set, D_3, is constructed by collecting data to which the two classifiers, C_1 and C_2, output different classes. Let a classifier constructed from D_3 be denoted by C_3. A boosting algorithm combines the three weak classifiers, C_1, C_2, and C_3, to construct a strong classifier.

There are many boosting algorithms. The AdaBoost algorithm is one of the most popular ones. In the AdaBoost algorithm, weak classifiers can continue to be added until the performance becomes sufficiently strong. The resulting strong classifier is represented by a weighted sum of weak classifiers such that

$$f(x) = \sum_k \alpha_k h_k(x), \tag{2.111}$$

where α_k is a scholar and denotes the weight and $h_k(x)$ denotes a weak classification function constructed by the given algorithm. The k-th weak classifier is constructed from a set of data obtained by sampling from the training data according to the weight of each dataset, and the weight is computed according to the classification results of the $k-1$-th classifier: If the $k-1$-th classifier classifies the j-th training data, x_j, correctly/incorrectly, then its weight, w_j, is decreased/increased, respectively. It is known that the total training error of the resultant strong classifier can be arbitrarily lowered by combining a large number of weak classifiers.

2.3 Computational Anatomical Model

Hidekata Hontani, Yasushi Hirano, Xiao Dong, Akinobu Shimiz, and Shohei Hanaoka

2.3.1 *Models for Segmentation*

One of the main objectives in medical image analysis is to segment given medical images into organs. One can segment the images by determining boundaries of the organs in the images. Organ boundaries mainly consist of edges. It is pointed out in [50] that the concept of determining edges in given images is an ill-posed problem. A problem is well posed if the following three conditions are satisfied:

1. The solution exists,
2. The solution is unique, and
3. The solution depends continuously on the initial data.

One can convert ill-posed problems to well-posed ones by introducing some prior knowledge of the problems in order to restrict the class of admissible solutions. A major technique for the restriction is regularization [50, 51].

Assuming that the objective is to detect the boundary of a target organ in a given image, let D denote a set of edge points detected from the given image as the candidates for points on the boundary, and let the parameter of a model representing the surface be denoted by $\boldsymbol{\theta}$. $L_{\text{Data}}(\boldsymbol{\theta}\,|D)$ describes a cost function that defines the error between the edge points, D, and the model surface $\boldsymbol{\theta}$. In general, edge detection often fails to detect some portions of the boundary and detects many false candidates. It is hence very difficult to design a cost function, $L_{\text{Data}}(\boldsymbol{\theta}\,|D)$, such that the boundary of the target organ is delineated correctly only by minimizing it.

In the *regularization technique*, prior knowledge introduces some regularity of the sizes and shapes of surfaces by adding new terms, *regularization terms*, to the cost function, allowing detection of the boundary by minimizing the new cost function, $L(\boldsymbol{\theta};D) = L_{\text{Data}}(\boldsymbol{\theta};D) + \lambda L_{\text{reg}}(\boldsymbol{\theta})$, where λ controls the compromise between the strength of the regularization and the closeness of the model to the data. For example, an Active Contour Model (ACM) detects contours in given images by minimizing a cost function with regularization terms that restrict the curves represented by the model to short and smooth ones [52]. The regularization terms are manually determined based on geometrical features of the boundaries of target regions, and the regularization techniques work well only if all of the boundaries have common geometric features that can be represented by the regularization terms. It is difficult in general to find such features and to determine appropriately the regularization terms.

More powerful strategies convert ill-posed problems to well-posed ones by introducing a framework of the Bayes' estimation or of the *maximum a posteriori*

(MAP) estimation [53, 54], in which the class of admissible solutions is restricted by prior probability distributions of targets. In medical image analysis, the regions or their boundaries are determined by maximizing the posterior probability distributions which consist of the prior distributions and the likelihood distributions. Assuming again that the objective is to detect the boundary surface of a target organ in a given image and letting the posterior probability distribution be denoted by $p(\boldsymbol{\theta}|D)$, in the framework of the MAP estimation, the boundary can be detected by maximizing $p(\boldsymbol{\theta}|D)$. As described in ***, $p(\boldsymbol{\theta}|D) \propto p(\boldsymbol{\theta})p(D|\boldsymbol{\theta})$, where $p(\boldsymbol{\theta})$ is the prior probability distribution of the model parameters and $p(D|\boldsymbol{\theta})$ denotes the likelihood of the parameters with respect to the given data, D. The prior distribution, $p(\boldsymbol{\theta})$, is introduced to restrict the admissible solutions and can be constructed by learning from sets of training data. Assuming that these probability distributions can be represented using exponential functions such as $p(\boldsymbol{\theta}) = \exp(F_{\text{pri}}(\lambda\boldsymbol{\theta}))$ and $p(D|\boldsymbol{\theta}) = \exp(F_{\text{like}}(\boldsymbol{\theta}; D))$, the posterior probability distribution can be maximized by maximizing $\log p(\boldsymbol{\theta}|D) = F_{\text{like}}(\boldsymbol{\theta}; D) + \lambda F_{\text{pri}}(\boldsymbol{\theta})$, the form of which is analogous to the cost function with the regularization terms.

There are many models that represent regions or boundaries in medical images: For example, regions may be represented by labeling voxels in the images, and boundaries may be represented by using parametric surface models. The statistical models of organs represent the varieties observed among patients. Even when a single set of training data is given for constructing an SSM, the resultant SSM would vary with respect to the representation of the regions or of the boundaries employed, and the performance of the image segmentation would largely depend on the SSM. In the following section, the representation and the corresponding segmentation methods will be described.

2.3.2 Geometrical Representation

In this section, some representations of geometrical figures such as points, curves, surfaces, and regions (volumes) in a three-dimensional space, are described. As described in the previous subsection, representation methods of geometric figures in given images can be roughly classified into two categories. In the first category, targets are represented using functions defined on an image space. In the second category, they are represented using functions defined on the parameter space, Sect. 2.3.2.1, with the resulting representation depicting the organ. Figure 2.11 shows some examples of different representations of an organ. An organ can be represented by labeling the voxels inside the organ, as will be described in Sect. 2.3.2.1, and the resultant representation represents the region with a set of the labeled voxels as shown in the top left panel in Fig. 2.11. Another method to represent an organ is by assigning negative values to the voxels inside the organ volume and positive values to the outside, using a level-set function, which implicitly represents the boundary of the region with its zero-crossing (see the bottom left panel in Fig. 2.11); this level-set representation will also be described

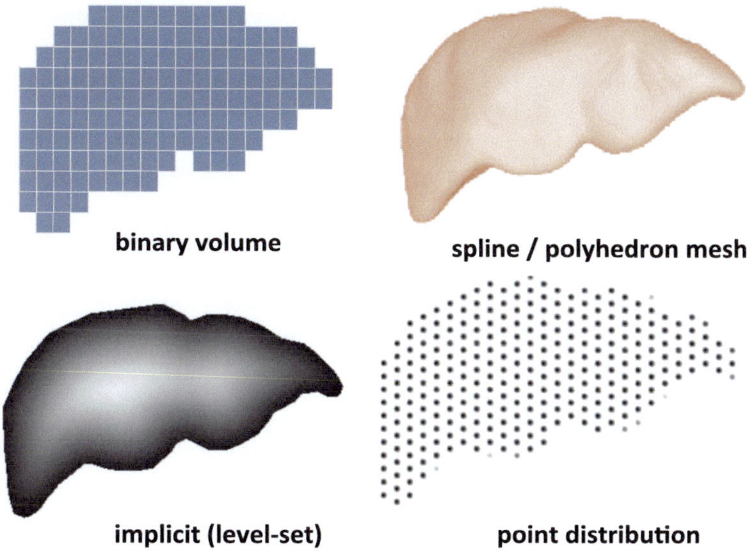

Fig. 2.11 Different representations of a target organ

in Sect. 2.3.2.1. The boundary of an organ can also be explicitly represented by parametric functions such as spline surfaces. These parametric representations will be described in Sect. 2.3.2.2. A set of points distributed on a target surface (see the bottom right panel of Fig. 2.11) is also widely employed for the representation of an organ's surface. The distribution of the points is described by the coordinates of each of the points, and those coordinates are the parameters of the representation. Such a representation with the set of the points will also be described in Sect. 2.3.2.2.

2.3.2.1 Representation Using Functions of Voxels

Target regions and boundaries in images can be represented by using functions defined on an image space. As will be described below, one of the advantages of this category of representations is that they can straightforwardly depict topological changes in the figures, and one of the disadvantages is that the representations are largely redundant.

Letting $\boldsymbol{x}$ denote the coordinates of a voxel in a three-dimensional image space, where $\boldsymbol{x} = [x, y, z]^T$ denotes the coordinates of a voxel in a given image, the figures in the images can be explicitly represented by using discrete functions, $f_{\text{label}}(\boldsymbol{x}) : \mathbb{R}^3 \mapsto \mathbb{N}$. A value of $f_{\text{label}}(\boldsymbol{x})$ at $\boldsymbol{x}$ describes a label corresponding to each figure in an image. Letting $k = 1, 2, \ldots, K$ denote the label that identifies the figures, where K is the number of target figures to be represented, a labeling function explicitly represents the figures as

$$f_{\text{label}}(\boldsymbol{x}) = k(\boldsymbol{x}), \tag{2.112}$$

where $k(\boldsymbol{x})$ denotes the label of the figure to which the voxel, $\boldsymbol{x}$, belongs. Letting a binary function, $f_k(\boldsymbol{x})$, denote if a voxel, [x], belongs to the k-th figure or not, the figures in a given image can be represented using following vector labeling function, $f_{\text{vec}}(\boldsymbol{x}) : \mathbb{R}^3 \mapsto \{0, 1\}^K$ such that

$$f_{\text{vec}}(\boldsymbol{x}) = \boldsymbol{k}(\boldsymbol{x}), \tag{2.113}$$

where $\boldsymbol{k} = [f_1(\boldsymbol{x}), f_2(\boldsymbol{x}), \ldots, f_K(\boldsymbol{x})]^T$. When each voxel belongs to a single figure, then $\sum_k f_k(\boldsymbol{x}) = 1$ should be satisfied.

Continuous functions, $f_{\text{cont}}(\boldsymbol{x}) : \mathbb{R}^3 \mapsto \mathbb{R}$, defined on an image space can represent figures implicitly. For example, regions, $\mathscr{R}$, in a given figure can be represented as follows:

$$f_{\text{cont}}(\boldsymbol{x}) \begin{cases} < T, \text{ if } \boldsymbol{x} \in \mathscr{R}, \\ \geq T, \text{ otherwise}, \end{cases} \tag{2.114}$$

where $T \in \mathbb{R}$ is a threshold. Then, the closed boundary surfaces of the regions, $\mathscr{R}$, can be represented as the voxels where the value of $f_{\text{cont}}(\boldsymbol{x})$ passes through T. For example, zero-crossings of the Laplacian of a (smoothed) image are widely employed for the detection of edge points [55–57].

Level-set representation [58] is also widely employed for implicitly representing surfaces in images. Let a level-set function be denoted by $f_{\text{level}}(\boldsymbol{x}) : \mathbb{R}^3 \mapsto \mathbb{R}$. Surfaces in an image can be implicitly represented by the zero-crossings of $f_{\text{level}}(\boldsymbol{x})$ where the sign of the function changes. Different from the representations using the discrete functions, you can differentiate the level-set functions and can compute some geometric properties of the surfaces from the differential coefficients. Let a closed surface represented by $f_{\text{level}}(\boldsymbol{x})$ be denoted by S and assume that $f_{\text{level}}(\boldsymbol{x}) < 0$ is satisfied inside S. Let the differential coefficients of $f_{\text{level}}(\boldsymbol{x})$ be denoted as $\phi_x = \partial f_{\text{level}}/\partial x$ or as $\phi_{zx} = \partial^2 f_{\text{level}}/\partial z \partial x$. Then, for example, the outward unit normal vector at $\boldsymbol{x}$ on S, $\boldsymbol{n}(\boldsymbol{x})$, can be obtained as follows:

$$\boldsymbol{n}(\boldsymbol{x}) = \frac{\nabla f_{\text{level}}(\boldsymbol{x})}{\|\nabla f_{\text{level}}(\boldsymbol{x})\|} = \frac{1}{\sqrt{\phi_x^2 + \phi_y^2 + \phi_z^2}} \begin{bmatrix} \phi_x \\ \phi_y \\ \phi_z \end{bmatrix}. \tag{2.115}$$

The mean curvature, κ_M, and the Gaussian curvature, κ_G, can be computed as follows:

$$\kappa_M = \nabla \cdot \frac{\nabla f_{\text{level}}(\boldsymbol{x})}{\|\nabla f_{\text{level}}(\boldsymbol{x})\|} = \frac{\left\{ \begin{array}{c} (\phi_{yy} + \phi_{zz})\phi_x^2 + (\phi_{xx} + \phi_{zz})\phi_y^2 + (\phi_{xx} + \phi_{yy})\phi_z^2 \\ -2\phi_x\phi_y\phi_{xy} - 2\phi_y\phi_z\phi_{yz} - 2\phi_z\phi_x\phi_{zx} \end{array} \right\}}{\left(\phi_x^2 + \phi_y^2 + \phi_z^2\right)^{3/2}}, \tag{2.116}$$

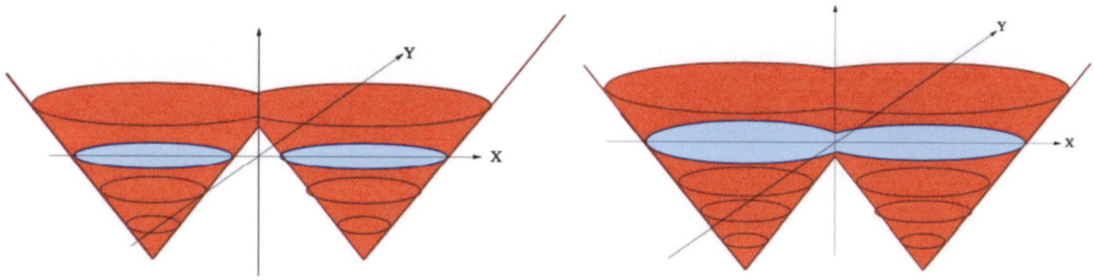

Fig. 2.12 An example of level-set representations corresponding to topological changes. Changing a level-set function continuously enables representation of topological changes of the target figures (These figures are appeared in [58])

$$\kappa_G = \frac{\left\{\begin{array}{c} \phi_x^2\left(\phi_{yy}\phi_{z}z - \phi_{y}z^2\right) + \phi_y^2\left(\phi_{xx}\phi_{zz} - \phi_{xz}^2\right) + \phi_z^2\left(\phi_{xx}\phi_{yy} - \phi_{xy}^2\right) \\ +2[\phi_x\phi_y(\phi_{xz}\phi_{yz} - \phi_{xy}\phi_{zz}) + \phi_y\phi_z(\phi_{xy}\phi_{xz} - \phi_{yz}\phi_{xx}) \\ +\phi_x\phi_z(\phi_{xy}\phi_{yz} - \phi_{xz}\phi_{yy})] \end{array}\right\}}{\left(\phi_x^2 + \phi_y^2 + \phi_z^2\right)^2}. \quad (2.117)$$

This property of level-set representation enabling direct computation of the geometric properties with the differential coefficients of the level set function is used in the level methods for propagating surfaces [58, 59]. One of the strongest advantages of level-set representations is the ability to represent the topological changes of the surfaces (e.g., a single closed surface split into two surfaces) straightforwardly as shown in Fig. 2.12.

It is often necessary to measure quantitatively the distance between figures. This measurement is performed by comparing the values of the functions that represent the figures at each voxel. For example, assume two regions, R_α and R_β, are represented by the binary functions, $f_\alpha(\boldsymbol{x})$ and $f_\beta(\boldsymbol{x})$, where α and β are the labels of the regions, respectively. Then, the Jaccard index (JI), $d_J(R_\alpha, R_\beta)$, defined below can be employed for measuring the distance:

$$d_J(R_\alpha, R_\beta) = \frac{|R_\alpha \cap R_\beta|}{|R_\alpha \cup R_\beta|}, \quad (2.118)$$

where $|R|$ denotes the number of voxels belonging to the region, R. $|R_\alpha \cap R_\beta$ denotes the number of voxels at which $f_\alpha(\boldsymbol{x}) = 1$ and $f_\beta(\boldsymbol{x}) = 1$ are satisfied and $|R_\alpha \cup R_\beta|$ denotes the number of voxels where $f_\alpha(\boldsymbol{x}) = 1$ or $f_\beta(\boldsymbol{x}) = 1$ are satisfied. When the regions are represented by the level-set functions, $f_{\text{level}}^{(\alpha)}(\boldsymbol{x})$ and $f_{\text{level}}^{(\beta)}(\boldsymbol{x})$, then the distance $d_{\text{level}}(R_\alpha, R_\beta)$ can be expressed as

$$d^2(R_\alpha, R_\beta) = \sum_{\boldsymbol{x}\in\Omega} \| f_{\text{level}}^{(\alpha)}(\boldsymbol{x}) - f_{\text{level}}^{(\beta)}(\boldsymbol{x}) \|^2, \quad (2.119)$$

where Ω denotes the region of interest in which the two regions are included, under the assumption that a unique level-set function that represents a given region is obtained.

2.3.2.2 Representation Using Parametric Functions

Parametric functions, whose values denote the coordinates in an image space, are also widely employed for representing figures in images.

2.3.2.3 Curves

Let s denote a scholar parameter that indicates the location along a given curve, C. Then, the curve, C, can be explicitly represented as $\boldsymbol{x}(s) = (x(s), y(s), z(s))^T : \mathbb{R} \mapsto \mathbb{R}^3$. Different from the representations described in the previous subsection, the functions $x(s)$, $y(s)$, and $z(s)$ are functions of a parameter, and the values of the functions denote the coordinates in the image. Here, assume that $x(s)$, $y(s)$, and $z(s)$ are represented by linear combinations of basis functions, $C_i(s)$ $(i = 1, 2, \ldots, N_B)$, as follows:

$$x(s) = \sum_i^{N_B} \alpha_i C_i(s), \ y(s) = \sum_i^{N_B} \beta_i C_i(s), \text{ and } z(s) = \sum_i^{N_B} \gamma_i C_i(s), \tag{2.120}$$

where α_i, β_i, and γ_i are scalar coefficients of the basis functions. Then, a curve with explicit functions, $\boldsymbol{x} = \boldsymbol{x}(s; \boldsymbol{\theta}) canberepresentedthus : \mathbb{R}^{3N_B} \mapsto \mathbb{R}^3$, where $\boldsymbol{\theta} = [\alpha_1, \beta_1, \gamma_1, \alpha_2, \cdots, \gamma_{N_B}]^T$.

One of the most popular basis functions employed for representing curves is B-spline functions ("B" comes from *basis*), which are piecewise polynomials. Let the i-th B-spline basis function of degree k in the variable, s, be denoted by $B_i^k(s)$. The B-spline functions, $B_i^k(s)$, have the following properties: first, the functions are not negative, $B_i^k(s) \geq 0$, have a compact support $[u_i, u_{i+k+1}]$, and are $k-1$ times continuously differentiable. In addition, by convention, the B-spline functions are constructed in a way such that they sum to 1 at all points:

$$\sum_{n=0}^{N_B - 1} B_i^k(s) = 1 \text{ for all } s. \tag{2.121}$$

A variety of functions can have the properties described above, and a simple example of the B-spline bases for representing closed curves is described below. More details of the spline representations can be found in [60].

Using the de Boor's algorithm, the k-th order B-spline function can be constructed from the $k-1$-th order functions ($k \geq 1$) as follows:

$$B_i^k(s) = \frac{x - u_i}{u_{i+k} - u_i} B_i^{k-1}(s) + \frac{u_{i+k+1} - x}{u_{i+k+1} - u_{i+1}} B_{i+1}^{k-1}(s), \tag{2.122}$$

and

$$B_i^0 = \begin{cases} 1, \text{ if } s \in [u_i, u_{i+1}], \\ 0, \text{ otherwise}. \end{cases} \tag{2.123}$$

For example, the second-order ($k = 2$) function, $B_0^2(s)$ has the form as follows when $u_i = i$:

$$B_0^2(s) = \begin{cases} s^2/2, & \text{if } 0 \leq s < 1, \\ 3/4 - (s - 3/2)^2, & \text{if } 1 \leq s < 2, \\ (s - 3) - 2/2, & \text{if } 2 \leq s < 3, \\ 0, & \text{otherwise}. \end{cases} \tag{2.124}$$

The graphs of the second-order functions are demonstrated in Fig. 2.13. A closed curve can be represented by a periodic function, which satisfies $\boldsymbol{x}(s) = \boldsymbol{x}(s + qL)$ for any integer q where L denotes the period, and $L = u_{N_B}$ when the closed curve is represented by N_B B-spline basis functions. The basis functions should be appropriately wrapped as shown in Fig. 2.12 for representing closed curves. Using these N_B wrapped basis functions enables the representation of closed curves as

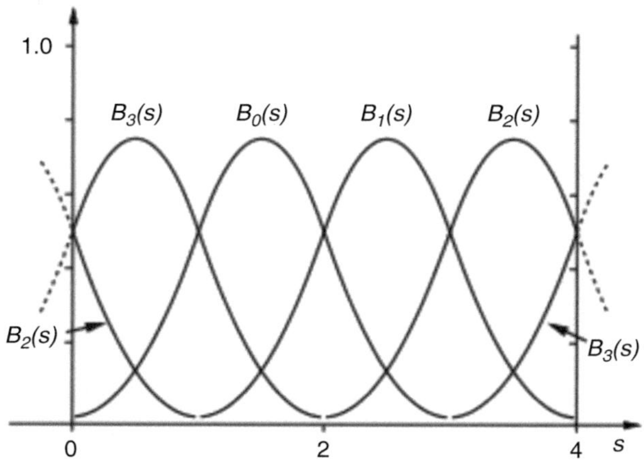

Fig. 2.13 An example of the second-order B-spline functions (This figure comes from the website of active contours)

follows:

$$\boldsymbol{x}(s;\boldsymbol{\theta}) = \sum_{i=0}^{N_B-1} \boldsymbol{\theta}_i B_i^k(s), \tag{2.125}$$

where $\boldsymbol{\theta}_i$ $(i = 0, 1, \ldots, N_B)$ are three-vector, each denoting the coordinates in an image space. The N_B points whose coordinates are represented by $\boldsymbol{\theta}_i$ are called *control points*, and their coordinates, $\boldsymbol{\theta} = \left[\boldsymbol{\theta}_1^T, \boldsymbol{\theta}_2^T, \ldots, \boldsymbol{\theta}_{N_B-1}^T\right]^T \in \mathbb{R}^{3N_B}$, are the parameters for the representation of the curves. Changing the locations of the control points allows changing of locations, sizes, and shapes of the curves. It should be noted that, by changing the location of a control point, the local portion of the curve can be deformed because each of the B-spline functions has a compact support. Let two curves be denoted by $\boldsymbol{x}(s;\boldsymbol{\theta}^\alpha)$ and $\boldsymbol{x}(s;\boldsymbol{\theta}^\beta)$. The distance between the two curves, $d(\boldsymbol{x}(s;\boldsymbol{\theta}^\alpha), \boldsymbol{x}(s;\boldsymbol{\theta}^\beta))$, can be defined as follows:

$$d^2(\boldsymbol{x}(s;\boldsymbol{\theta}^\alpha), \boldsymbol{x}(s;\boldsymbol{\theta}^\beta)) = \frac{1}{L}\int_0^L \|\boldsymbol{x}(s;\boldsymbol{\theta}^\alpha) - \boldsymbol{x}(s;\boldsymbol{\theta}^\beta)\|^2 ds, \tag{2.126}$$

where $[0, L]$ denotes the domain of s. Substituting (2.125) with (2.126) results in

$$d^2(\boldsymbol{x}(s;\boldsymbol{\theta}^\alpha), \boldsymbol{x}(s;\boldsymbol{\theta}^\beta)) = \frac{1}{L}\int_0^L \left\| \sum_{i=0}^{N_B-1} (\boldsymbol{\theta}_i^\alpha - \boldsymbol{\theta}_i^\beta) B_i^k(s) \right\|^2 ds. \tag{2.127}$$

Let

$$b_{ij} = \frac{1}{L}\int_0^L B_i^k(s) B_j^k(s) ds. \tag{2.128}$$

Then, Eq. (2.127) can be written as

$$d^2(\boldsymbol{x}(s;\boldsymbol{\theta}^\alpha), \boldsymbol{x}(s;\boldsymbol{\theta}^\beta)) = \sum_{i,j} b_{ij} \left\| \boldsymbol{\theta}_i^\alpha - \boldsymbol{\theta}_j^\beta \right\|^2. \tag{2.129}$$

The distance between the two curves is defined as the distance between the two corresponding parameter vectors with a metric defined by b_{ij}. It should be noted that the metric, b_{ij}, can be determined only by the basis functions and can be computed before figures are given. It should also be noted that the distance is defined based on the distance between two points whose locations are indicated by the same value of s on both curves. This means that the locations, sizes, and orientations (poses) of curves should be normalized *before* the distances among them are measured and that each of the curves should be appropriately parameterized so that corresponding points on the curves have the same values of s.

A Fourier series can also represent a closed curve in an explicit way. Again, letting a positive scalar, L, denote the period of $\boldsymbol{x}(s)$, where $\boldsymbol{x}(s) = [x(s), y(s), z(s)]^T = [x(s+L), y(s+L), z(s+L)]^T$ for any s, allows the representation of $x(s)$, $y(s)$, and $z(s)$ by linear combinations of the Fourier basis functions:

$$\begin{aligned} x(s) &= u_0 + \sum_{i=1}^{N_B-1} u_i \cos\left(\frac{2\pi si}{L}\right) + u_i' \sin\left(\frac{2\pi si}{L}\right), \\ y(s) &= v_0 + \sum_{i=1}^{N_B-1} v_i \cos\left(\frac{2\pi si}{L}\right) + v_i' \sin\left(\frac{2\pi si}{L}\right), \\ z(s) &= w_0 + \sum_{i=1}^{N_B-1} w_i \cos\left(\frac{2\pi si}{L}\right) + w_i' \sin\left(\frac{2\pi si}{L}\right). \end{aligned} \tag{2.130}$$

Letting the Fourier coefficients be denoted by $\boldsymbol{\theta} = [\alpha_0^{(x)}, \alpha_0^{(y)}, \ldots, \beta_{L-1}^{(z)}]^T$, an explicit representation of a closed curve can be obtained such that $\boldsymbol{x} = \boldsymbol{x}(s; \boldsymbol{\theta})$. Changing the value of each of the Fourier coefficients can change the amount of the component of the corresponding frequency and can deform the whole shape of the target curve. For example, setting the values of the coefficients corresponding to higher frequencies to zero can smooth the target curve.

The same definition (Eq. (2.126)) of the distances between figures can be employed, and the distances can be computed in the same way (Eq. (2.129)) even when using different basis functions for the representation. A difference appears in the metric tensor, $\mathbf{B}$. Because of the unity and the orthogonality of the sinusoidal basis functions, an identity matrix for the metric tensor is obtained, and hence the following equation holds:

$$d^2(\boldsymbol{x}(s; \boldsymbol{\theta}^\alpha), \boldsymbol{x}(s; \boldsymbol{\theta}^\beta)) \propto \|\boldsymbol{\theta}^\alpha - \boldsymbol{\theta}^\beta\|^2. \tag{2.131}$$

A PDM of a curve [61], C, is another example of explicit representations of curves, where it represents a curve with a series of the coordinates of L points. Letting i ($i = 0, 2, \ldots, L-1$) denote an index number for each of the points along the curve, and letting $\boldsymbol{x}(i)$ denote the coordinates of the i-th point, $\boldsymbol{x} = \boldsymbol{x}(i)$ is an explicit representation of a curve. Letting a $3L$-vector, $\boldsymbol{u}_C = [\boldsymbol{x}^T(0), \boldsymbol{x}^T(1), \ldots, \boldsymbol{x}^T(L-1)]^T$ denote the coordinates of the L points and represent the curve, the distance between two curves represented by a PDM can be defined as

$$d^2(\boldsymbol{x}^\alpha(i), \boldsymbol{x}^\beta(i)) = \frac{1}{L}\sum_{i=0}^{L-1} \|\boldsymbol{x}^\alpha(i) - \boldsymbol{x}^\beta(i)\|^2. \tag{2.132}$$

The distances between two corresponding points with identical numbers are used to define the distance between the two curves.

A PDM can also be represented by a linear combination of basis functions. Letting a $3L$-vector, $\boldsymbol{x}$, denote a representation of some curves and letting $3L$-vectors, $\boldsymbol{u}_i$ ($i = 0, 1, \ldots, N_B - 1$), denote basis functions, where $N_B \leq 3L$, allows representation of a curve with a linear combination of the basis functions as follows:

$$\boldsymbol{x}(\boldsymbol{\theta}) = \sum_{i=0}^{N_B-1} \theta_i \boldsymbol{u}_i, \tag{2.133}$$

where θ_i is a scalar and $\boldsymbol{\theta} = [\theta_0, \theta_1, \ldots, \theta_{N_B}]^T$. Changing the values of the parameters, $\boldsymbol{\theta}$ causes deformation of the shape of the curve. In many applications, the basis functions are constructed from a set of training data, $\mathscr{D} = \{\boldsymbol{x}^{(j)} | j = 1, 2, \ldots, M\}$, where M denotes the number of the data: A PCA is applied to the data set, $\mathscr{D}$, and the mean of $\mathscr{D}$ and the eigenvectors corresponding to the largest N_B eigenvalues are selected for the basis functions. Then the variety of a target curve can be represented by an N_B-dimensional linear space as follows:

$$\boldsymbol{x}(\boldsymbol{\theta}) = \bar{\boldsymbol{x}} + \sum_{i=0}^{N_B-1} \theta_i \boldsymbol{u}_i, \tag{2.134}$$

where $\bar{\boldsymbol{x}}$ is the mean vector and $\boldsymbol{u}_i$ denotes the eigenvectors. As described in Sect. 2.2.3, the Mahalanobis distance between the mean shape, $\bar{\boldsymbol{x}}$, and a described curve, $\boldsymbol{x}(\boldsymbol{\theta})$, is given as

$$d(\bar{\boldsymbol{x}}, \boldsymbol{x}(\boldsymbol{\theta})) = \sum_{i=0}^{N_B-1} \frac{\theta_i}{\lambda_i}, \tag{2.135}$$

where λ_i are the eigenvalues corresponding to $\boldsymbol{u}_i$.

2.3.2.4 Surfaces

Let s and t denote scalar parameters that indicate the location along a given surface, S. The surface, S, can be represented explicitly as $\boldsymbol{x}(s,t) = [x(s,t), y(s,t), z(s,t)]^T : \mathbb{R}^2 \mapsto \mathbb{R}^3$. Assume that $x(s,t)$, $y(s,t)$, and $z(s,t)$ are represented by linear combinations of basis functions, $C_i(s,t)$, as follows:

$$x(s,t) = \sum_{i=1}^{N_B-1} u_i C_i(s,t), \quad y(s,t) = \sum_i v_i C_i(s,t), \text{ and } z(s,t) = \sum_i w_i C_i(s,t), \tag{2.136}$$

where u_i, v_i, and w_i are scalar coefficients of the basis functions. Then, the surface can be represented in an explicit manner as $\boldsymbol{x} = \boldsymbol{x}(s, t; \boldsymbol{\theta})$, where $\boldsymbol{\theta} = [u_0, v_0, w_0, u_1 \ldots, v_{N_B-1}, w_{N_B-1}]^T$. B-spline or Fourier functions can be employed for the basis functions. Here, the NURBS (nonuniform rational basis spline) functions are described because they are also widely employed for the surface representation. Letting a B-spline function be denoted by $B_i^k(s)$, a NURBS function, $b_i^k(s)$, is defined as

$$b_i^k(s) = \frac{w_i B_i^k(s)}{\sum_{j=0}^{N_B-1} w_j B_j^k(s)}, \tag{2.137}$$

where w_j is a positive weighting coefficient, which determines the weight of the j-th control point. $B_i^k(s) = b_i^k(s)$ when $w_j = 1$ for all j. The NURBS functions in (2.137) can be used as basis functions for representing curves. NURBS basis functions for surfaces can be constructed thus:

$$b_{i,j}^k(s,t) = \frac{w_{i,j} B_i^k(s) B_j^k(t)}{\sum_{i'}\sum_{j'} w_{i',j'} B_{i'}^k(s) B_{j'}^k(t)}. \tag{2.138}$$

Letting the coordinates of the (i,j)-th control point be denoted by $\boldsymbol{\theta}_{i,j}$ and letting $\boldsymbol{\theta} = [\boldsymbol{\theta}_{0,0}^T, \boldsymbol{\theta}_{0,1}^T, \ldots, \boldsymbol{\theta}_{N_B-1,N_B-1}^T]^T$, a can be represented using the NURBS functions as follows [62–64]:

$$\boldsymbol{x}(s,t;\boldsymbol{\theta}) = \sum_i \sum_j b_{i,j}^k(s,t)\boldsymbol{\theta}_{i,j} = \frac{\sum_i \sum_j w_{i,j} B_i^k(s) B_j^k(t) \boldsymbol{\theta}_{i,j}}{\sum_i \sum_j w_{i,j} B_i^k(s) B_j^k(t)}. \tag{2.139}$$

Analogous to Eq. (2.126), the distance between two surfaces, $\boldsymbol{x}(s,t;\boldsymbol{\theta}^\alpha)$ and $\boldsymbol{x}(s,t;\boldsymbol{\theta}^\beta)$, can be defined as follows:

$$d^2(\boldsymbol{x}(s,t;\boldsymbol{\theta}^\alpha), \boldsymbol{x}(s,t;\boldsymbol{\theta}^\beta) = \frac{1}{S}\iint \|\boldsymbol{x}(s,t;\boldsymbol{\theta}^\alpha) - \boldsymbol{x}(s,t;\boldsymbol{\theta}^\beta)\|^2 ds dt, \tag{2.140}$$

where S is a coefficient for the normalization with respect to the size of the domain of (s,t). Equation (2.140) can be written thus:

$$d^2(\boldsymbol{x}(s,t;\boldsymbol{\theta}^\alpha), \boldsymbol{x}(s,t;\boldsymbol{\theta}^\beta)) = \sum_{i,j}\sum_{l,m} b_{i,j,l,m} \|\boldsymbol{\theta}_{i,j} - \boldsymbol{\theta}_{k,l}\|^2 = (\boldsymbol{\theta}^\alpha - \boldsymbol{\theta}^\beta)^T \mathbf{B} (\boldsymbol{\theta}^\alpha - \boldsymbol{\theta}^\beta), \tag{2.141}$$

where

$$b_{i,j,l,m} = \iint b_{i,j}^k(s,t) b_{l,m}^k(s,t) ds dt, \tag{2.142}$$

and the component of the metric tensor, $\mathbf{B}$, is determined by $b_{i,j,l,m}$.

When lines/surfaces are represented using the B-splines or NURBS, the coordinates of each point on the lines/surfaces are represented in a Cartesian coordinate system. Other coordinate systems can also be employed for representation. One of

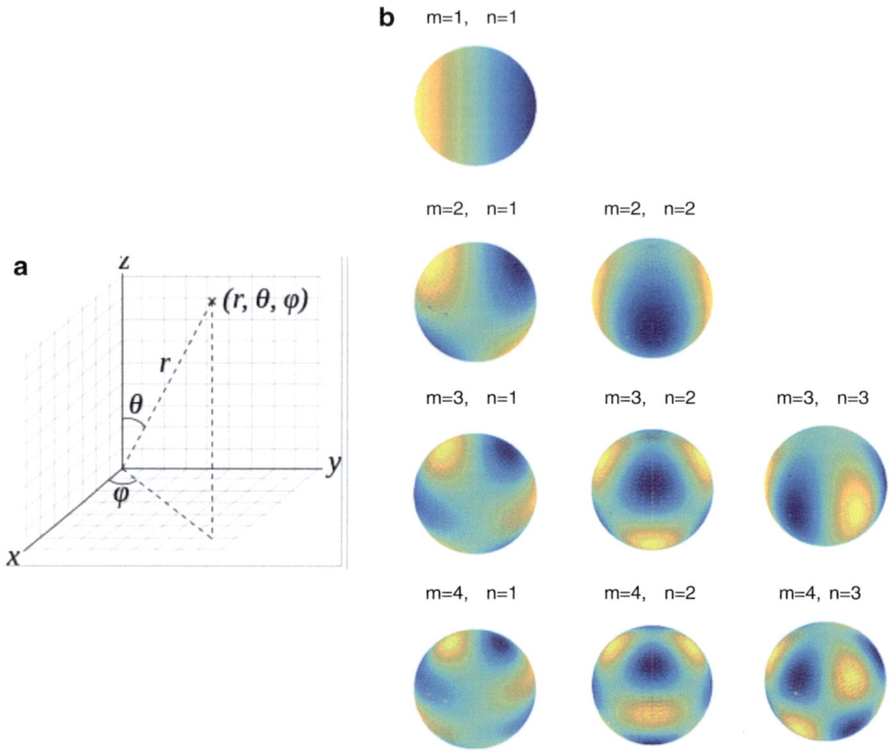

Fig. 2.14 (**a**) The spherical coordinates. (**b**) Spherical harmonics functions corresponding to $l = 0, 1, 2, 3$, and 4

the most popular coordinate systems for non-Cartesian representation is a spherical coordinate system, in which the coordinates of a point in a three-dimensional space are represented with three parameters, the radial distance r, the polar angle θ, and the azimuth angle ψ as shown in Fig. 2.14a. Spherical harmonics form a complete set of orthonormal functions defined on the unit sphere, on which the location of a point can be represented by the angular components, (θ, ψ): Analogous to the Fourier series that are a complete set of orthonormal functions defined on the unit circle, the spherical harmonics can expand any square-integrable function as a linear combination of these functions. Letting a given square-integrable function be denoted by $f(\theta, \psi) : \mathbb{R}^2 \mapsto \mathbb{R}$ and letting the spherical harmonics be denoted by $Y_l^m(\theta, \psi)$, where l is a nonnegative integer and m is an integer satisfying $-l \leq m \leq l$ enable representation of the given function as

$$f(\theta, \psi) = \sum_{l=0}^{\infty} \sum_{m=-l}^{+l} f_l^m Y_l^m(\theta, \psi), \tag{2.143}$$

where f_l^m denotes the complex expansion coefficients. The spherical harmonics functions are given as follows:

$$Y_l^m(\theta, \psi) = (-1)^{(m+|m|)/2} \sqrt{\frac{2l+1}{4\pi} \frac{(l-|m|)!}{(l+|m|)!}} P_l^{|m|}(\cos\theta) e^{im\psi}, \tag{2.144}$$

where $P_l^{|m|}$ is an associated Legendre polynomial:

$$P_l^m(x) = \frac{(-1)^m (1-x^2)^{m/2}}{2^l l!} \frac{d^{l+m}}{dx^{l+x}} (x^2-1)^l. \tag{2.145}$$

Given a function defined on the unit sphere, $f(\boldsymbol{\eta})$, where $\boldsymbol{\eta} = (\theta, \psi)^T$, allows computation of the coefficients as follows:

$$f_l^m = \int_{\eta \in S^2} f(\eta) \bar{Y}_l^m(\eta) d\eta, \tag{2.146}$$

where $\bar{Y}_l^m$ is the complex conjugate of Y_l^m. Some examples of the spherical harmonics are shown in Fig. 2.14b. Setting a two-dimensional coordinate system (θ, ψ) on a given simple[1] closed surface allows representation of the coordinates of each point on the surface in a Cartesian coordinate system by functions of the angular components, (θ, ψ), as $\boldsymbol{x} = \boldsymbol{x}(\theta, \psi) = [x(\theta, \psi), y(\theta, \psi), z(\theta, \psi)]^T$. Expanding these functions, x, y, and z, to linear combinations of the spherical harmonics functions results in representation of the surface as

$$\boldsymbol{x}(\theta, \phi; \boldsymbol{c}) = \sum_{l=0}^{\infty} \sum_{m=-1}^{l} \boldsymbol{c}_l^m Y_l^m(\theta, \psi), \tag{2.147}$$

where $\boldsymbol{c} = \{\boldsymbol{c}_l^m | l \in \mathbb{N}, -l \leq m \leq l\}$ are the coefficients, each of which is a three-vector, $\boldsymbol{c}_l^m = (x_l^m, y_l^m, z_l^m)^T$, where each of its components is computed from $x(\theta, \psi)$, $y(\theta, \psi)$, and $z(\theta, \psi)$, respectively, as shown in Eq. (2.146). Changing the values of the coefficients, $\boldsymbol{c}_l^m$, allows deformation of the surface. It should be noted that the coefficients with larger values of l correspond to the higher frequencies of the surface and to the smaller details of its shape.

A PDM [8, 65–68] represents a surface using the coordinates of the points distributed on it: $\boldsymbol{x}(i) : \mathbb{N} \mapsto \mathbb{R}^3$, where i denotes an index of each of the points on the surface. The distance between two surfaces represented by PDMs is defined as (2.132). Examples of liver surfaces represented by PDMs are shown in Fig. 2.15. Analogous to PDM-based curve representations shown in (2.134), surfaces can be represented using linear combinations of eigenvalues obtained by applying a PCA

[1] A simple surface does not have a self-crossing point.

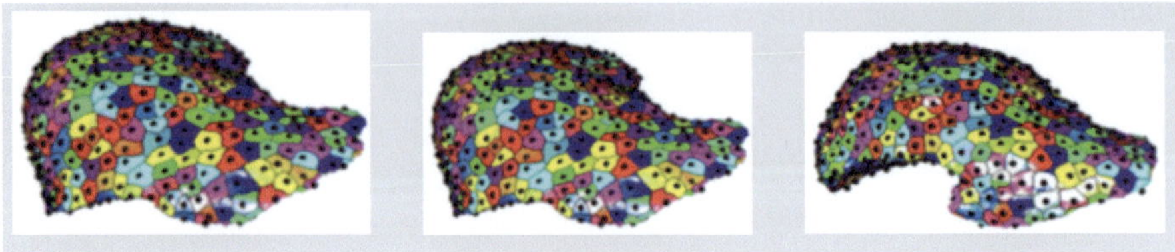

Fig. 2.15 Examples of liver surfaces represented by PDMs

to a set of training surfaces for targets:

$$\boldsymbol{x}(\boldsymbol{\theta}) = \bar{\boldsymbol{x}} + \sum_{i=0}^{N_B-1} \theta_i \boldsymbol{u}_i, \tag{2.148}$$

where $\bar{\boldsymbol{x}}$ and $\boldsymbol{u}_i$ denote the mean of the training surfaces and the eigenvalues corresponding to the N_B largest eigenvalues, respectively.

2.3.2.5 Registration Required Before Measurement or Analysis

Target figures such as curves or surfaces should be registered or normalized appropriately before measuring the distances between them or analysis of their the statistics. If you represent figures in different images which are represented using functions of voxels, $f(\boldsymbol{x})$, the values must be compared at the corresponding voxels; i.e., each voxel in one of the images should be mapped to a voxel in each of the other images before the comparison. Registration of given images is hence required, and use of anatomical landmarks is vital. Medical images can be registered by detecting anatomical landmarks and by deforming the images so that the detected corresponding landmarks have (approximately) identical coordinates. Landmark detection will be described in Sect. 2.3.3.

For statistical analysis of figures, if the figures are represented using parametric functions, $\boldsymbol{x}(s, t; (\theta))$, the locations of points on the figures that have identical values of (s, t) that indicate the locations on the figures must be compared. For example, when using the NURBS functions to represent surfaces, the distances among the figures are calculated as shown in (2.140), in which the distance between two points on different figures that have identical values of the parameters, (s, t), is integrated. The computed distance, therefore, is plausible from the point of view of medical imaging only when the parameters, (s, t), that indicate the locations are appropriately set for each of the given figures; i.e., the points that are on different figures and that have identical values of the parameters, (s, t), should appropriately correspond. The parameters, (θ, ψ), should correspond when employing spherical harmonics for the surface representation, and the index numbers, i, of the points should correspond when PDMs are used for the representations.

When parametric functions are employed for the representation, correspondences among given figures must be made in order to compare them and to analyze their statistics. Several approaches for making these correspondences can be utilized [69]. Diffeomorphism-based frameworks figure prominently in this methodology and will be described in Sect. 2.3.4.

2.3.3 *Image Features and Landmarks*

Local image features, like protuberances, ridges, and edges, are often detected at early stages of image processing and are used at higher stages to arrive at compact descriptions of targets, for example, for identifying the anatomical structures of a patient captured in a given image. Many local image features are defined for general purposes: edges, for example, are the locations at which the brightness changes sharply and are detected for a variety of purposes. Many different features have been proposed for higher-level image processing, e.g., image segmentation, image recognition, and for landmark detection. In many medical image processing programs, anatomical landmarks are detected first to determine the locations of the voxels in given images. Only after detecting the landmarks can the body sizes and shapes of the patients be normalized, allowing introduction of a coordinate system appropriate for using computational models. For the landmark detection, local image features that are specifically observed at each landmark location must be designated. This may pose some difficulty if corresponding landmarks in different patients have different appearances and because multiple locations can have appearances similar to those of some landmarks.

2.3.3.1 Anatomical Landmarks

Anatomical landmarks are the anatomical structures that can be reliably detected on given images of any patients and that can be used for uniquely identifying the location in each given image [70]. It is very important for the medical image segmentation to detect anatomical landmarks accurately because they are useful for anatomically identifying body regions in the images and for registering organ models to the images (see Sect. 2.3.5). One of the difficulties of automatic landmark detection is finding a set of local structures of whose features are similar from patient to patient. Many local structures have similar appearances, and it is difficult to find a set of landmarks each of which can signify a single location. One approach for handling this difficulty is to detect a set of landmarks simultaneously by using not only the local image features of each landmark but also a model of their spatial distributions. In the following, the methodology of this approach is described [70].

The algorithm of a method [70] that detects a set of anatomical landmarks consists of two steps: A set of candidates for each of the anatomical landmarks is first detected by using only its local image features and then false candidates are rejected using the model of the relative locations between the landmarks at the second step. At the first step, a detector for each of the landmarks is constructed by using a set of training data. Let the location of j-th landmark ($j = 1, 2, \ldots, N$) in the i-th training image, I_i ($i = 1, 2, \cdots, M$) be denoted by $\boldsymbol{p}_i^j$, which is entered manually.

Let I_i^j denote a local image pattern observed in a small region centered at $\boldsymbol{p}_i^j$ in I_i. In cases where the region is a cube and its size is $L \times L \times L$, then I_i^j is represented by a L^3-vector. The detector for the j-th landmark is constructed from a set of the local images, $\mathscr{S}^j = \{I_i^j | i = 1, 2, \cdots, M\}$, by applying a PCA to $\mathscr{S}^j$. Let a set of eigenvectors obtained by the PCA be denoted by $\{\boldsymbol{v}_k^j | k = 1, 2, \ldots, T^j\}$, which consists of T^j largest eigenvalues, $\{\lambda_1^j \geq \lambda_2^j \geq \ldots \geq \lambda_{T^j}^j\}$. Let a L^3-vector, $\boldsymbol{I}(\boldsymbol{p})$, denote a local appearance observed around a point $\boldsymbol{p}$ in a given new image. Then the log likelihood of the location $\boldsymbol{p}$ for the j-th landmark, $p(I|\boldsymbol{p}) = L^j(\boldsymbol{p})$, can be computed:

$$L^j(\boldsymbol{p}) = -\sum_{k=1}^{T^j} \frac{\left[(\boldsymbol{v}_k^j)^T(\boldsymbol{I}(\boldsymbol{p}) - \bar{\boldsymbol{I}}^j)\right]^2}{\lambda_k^j}, \tag{2.149}$$

where $\bar{\boldsymbol{I}}^j$ is the average of $\{\boldsymbol{I}_i^j | i = 1, 2, \cdots, M\}$: $\bar{\boldsymbol{I}}^j = \sum_i \boldsymbol{I}_i^j / M$. The likelihood of $\boldsymbol{p}$, $l(\boldsymbol{p})$, is proportional to $\exp\{L^j(\boldsymbol{p})\}$:

$$l^j(\boldsymbol{p}) \propto \exp\{L^j(\boldsymbol{p})\}. \tag{2.150}$$

This computation of the likelihood can be interpreted as the extraction of a feature that is observed around the landmark, $\boldsymbol{p}$. Unfortunately, the likelihood distribution cannot identify the location of the j-th landmark in many cases because one can find other anatomical structures that have the appearances similar to that of the target. A set of multiple candidates for the j-th landmark, hence, would be obtained by detecting the local maximums of $L^j(\boldsymbol{p})$ (or of $l^j(\boldsymbol{p})$). Let the detected candidates be denoted by $\boldsymbol{c}_k^j$ ($k = 1, 2, \ldots, K^j$).

At the second step, a true point is selected from the candidates $\{\boldsymbol{c}_k^j | k = 1, 2, \ldots, K^j\}$ in the given new image by using a model that represents the relative relationships between the anatomical landmarks. Let a true position of the j-th landmark be denoted by a three-vector, $\boldsymbol{p}^j$, and the distance between the j-th landmark and the j'-th landmark be denoted by $d^{jj'}$. In this method [70], a model of $d^{jj'}$ is constructed for each pair of the landmarks by using the training set of the landmarks manually extracted from all of the training images, $\{\boldsymbol{p}_i^j | i = 1, 2, \ldots, M, \quad j = 1, 2, \ldots, N\}$. Let $d_i^{jj'} = \|\boldsymbol{p}_i^j - \boldsymbol{p}_i^{j'}\|$ ($j \neq j'$). Then, it is straightforward to construct a Gaussian model of $d^{jj'}$ in (2.151) from the training set:

$$p(d^{jj'}) = \frac{1}{\sqrt{2\pi}\sigma^{jj'}} \exp\left\{-\frac{1}{2}\frac{(d^{jj'} - \bar{d}^{jj'})^2}{(\sigma^{jj'})^2}\right\}, \tag{2.151}$$

where $\bar{d}^{jj'}$ and $\sigma^{jj'}$ are the average and the standard deviation of $\{d_i^{jj'}|i = 1, 2, \cdots, M\}$, respectively. A true point, $\hat{\boldsymbol{c}}^j$, is selected for each j ($j = 1, 2, \cdots, N$) from the candidate set, $\{\boldsymbol{c}_k^j | k = 1, 2, \ldots, K^j\}$, by finding the number $\hat{k}^j$ for each j that maximizes the following simultaneous probability:

$$p(\boldsymbol{c}_{k^1}^1, \boldsymbol{c}_{k^2}^2, \ldots, \boldsymbol{c}_{k^N}^N) \propto \prod_j l^j(\boldsymbol{c}_{k^j}^j) \prod_{j \neq j'} p(d^{jj'}), \tag{2.152}$$

where $d^{jj'} = \|\boldsymbol{c}_{k^j}^j - \boldsymbol{c}_{k^{j'}}^j\|$. The final estimated location of the j-th landmark can be obtained as $\hat{\boldsymbol{c}}^j = \boldsymbol{c}_{\hat{k}^j}^j$. The model shown in (2.152) is a Markov random field (MRF) model that can be used to estimate the arguments that maximize the probability.

2.3.3.2 Keypoints

Keypoints are points with local image features that can be used for matching multiple images, in other words, for finding pairs of corresponding points in the images. As opposed to anatomical landmarks, it is not necessary to have information about the correspondence among different keypoints detected from an image. Keypoints used for image matching are determined after the target images are acquired, though which anatomical landmarks that are to be detected are determined in advance. Scale-invariant feature transform (SIFT) is one of the most popular algorithms for detecting keypoints and for describing their features. The details of SIFT can be found in the literature [71, 72].

The SIFT algorithm consists of two steps: Keypoints are detected at the first step and then local image features are described for each of the detected keypoints. At the first step, not only the spatial location but also the scale are detected for each keypoint. Here, the scale represents the size of the local structure around the keypoint: For example, keypoints are often detected on center lines of blood vessels, and the scale of each keypoint detected on the center line will approximately equal to the radius squared of the blood vessel. The theoretical foundation of the keypoint detection comes from the research field of the scale-space analysis [73]. At the second step, a histogram of the gradient directions observed in a local region around each keypoint is computed as the image feature. It should be noted that the histogram of the gradient directions includes some information about the curvatures of constant-level surfaces around each keypoint because the gradients are perpendicular to the level surfaces. Analogous to SIFT features, many other image features that are computed based on image gradients, including histograms of oriented gradients (HOG) [74] and Speeded Up Robust Features (SURF) [75], can be used for image description.

2.3.3.3 Edges and Ridges

In medical image analysis, local image features are detected from given images in order to constrain degrees of freedom (DoF) of the locations of anatomical structures. Though one needs to constrain three DoF for uniquely identifying the location of each structure, image features that can constrain only one or two DoFs are also useful for accurately segmenting organ regions in given images. Local appearances that are invariant with respect to translation cannot constrain three DoF of their locations. For example, a local appearance around a point on a smooth surface is invariant with respect to translation along tangential directions and can constrain the spatial location only along the direction perpendicular to the surface. In this subsection, edge and ridge features are described. Edge features constrain one DoF in given 3D images and ridge features constrain two DoFs. In this subsection, the functions with subscripts denote the derivatives of the functions with respect to the variables indicated by the subscripts. For example, I_x and I_{xy} denote $I_x = \partial I/\partial x$ and $I_{xy} = \partial^2 I/\partial x \partial y$, respectively.

Edges are defined as the locations where the voxel values change rapidly in one direction and are detected because they are often observed at the boundaries of organs. The rapid changes of voxel values can be detected by finding the locations where the second derivatives of images are equal to zero as the magnitudes of the first derivatives are maximal. Roughly, there are two approaches for finding such locations. In one approach, the Laplacian of a given image is computed, and its zero-crossings, at which the following equation holds, are detected as edges:

$$I_{xx} + I_{yy} + I_{zz} = 0. \tag{2.153}$$

In the other approach, the local maxima of the gradient magnitudes in directions of the gradients are detected as edges. Let a unit vector parallel to the gradient be denoted by $\boldsymbol{\xi}(x, y, z)$, where

$$\boldsymbol{\xi}(x, y, z) = \nabla I(x, y, z)/\|\nabla I(x, y, z)\|. \tag{2.154}$$

Then, at the local maxima, the second derivatives along the gradient direction, $\boldsymbol{\xi}$, should be zero:

$$\boldsymbol{\xi}(x, y, z) \cdot \nabla\left(\|\nabla I(x, y, z)\|\right) = 0, \tag{2.155}$$

where $\nabla(\|\nabla I(x, y, z)\|)$ denotes the gradient of the gradient magnitude. Computing its inner product with $\boldsymbol{\xi}$ results in the directional second derivative of the gradient magnitude. The Eq. (2.155) can be rewritten as follows:

$$\frac{I_x^2 I_{xx} + I_y^2 I_{yy} + I_z^2 I_{zz} + 2I_x I_y I_{xy} + 2I_y I_z I_{yz} + 2I_z I_x I_{zx}}{(I_x^2 + I_y^2 + I_z^2)^2} = 0. \tag{2.156}$$

The edge locations detected by (2.153) and by (2.156) are identical only when the constant-level surface of $I(x, y, z)$ is flat. One difficulty of the edge detection is that the values of the spatial derivatives generated by these two equations are susceptible to image noise because spatial derivative operations enhance high-frequency components of signals as described in Sect. 2.2.1.5, and many false edges are detected on noisy images. Therefore, given images are smoothed to suppress noise, usually using a Gaussian filter, $g(\boldsymbol{x}|t)$. Once again, letting $L(\boldsymbol{x}|t) = I(\boldsymbol{x}) * g(\boldsymbol{x}|t)$ and replacing I_* and I_{**} in (2.153) and in (2.156) with L_* and L_{**}, results in increased resistance to noise in the edge detectors. One drawback of this Gaussian smoothing is that the locations of the detected edges become inaccurate at the locations where the constant-level surfaces have high curvatures. Nonlinear image smoothing or edge-preserving algorithms [58, 76, 77] are often employed to remedy this.

Ridges (or "bright tubes") are defined as locations where the voxel values are constant in one direction, $\boldsymbol{u}$, and are maximum along planes perpendicular to $\boldsymbol{u}$. Ridges are detected on medical images because they are often associated with curvilinear structures such as vessels and bronchi. A Hessian matrix can be used to detect ridges:

$$H(\boldsymbol{x}) = \begin{bmatrix} L_{xx}(\boldsymbol{x}|t) & L_{xy}(\boldsymbol{x}|t) & L_{xz}(\boldsymbol{x}|t) \\ L_{yx}(\boldsymbol{x}|t) & L_{yy}(\boldsymbol{x}|t) & L_{yz}(\boldsymbol{x}|t) \\ L_{zx}(\boldsymbol{x}|t) & L_{zy}(\boldsymbol{x}|t) & L_{zz}(\boldsymbol{x}|t) \end{bmatrix}. \tag{2.157}$$

Letting the eigenvalues of $H(\boldsymbol{x})$ be denoted by $\lambda_1 \geq \lambda_2 \geq \lambda_3$ and letting the corresponding eigenvectors be denoted by $\boldsymbol{e}_1$, $\boldsymbol{e}_2$, and $\boldsymbol{e}_3$ (it should be noted that these eigenvectors are orthogonal to each other because $H(\boldsymbol{x})$ is symmetric), the Taylor expansion of $L(\boldsymbol{x})$ is

$$L(\boldsymbol{x} + \boldsymbol{\delta}) \simeq L(\boldsymbol{x}) + \nabla L(\boldsymbol{x}) \cdot \boldsymbol{\delta} + \frac{1}{2}\boldsymbol{\delta}^T H(\boldsymbol{x})\boldsymbol{\delta} + O(\|\boldsymbol{x}\|^3). \tag{2.158}$$

The last term on the right side can be rewritten as

$$\frac{1}{2}\boldsymbol{\delta}^T H(\boldsymbol{x})\boldsymbol{\delta} = \frac{1}{2}(\lambda_1 a^2 + \lambda_2 b^2 + \lambda_3 c^2), \tag{2.159}$$

where a, b, and c denote the small disturbances along $\boldsymbol{e}_1$, $\boldsymbol{e}_2$, and $\boldsymbol{e}_3$, respectively, such that $a = \boldsymbol{e}_1^T\boldsymbol{\delta}$, $b = \boldsymbol{e}_2^T\boldsymbol{\delta}$, and $c = \boldsymbol{e}_3^T\boldsymbol{\delta}$. In other words, the eigenvalues λ_1, λ_2, and λ_3 are the second derivatives in the directions of $\boldsymbol{e}_1$, $\boldsymbol{e}_2$, and $\boldsymbol{e}_3$, respectively. At ridge points, the largest eigenvalue is equal to zero, $\lambda_1 = 0$, and the corresponding eigenvector, $\boldsymbol{e}_1$, is parallel to the tangential direction along which the voxel values are constant. The other two eigenvalues, λ_2 and λ_3, are negative and are locally minimum along $\boldsymbol{e}_2$ and $\boldsymbol{e}_3$ because, at ridge points, the voxel values are maximum along the plane spanned by $\boldsymbol{e}_2$ and $\boldsymbol{e}_3$. Hence, the conditions ridges should satisfy are

$$\lambda_1 \simeq 0, \tag{2.160}$$

$$\lambda_2 \simeq \lambda_3 \ll 0, \tag{2.161}$$

$$\boldsymbol{e}_2 \cdot \nabla L(\boldsymbol{x}|t) = 0, \tag{2.162}$$

$$\boldsymbol{e}_3 \cdot \nabla L(\boldsymbol{x}|t) = 0. \tag{2.163}$$

The last two conditions reveal the locations of ridges. The scale, t, of the Gaussian applied to smooth the input images should be determined based on the radii of the local curvilinear structures. Assuming that a radius of a curvilinear structure is equal to r, the scale should be equal to $t \simeq r^2$, and some multiscale methods can determine an appropriate scale for each curvilinear structure in a given image [73, 78].

2.3.4 *Diffeomorphism Frameworks*

CA [79] aims to develop models to understand the anatomical variability of organs and tissues, including *(a) automated construction of anatomical manifolds, points, curves, surfaces, and subvolumes; (b) comparison of these manifolds; and (c) the statistical codification of the variability of anatomy via probability measures allowing for inference and hypothesis testing of disease states* [80]. An applicable framework to achieve these goals should provide: (a) a mathematical model to describe the space and variability of anatomy, i.e., the shape space and the transformations between shapes; (b) a computable distance metric to measure the difference between shapes; and (c) statistical analysis tools for the shape space.

In the last decade, a diffeomorphism-based CA framework developed by A. Trouvé [81], M. Miller [82], L. Younes [83], X. Pennec [84], D. Holm [85], S. Durrleman [86], and their collaborators has undergone tremendous progress to reach these goals. Existing computational frameworks developed along this road map fall into two categories: the Riemannian manifold solution and the Lie group solution with their corresponding terms large deformation diffeomorphic metric mapping (LDDMM) well-known representatives [81–83, 85, 87–89] and stationary vector fields (SVF) [90–93], respectively. Of the two, the LDDMM framework is more mathematically fundamental and general. The SVF can be regarded as a simpler alternative which provides better computational efficiency at the cost of more theoretical limitations.

Taking LDDMM as a general framework of diffeomorphism-based CA, its key components to solve the abovementioned tasks include:

- Accommodation of variation in the shape space by introducing groups of diffeomorphic transformations carrying individual elements from one to another.
- Defining a Riemannian metric defined on the shape space to measure continuous deformations of shapes, that is, paths in shape space. The distance between

shapes can be defined as the length of the shortest path, in other words, the geodesic that connects two shapes.
- Providing a Riemannian exponential map that generates the geodesic allowing linearization of shape space. When shapes are represented as initial velocity fields of geodesics connecting them with a fixed reference shape, one effectively works in the linear tangent space over the reference shape. The exponential map allows calculation of statistics in shape space.

In practice, finding the distance and the corresponding *optimal* curve connecting shapes in the shape space to realize this distance is formalized as a registration problem. Once the optimal deformation curves are found, their linearized representations can be used for statistical analysis on the shape manifold.

Sections 2.3.4.1 and 2.3.4.2 explain the LDDMM- and SVF-based image registration and their underlying geometry. Section 2.3.4.3 sketches the basic ideas of statistical analysis on shape space. Typical applications and future directions of diffeomorphism-based CA are addressed in Sect. 2.3.4.4.

2.3.4.1 LDDMM Framework for Registration

General Setting

The LDDMM approach models CA as a deformation of an initial template image $I \in V$ by diffeomorphic transformations $g \in G$, where $V = \{I : \Omega \to \mathcal{M}\}$ is a vector space of images with domain Ω and G is a Lie subgroup of the diffeomorphism group $Diff(\Omega)$ on Ω with Lie algebra $\mathfrak{g}$. The deformation of an image $I \in V$ by a diffeomorphic transformation $g \in G$ is defined by a smooth map:

$$\mathfrak{l} : G \times V \to V, (g, I) \to gI \tag{2.164}$$

Inner products of V and g are also defined as V and $\mathfrak{g}$ as $< \cdot, \cdot >_V = (\cdot, \cdot)_{V^* \times V}$ and $< \cdot, \cdot >_{\mathfrak{g}} = (\cdot, \cdot)_{\mathfrak{g}^* \times \mathfrak{g}}$ [94, 95].

LDDMM Diffeomorphic Registration

Given two images $I_0, I_1 \in V$, the objective of the diffeomorphic registration is to find a curve $t \mapsto u_t \in \mathfrak{g}$ that minimizes the energy

$$E(u_t) = \frac{1}{2} \int_0^1 < u_t, u_t >_{\mathfrak{g}} dt + \frac{1}{2\sigma^2} < g_1 I_0 - I_1, g_1 I_0 - I_1 >_V \tag{2.165}$$

where g_1 is the endpoint of the flow of u_t given by

$$\partial_t g_t = u_t g_t, g_0 = Id, g_t \in G \tag{2.166}$$

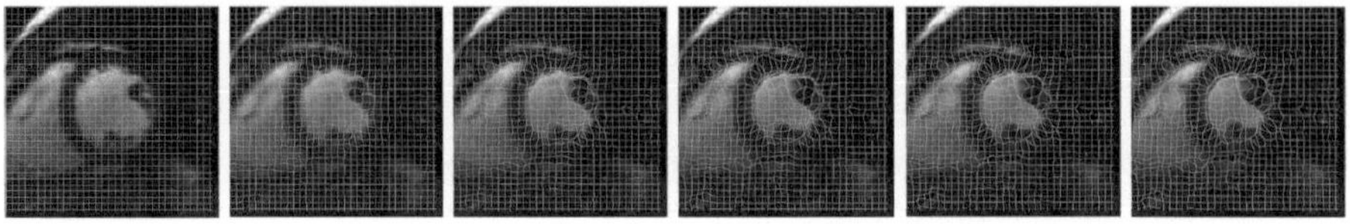

Fig. 2.16 Deformation fields g_t and deformed images g_tI_0 at time slot $t = 0, 0.2, 0.4, 0.6, 0.8, 1.0$ in a 2D LDDMM image registration

This means I_0 is smoothly deformed by $g_t, t \in [0, 1]$ to I_1. Figure 2.16 shows an example of a sequence of diffeomorphic registration.

Computing the derivative of the matching energy $E(u_t)$ should result in the optimal u_t satisfying

$$u_t^\flat + g_tI_0 \diamond g_tg_1^{-1}\pi = 0 \tag{2.167}$$

where $\pi = \frac{1}{\sigma^2}(g_1I_0 - I_1)^\flat$ and the $\flat$-map on a vector space V is defined by

$$\flat : V \to V^*, (u^\flat, v)_{V^* \times V} =< u, v >_V \tag{2.168}$$

The *momentum map* $\diamond : T * V \to \mathfrak{g}^*$ is defined by $(I \diamond \pi, \mu)_{\mathfrak{g}^* \times \mathfrak{g}} = (\pi, \zeta_\mu I)_{V^* \times V}$, and $\zeta_\mu : V \to TV$ is the fundamental vector field generated by $\mu \in \mathfrak{g}$ [85, 94].

Geometry of LDDMM Registration

Hidden behind the complex formulae (2.164), (2.165), (2.166), and (2.167) is an intuitive geometric picture, which helps to understand the key characteristics of the LDDMM framework.

Riemannian structure The Riemannian structure of LDDMM can be understood from the abstract formulation of the diffeomorphic deformation in (2.164).

- **Images** We can think of the vector space V as containing different types of images encountered in CA by selecting $\Omega, \mathscr{M}$. The inner product $< \cdot, \cdot >_V$ can be used to measure the difference between two images. For example, brain MRI images can be represented as $V = \{I : \Omega \to R\}, \Omega \in R^3$, and the difference between I_0 and I_1 is given by $\|I_0 - I_1\|^2 =< I_0 - I_1, I_0 - I_1 >_V= \int_\Omega |I_0(x) - I_1(x)|^2 dx$.
- **Transformation** Diffeomorphic transformations on V is a Lie group $G \in \mathit{Diff}(\Omega)$ essentially determined by V through the image domain Ω. A transformation $g \in G$ deforms $I \in V$ by the action $\mathfrak{l}$ given in (1). For intensity MRI images, e.g., T1-weighted images, the action is $gI = I \circ g^{-1}$. For more information about how transformations act on different images, the reader is directed to [95, 96].

- **Riemannian manifold** Lie group $G \in Diff(\Omega)$ with the inner product $< \cdot, \cdot >_{\mathfrak{g}}$ on its Lie algebra $\mathfrak{g}$ define a Riemannian structure on G with a right invariant metric [81, 83]. Obviously by selecting different G and $<, >_{\mathfrak{g}}$, we are in fact working on different Riemannian manifolds, which may lead to different registration results as will be explained later.

Geodesics The objective of LDDMM registration (2.165) and (2.166) is to find the optimal path $g_t \in G$ continuously parametrized by time t that smoothly deforms I_0 through $I_t = g_t I_0$. The optimal path is defined as the path that costs the least in time-integrated kinetic energy $\frac{1}{2}\int_0^1 < u_t, u_t >_{\mathfrak{g}} dt$ for a given error tolerance $\frac{1}{2\sigma^2} < g_1 I_0 - I_1, g_1 I_0 - I_1 >_V$. The optimal solution given by (2.166) and (2.167) is simply a geodesic connecting I_0 and a point in the near neighborhood of I_1 determined by (2.166) and (2.167), equivalent to the Euler-Poincaré equation [89]. It can be solved by gradient flow [87] or geodesic shooting methods [82, 97].

Riemannian exponential and shape space linearization Besides a distance metric on shape space, by modeling optimal transformations as geodesics, LDDMM also linearizes the shape space to validate statistical analysis on it. Any geodesic fulfilling differential equations (2.167) and (2.168) is uniquely determined by I_0, I_1 and the initial momentum $u_0^{\flat} \in \mathfrak{g}^*$ or equivalently the initial vector field $u_0 \in \mathfrak{g}$. Taking I_0 as a common reference, any image $I \in V$ can be reached within an error tolerance by a geodesic determined by a vector field $u_0(I)$ in the vector space $\mathfrak{g}$. So the Riemannian exponential map *exp* and its inverse *log* linearize the shape space by representing any image $I = gI_0 \in G$ with $u_0(I) = log(g) \in \mathfrak{g}$ [98].

Further considerations To consolidate the foundations of the LDDMM framework, the following questions also need to be answered:

- *When does the geodesic provide a mathematically and physically valid distance metric between shapes, for example, satisfying the triangle relationship?*
- *Is the distance metric between shapes a smooth function? Does a small change in images lead to a small change of the distance between them?*
- *Does a geodesic with a limited length always exist between two shapes? Is it unique?*
- *How reliable is the linearized representation of the shape space? For example, starting from a reference shape I_0, does a small deviation of the initial vector field u_0 result in a small change of the correspondent destination shape $g_1 I_0$?*

Answers to these questions lie in the Riemannian structure on G defined by G, Ω, and $< \cdot, \cdot >_{\mathfrak{g}}$, especially the curvature of the Riemannian manifold. A few interesting facts about these questions include: [99, 100]

- In the standard LDDMM framework, the group G is set as the flow of all time-dependent vector fields u_t defined on an admissible Banach space V. The distance metric defined there is a valid distance on G_V.

- Right-invariant L^2 metric on the full diffeomorphic group $\mathit{Diff}(\Omega)$ leads to vanishing geodesic distance so that any two shapes can be deformed into each other by a deformation that is arbitrarily small with this metric.
- An H^1 metric on $\mathit{Diff}(\Omega)$ introduces nondegenerate geodesic distance.
- For a unit n-dimensional cube M^n in R^n, the diameter (maximal geodesic distance on the manifold) of the smooth volume-preserving diffeomorphism group $\mathit{SDiff}(M^n)$ is finite with an L^2 right-invariant metric for $n \geq 3$. And the diameter is infinite for $n = 2$.

These are only a few examples to remind the readers how complex the situation can be in the LDDMM framework with different Riemannian structures.

For more information about related topics on the LDDMM framework, such as the metamorphosis, currents, inner and outer metrics, and curvature of the shape space, the reader is referred to [85, 86, 88, 99].

2.3.4.2 SVF Framework

The SVF framework of CA was developed by Arsigny, Pennec [84, 90, 93, 101] as an alternative to the LDDMM framework.

Basic Setting

Similar to LDDMM, SVF framework also works on a vector space of images V and a Lie group of diffeomorphic transformation G with inner products on V and $\mathfrak{g}$. The action of $g \in G$ on $I \in V$ is exactly as given in (2.164).

SVF Diffeomorphic Registration

The registration of two images I_0, I_1 is formulated in SVF framework as

$$E(v) = \frac{1}{2} < v, v >_{\mathfrak{g}} + \frac{1}{2\sigma^2} < g_1 I_0 - I_1, g_1 I_0 - I_1 >_V \tag{2.169}$$

where

$$\partial_t g_t = DR(g_t)(v) = v \circ g_t \Rightarrow g_t = Exp(vt) \tag{2.170}$$

and $Exp : \mathfrak{g} \rightarrow G$ is the Lie group exponential map.

The goal of the registration formulation is to find the optimal $v \in \mathfrak{g}$ that generates a smooth transformation g_t to deform I_0 to I_1. The distance between shapes is given by the norm $< v, v >_{\mathfrak{g}}$.

SVF vs. LDDMM

By comparing (2.169) and (2.170) with (2.165) and (2.166), we can observe that:

Lie group structure The underlying space of the SVF framework is not a Riemannian manifold and there is no Riemannian metric, geodesic, or connection involved. Instead, SVF works on the Lie group structure of G.

One-parameter subgroup The optimal curve $g_t I_0$ connecting I_0 and I_1 is formulated by $g_t = Exp(vt)$, which is a one-parameter subgroup of G. Different from LDDMM, which constructs g_t by integrating a time-dependent vector field u_t specified by the Riemannian metric on G, the g_t of SVF is an integral curve of a stationary vector field (the source of the term) $v \in \mathfrak{g}$ (2.169). Finding the optimal v to register I_0 with I_1 can also be achieved by gradient flow or geodesic shooting algorithms as in LDDMM, but in practice, the most commonly used algorithm is the Log-Demons algorithm [91, 92].

Lie group exponential and shape space linearization Similar to LDDMM, SVF assumes that every image $I \in V$ can be reached from a template image I_0 by a Lie group exponential map $Exp(v(I))$. The linearization of the shape space is achieved by the Lie group exponential *Exp* instead of the Riemannian exponential map *exp* in LDDMM, i.e., representing $I = Exp(v(I))I_0 \in G$ by $v(I) \in \mathfrak{g}$.

So, to a first approximation, SVF framework can be regarded as a simplified alternative to LDDMM which (a) works on stationary vector fields on the tangent space of $Diff(\Omega)$, equivalently vectors in $\mathfrak{g}$ instead of time dependent vector fields in LDDMM and (b) replaces the Riemannian exponential with the Lie group exponential as explained in Fig. 2.17a, b.

The main advantage of SVF over LDDMM is that both the Lie group exponential *Exp* and its inverse *Log* can be computed with a higher efficiency as matrix operations than the Riemannian *exp* and *log* [92]. Another advantage of using SVF is that the parallel transport operation for longitudinal data analysis can also be carried

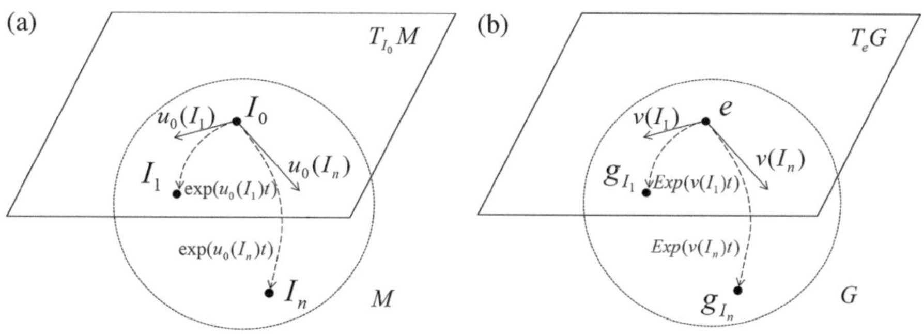

Fig. 2.17 Geometry of diffeomorphic registration in LDDMM and SVF frameworks (**a**) LDDMM registration between images $I_0 \rightarrow I_1$, $I_0 \rightarrow I_n$ and their tangent space representations on $T_{I_0}M$ by Riemannian exponential map. (**b**) SVF registration between images and their tangent space representations on $T_eG = \mathfrak{g}$ by Lie group exponential map, where $g_{I_i} \in G$ and $g_{I_i}I_0 = I_i$

out much more easily than LDDMM when the Cartan connections are selected to define the parallel transport operation on the manifold G [93].

The main difficulty of SVF appears when we consider the existence, and properties of the *optimal* transformation between images are considered, as explained in the LDDMM case. For SVF framework, we know that generally:

- The *Exp* map is usually not subjective, which means that not all images in G can be reached by an $Exp(v)$ from a template I_0.
- The optimal transformation is not smooth with regard to the images, so that a small change in images may lead to a large change of the curve connecting them.

But even with these known theoretical obstacles, SVF framework still shows promising performance in practical applications and plays an important role in CA computation [101, 102]. The reader is referred to [100, 103] for more information about the mathematical foundation of SVF.

2.3.4.3 Statistical Analysis on Shape Manifold

Statistical analysis on anatomy shapes is one of the key goals of CA. Two of the most commonly used statistical tools are the:

Statistical shape atlas, which aims to build statistics of organ shapes across diseases, populations, species, or ages. Its key task is to estimate representative organ anatomies and the intersubject shape variability.

Longitudinal shape data analysis is used to model the organ development across time, i.e., statistical analysis of the dynamic trajectories of organ shapes.

Tangent Space Statistical Shape Atlas

The abovementioned diffeomorphic registration frameworks provide two essential components for the statistical analysis on the shape space: the distance metric between anatomic shapes and the linearized representation of shape difference on a vector space.

We take one of the most commonly used statistical tools, the PCA atlas of construction of anatomic structures, as an example to explain how statistical analysis can be carried out. Usually, it is called "tangent space PCA," since the underlying vector space is essentially the tangent space at the mean shape of the PCA model [86, 104].

Given a collection of anatomic shapes $S = \{I_0, I_1, \ldots, I_{N-1}\}$, the tangent space PCA atlas consists of a *mean* model $\bar{I}$ and a covariance matrix of the deviations of the shapes from $\bar{I}$. The PCA model can be constructed thus:

1. The mean model $\bar{I}$ is computed from S.

2. The linearized shape deviation $\boldsymbol{x}^n$ of I_n, represented by the initial vector field u_0^n or its dual $u_0^{\flat n}$ for LDDMM or the stationary vector field $v^n for SVF$, of each shape $I_n, n = 0, 1, \ldots, N-1$ is computed by registering $\bar{I}$ to I_n.
3. The covariance matrix of $\{\boldsymbol{x}^n\}$ is computed.

In CA, the most commonly used (weighted) mean shape $\bar{I}$ of S is the Fréchet mean or its local version, the Karcher mean with weighting factors $\{w_n\}$ defined as

$$\bar{I} = \min_{I} \sum_{n} w_n dist^2(I, I_n) \tag{2.171}$$

where $dist(I, I_n)$ is a distance metric between I and I_n. $\bar{I}$ in (2.171) can be found using an iterative fixed point algorithm [90, 105, 106]. Given the mean model $\bar{I}$, $\{\boldsymbol{x}^n\}$ and the covariance matrix can be easily computed to construct the generative PCA model from S as illustrated in Fig. 2.18a. For more details about the mean model such as its geometrical meaning, invariancy with group actions in LDDMM and SVF frameworks, and log-domain statistics, the reader is referred to [84, 107]. A methodological discussion on statistical atlas construction can be found in [86].

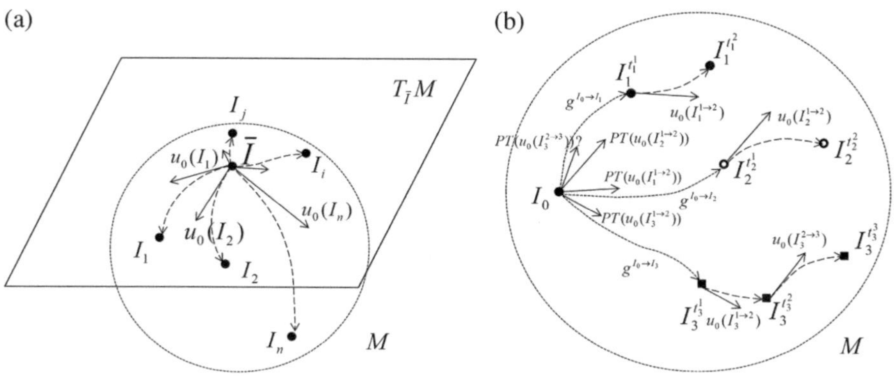

Fig. 2.18 Statistical analysis of diffeomorphism-based computational anatomy (CA) (**a**) Tangent space PCA model with LDDMM: $\bar{I}$ is the Fréchet/Karcher mean of the training data; each image I_i is represented in the vector space $T_{\bar{I}}M$ as $u_0(I_i)$, and the statistical analysis is carried out on that vector space. (**b**) Longitudinal data analysis in LDDMM: Three objects S_1, S_2, S_3 are observed at discrete time points represented as •, ∘, ■. Evolutionary trajectories are estimated by geodesics for S_1, S_2 and a piecewise geodesic for S_3. To compare trajectories of S_1 and S_2, the tangent space representations of their trajectories would undergo parallel transporting $u_0(I_1^{1\to2}), u_0(I_2^{1\to2})$ to a common reference I_0 as $PT(u_0(I_1^{1\to2})), PT(u_0(I_2^{1\to2}))$ along geodesics $g^{I_0\to I_1}, g^{I_0\to I_2}$ and then comparison in $T_{I_0}M$. How this parallel transport and comparison of the piecewise geodesic trajectory of S_3 is to be achieved remains open because the parallel transport is path-dependent [98] and there are different ways to transport $u_0(I_3^{2\to3})$ to $T_{I_0}M$

Longitudinal Shape Data Analysis

Longitudinal shape data analysis in CA analyzes the spatiotemporal variability of anatomic shapes to reveal dynamic development patterns of organs across diseases, disease course, ages, genders, etc. Its key task is to provide a model of how one individual's trajectory changes relative to those of other subjects.

Given a longitudinal data set for a collection of N objects $S = \{S_0, S_1, \ldots, S_{N-1}\}$ observed at different time slots. The observed data set for S_n includes T_n observations at time slots $\{t_n^0, t_n^1, \ldots, t_n^{T_n}\}$ as $I_n = \{I_n^{t_n^0}, I_n^{t_n^1}, \ldots, I_n^{t_n^{T_n}}\}$. Longitudinal data analysis should answer the following questions:

- How is the trajectory of each object evolution estimated and how is its shape predicted at any time t from limited observations?
- How are trajectories of different objects compared?
- How can a statistical atlas of evolutionary trajectories be constructed?

Trajectory Estimation

Trajectory estimation involves finding a continuous (smooth) curve in the spatiotemporal shape space that best fits the observed discrete time data of a single object. In diffeomorphism-based CA, existing solutions include:

- **Geodesic trajectory** Given the observed data of S_n, $I_n = \{I_n^{t_n^0}, I_n^{t_n^1}, \ldots, I_n^{t_n^{T_n}}\}$ at T_n time slots, the evolutionary trajectory can be computed as a piecewise geodesic (Riemannian geodesic or group geodesic for LDDMM and SVF, respectively) that connects successive observed data, which can be computed by diffeomorphic registration between successive observations by LDDMM and SVF. If there are only two observations at t_0 and t_1, the trajectory is just the geodesic connecting them. Also, such a trajectory holds a correspondent (piecewise) tangent space representation, which might be explored for the trajectory comparison and statistical analysis. The disadvantage is that such a greedy method may over-fit the observed data and lose the global smoothness of the trajectory [93, 108, 109].
- **Trajectory regression or interpolation** Another solution is to fit a smooth time-dependent curve simultaneously to all the observed data by kernel regression or the first- or second-order interpolations as in [86, 110]. Such methods result in smooth trajectories, but at a cost of losing the tangent space trajectory representation.

Trajectory Comparison

Depending on the trajectory estimation methods, trajectories can be compared with corresponding strategies:

- **Parallel transport** When the trajectories are just simple geodesics as explained above, comparison of trajectories can be achieved by the well-known *parallel transport* operation to translate their corresponding tangent space representatives to a common reference. But for piecewise geodesic trajectories, the problem remains open (see Fig. 2.18b). The reader is referred to [93, 108, 111] for more details on the parallel transport operation in LDDMM and SVF frameworks and their relationship.
- **Trajectory registration** For general interpolated trajectories, in [86] a trajectory registration strategy has been proposed to register spatiotemporally the trajectory of one object with the observed data of the other object, resulting in an atemporal spatial transformation ϕ and a time wrap ψ, which can be used to represent the difference between trajectories.

Spatiotemporal Atlas Construction

Similar to the statistical atlas of anatomic shapes, a spatiotemporal atlas of the evolutionary trajectories of anatomic shapes can also be constructed. In [86, 112] a subject-specific framework has been proposed to construct such an atlas. The basic assumptions are:

- All the individuals in the population share a common mean evolutionary trajectory $M(t) = \chi_t(M_0)$ with χ_t a time-dependent spatial transformation.
- The trajectory of object S_n is a deformation of $M(t)$ by a spatial morphological deformation ϕ_n and a time wrap φ_n, given by $I_n(t) = \phi_n(M(\varphi_n(t)))$.

Then the spatiotemporal atlas can be constructed as an optimization procedure to find the optimal $\{M_0, \chi_t, \{\phi_n\}, \{\varphi_n\}\}$ to fit the observed dataset. The reader is referred to [86, 112] for more algorithmic details and applications.

2.3.4.4 Applications and Future Works

As a computational framework on shape manifolds, diffeomorphism-based CA has been widely used for general image registration [91, 107, 111], morphology-based disease diagnosis [108, 113], SSA construction [90, 105, 114–116], and longitudinal data analysis [93, 102, 112, 117] even beyond the medical image processing field [118, 119].

Future work may be carried out on the following aspects:

- Extending the diffeomorphic registration framework of CA to various image modalities and multimodality image registration
- Extending the applications of SSA to achieve shape segmentation, registration, and classification

- Building longitudinal data analysis frameworks beyond the limitations of the framework of [112], which is essentially not a general and generative spatiotemporal model that can cover the variabilities of the evolutionary shape trajectories

2.3.5 Computational Anatomy and Registration

One of the most popular approaches employed for segmenting organ regions in given medical images is nonrigid registration of organ models to the images. There exist a huge number of model-to-image registration methods with several surveys of the registration methods in, e.g., [69, 120]. In this section, a very fundamental approach to registration, a MAP estimation, is described.

Many model registration methods first normalize the locations, shapes, and sizes of the patient bodies in given images (see also Fig. 2.6) and then register the organ models to the images. The normalization of the bodies in given images is required for decreasing the variation of the locations, sizes, and shapes of bodies of patients and for setting a common coordinate system to all of the training and test images. The organ models represent the statistics of the locations and/or of the shapes of target organs. In this section, two major statistical models of organs are explained: the probabilistic atlases and the SSMs.

2.3.5.1 Probabilistic Atlas

Letting the regions of target organs in given images be denoted by R_t ($t = 1, 2, \ldots, N$), where t denotes an index number or a label of each organ and N is the number of the organs to be detected, allows determination of the regions, R_t, in a given image, $I(\boldsymbol{x}) : \mathbb{R}^3 \mapsto \mathbb{R}$, by using a classifier that judges if each voxel in the image is inside of the t-th organ or not. The optimal classifier, which minimizes the expected error ratio, decides that a voxel is inside the region, R_t, only when the following condition is satisfied (Sect. 2.2.3):

$$p(\boldsymbol{x} \in R_t | I(\boldsymbol{x})) > p(\boldsymbol{x} \notin R_t | I(\boldsymbol{x})), \tag{2.172}$$

where $p(\boldsymbol{x} \in R_t | I(\boldsymbol{x}))$ and $p(\boldsymbol{x} \notin R_t | I(\boldsymbol{x}))$ denote the posterior probability distributions. These posterior probability distributions can be computed by multiplying the prior distributions by the likelihood distributions:

$$\begin{aligned} p(\boldsymbol{x} \in R_t | I(\boldsymbol{x})) &\propto p(\boldsymbol{x} \in R_t) p(I(\boldsymbol{x}) | \boldsymbol{x} \in R_t), \\ p(\boldsymbol{x} \notin R_t | I(\boldsymbol{x})) &\propto p(\boldsymbol{x} \notin R_t) p(I(\boldsymbol{x}) | \boldsymbol{x} \notin R_t). \end{aligned} \tag{2.173}$$

The prior probability distributions, $p(\boldsymbol{x} \in R_t)$, are called *the probabilistic atlas of target organs* [22, 121–123]. A probabilistic atlas of a target organ represents the probability of each voxel being inside of R_t and is constructed from a set of training images in which the target organs are manually labeled. Let M denote the number of the training images and let $I^i(\boldsymbol{x})$ $(i = 1, 2, \ldots, M)$ denote the training images in which the location, the size, and the shape of the body are normalized. Let $b_t^i(\boldsymbol{x})$ denote labeled images corresponding to $I^i(\boldsymbol{x})$ such that

$$b_t^i(\boldsymbol{x}) = \begin{cases} 1, & (\boldsymbol{x}) \in R_t^i, \\ 0, & (\boldsymbol{x}) \notin R_t^i, \end{cases} \tag{2.174}$$

where $t = 1, 2, \ldots$ denotes the ID of a target organ and R_t^i denotes the region of the target organ in $I^i(\boldsymbol{x})$. When the number of the training images, M, is large enough, the probabilistic atlas can be constructed as follows:

$$p(\boldsymbol{x} \in R_t) = \frac{1}{M} \sum_{i=1}^{M} b_t^i(\boldsymbol{x}). \tag{2.175}$$

When M is not large enough, though, a probabilistic atlas estimated by (2.175) over-fits to the training images. For example, it often happens that the estimated prior probability distributions have many zero values at inappropriate locations. One of the techniques for avoiding this over-fitting blurs $b^i(\boldsymbol{x})$ and estimates $p(\boldsymbol{x} \in R_t)$ as follows:

$$p(\boldsymbol{x} \in R_t) = \frac{1}{M} \sum_{i=1}^{M} \sum_{\boldsymbol{v}} w(\boldsymbol{v}) b_t^i(\boldsymbol{x} - \boldsymbol{v}), \tag{2.176}$$

where $\boldsymbol{v} \in \mathbb{R}^3$ and $w(\boldsymbol{v}) \geq 0$ is a unimodal weighting function for the spatial blurring that satisfies

$$\sum_{\boldsymbol{v}} w(\boldsymbol{v}) = 1. \tag{2.177}$$

For example, a Gaussian function can be used for $w(\boldsymbol{v})$. Once $p((x, y, z) \in R_t)$ is estimated, the prior of the voxel being outside of R_t can be estimated as $p((x, y, z) \notin R_t) = 1 - p((x, y, z) \in R_t)$. Figure 2.19 shows an example of a probabilistic atlas of the liver.

For computing the posterior probabilities in (2.172), not only the prior probability distribution but also the likelihood distributions are needed, $p(I(\boldsymbol{x})|\boldsymbol{x} \in R_t)$ and $p(I(\boldsymbol{x})|\boldsymbol{x} \notin R_t)$. Using these likelihoods, one can rewrite Eq. (2.172) as follows:

$$p(\boldsymbol{x} \in R_t) p(I(\boldsymbol{x})|\boldsymbol{x} \in R_t) > p(\boldsymbol{x} \notin R_t) p(I(\boldsymbol{x})|\boldsymbol{x} \notin R_t). \tag{2.178}$$

The models of the conditional probability distributions, $p(I(\boldsymbol{x})|\boldsymbol{x} \in R_t)$ and $p(I(\boldsymbol{x})|\boldsymbol{x} \notin R_t)$, are estimated from the set of training images. For example, let $h_t(I|\boldsymbol{x} \in R_t)$ $(h_t(I|\boldsymbol{x} \notin R_t))$ denote the histogram of pixel values observed in (out

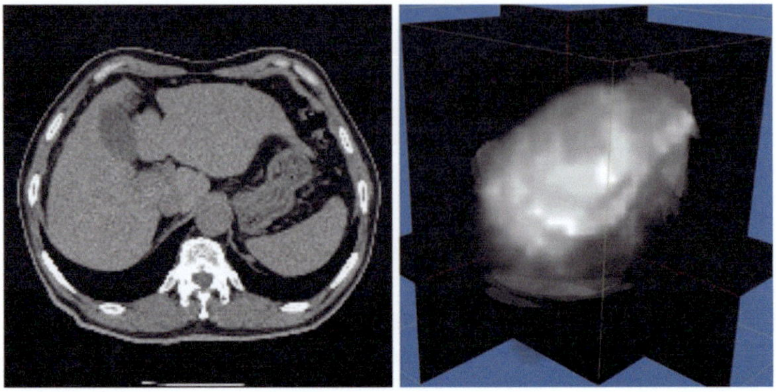

Fig. 2.19 An example of an input image and of the probabilistic atlas of the liver appeared in [124]

of) R_t^i $(i = 1, 2, \ldots, M)$. Then, the conditional distributions can be represented using, e.g., Gaussian mixture models, of which parameters are estimated by fitting the models to the histograms by means of an expectation/maximization (EM) algorithm. Once the representations of the conditional distributions are obtained, the likelihoods in (2.178) can be computed straightforwardly.

The segmentation method described above determines whether each voxel is included in the region of a target organ or not based only on the location of each voxel and on its voxel value. No image features that are obtained by observing multiple voxels, e.g., the continuity of the regions or the shapes of the region boundaries, are referred to for the segmentation. As a result, the boundaries of the segmented regions can have shapes different from those of target organs. Hence, SSMs of organs that represent the shapes of target organs are often employed for obtaining regions consistent with the targets.

2.3.5.2 SSMs

SSMs represent the prior distributions of the *boundary shapes* of target organs. An SSM is constructed from a set of training data, which are the descriptions of the target organs in the training images. Depending on the representation employed for the descriptions, different SSMs may result from identical training data. In this section, MAP estimation approaches of the registration are described for some of the representations.

SSM with Level-Set Representation

A level-set function, $f_t(\boldsymbol{x})$, represents organ regions in an implicit fashion [124–126], and SSMs based on the level-set functions [126, 127] represent the statistical varieties of the shapes of the target organs with probability distributions of the values of $f_t(\boldsymbol{x})$. The probability distributions are estimated from sets of the training data. Let R_t denote the region of an organ, where t ($t = 1, 2, \ldots, N$) denotes the label of organs. The region, R_t, can be represented by a level-set function, $f_t(\boldsymbol{x})$, as follows:

$$f_t(\boldsymbol{x}) \begin{cases} < 0, & \text{if } \boldsymbol{x} \in R_t, \\ \geq 0, & \text{otherwise.} \end{cases} \tag{2.179}$$

Given a region R_t in an image, such a level-set function (2.179) can be constructed by computing a signed distance function as described in [58]. For the construction of the statistical models, the level-set function defined in an image space is converted to a vector. Letting the number of voxels in an image be denoted by W, and letting the coordinates of the k-th voxel ($k = 1, 2, \ldots, W$) be denoted by $\boldsymbol{x}_k$, then the level-set function, $f_t(\boldsymbol{x})$, can be represented using a W-vector, $\boldsymbol{f}_t$ such that $\boldsymbol{f}_t = [f_t(\boldsymbol{x}_1), f_t(\boldsymbol{x}_2), \ldots, f_t(\boldsymbol{x}_W)]^T$. As described above, an SSM of a target organ is constructed from a set of training data. Letting a region of the target organ in the i-th training image, $I^i(\boldsymbol{x})$, be denoted by R_t^i ($i = 1, 2, \ldots, M$) and letting the corresponding level-set representation be denoted by $\boldsymbol{f}_t^i$, applying a PCA to the training set, $\mathscr{D}_t = \{\boldsymbol{f}_t^i | i = 1, 2, \ldots, M\}$ allows construction of a linear SSM as follows (see also Fig. 2.20):

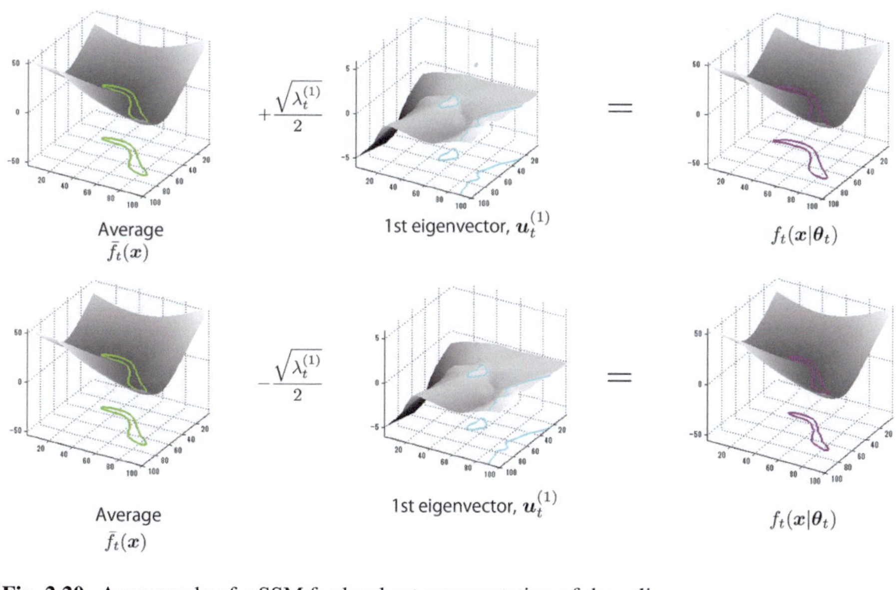

Fig. 2.20 An example of a SSM for level-set representation of the spline

Let the eigenvalues obtained by applying PCA to $\mathscr{D}_t$ be denoted by $\lambda_t^{(1)} \geq \lambda_t^{(2)} \geq \ldots \geq \lambda_t^{(W)}$, which are in decreasing order, and let the corresponding eigenvectors be denoted by $\boldsymbol{u}_t^{(s)}(\boldsymbol{x})$ $(s = 1, 2, \ldots, W)$. The following linear model is widely employed for an SSM:

$$\boldsymbol{f}_t(\boldsymbol{x}|\boldsymbol{\theta}_t) = \bar{\boldsymbol{f}}_t(\boldsymbol{x}) + \sum_{s=1}^{T_t} \theta_t^{(s)} \boldsymbol{u}_t^{(s)}(\boldsymbol{x}), \tag{2.180}$$

where $\bar{\boldsymbol{f}}_t$ is a mean vector of $\boldsymbol{f}_t^i$ $(i = 1, 2, \ldots, M)$, $T_t < W$ is a positive integer, and $\boldsymbol{\theta}_t = [\theta_t^1, \theta_t^2, \ldots, \theta_t^{T_t}]^T$ is a T_t-vector of the shape parameters.

As described in Sect. 2.2.3, the prior distribution of the shape parameters can be approximated using a multivariable Gaussian function such that

$$p(\boldsymbol{\theta}) = \frac{1}{Z} \exp\left\{ \sum_{s=1}^{T_t} \frac{\left(\theta_t^{(s)}\right)^2}{\lambda_t^{(s)}} \right\}. \tag{2.181}$$

The covariance matrix of the Gaussian distribution in (2.181) is diagonal, and the diagonal components consist of the eigenvalues, $\lambda_t^{(s)}$. The number of eigenvectors, T_t, in (2.180) can be determined as the minimum integer that satisfies $\sum_{s=1}^{T_t} \lambda_t^s / \sum_{s=1}^{W} \lambda_t^s > \gamma_t$, where γ_t is a threshold less than one (1), e.g., $\gamma_t = 0.95$. The linear model (2.180) constrains the level-set representation, $\boldsymbol{f}_t$, to an T_t-dimensional subspace.

Given an image, the region of the target organ can be segmented by estimating the values of $\boldsymbol{\theta}_t$ in (2.180), and the estimates can be obtained by maximizing the posterior probability distribution. Letting the MAP estimates be denoted by $\hat{\boldsymbol{\theta}}$, then,

$$\hat{\boldsymbol{\theta}}_t = \arg\max_{\boldsymbol{\theta}_t} \log p(\boldsymbol{\theta}_t | I(\boldsymbol{x})), \tag{2.182}$$

where $I(\boldsymbol{x})$ denotes a given image. It is difficult, though, to solve the problem (2.182) directly.

To make the problem tractable, a latent variable, $b_t(\boldsymbol{x})$, is introduced, where $b_t(\boldsymbol{x}) = 1$ if $\boldsymbol{x} \in R_t$ and $b_t(\boldsymbol{x}) = 0$ if $\boldsymbol{x} \notin R_t$. If not only an image, I, but also $b_t(\boldsymbol{x})$ are measured, then the problem of the estimation of $\boldsymbol{\theta}_t$ based on $p(\boldsymbol{\theta}_t, b_t(\boldsymbol{x}) | I(\boldsymbol{x}))$ is tractable. An EM algorithm is useful for the estimation.

In the EM algorithm, the parameters, $\boldsymbol{\theta}$, are estimated by maximizing the following MAP estimation problem [128]:

$$\hat{\boldsymbol{\theta}}_t = \arg\max_{\boldsymbol{\theta}} E_{b_t|I,\boldsymbol{\theta}} \left(\log p(\boldsymbol{\theta}\, b_t(\boldsymbol{x})|I)\right) \tag{2.183}$$

$$= \arg\max_{\boldsymbol{\theta}} \sum_{k=1}^{W} \sum_{b=0}^{1} p(b_t(\boldsymbol{x}_k) = b | I(\boldsymbol{x}_k), \boldsymbol{\theta}) \log p(\boldsymbol{\theta}, b_t(\boldsymbol{x}_k) = b | I(\boldsymbol{x})) \tag{2.184}$$

where $b \in \{0, 1\}$. The function to be maximized is the marginal distribution of $\log p(\boldsymbol{\theta}_t, b_t(\boldsymbol{x}_k)|I(\boldsymbol{x}))$ with respect to $b_t(\boldsymbol{x}_k)$. It should be noted that

$$\log p(\boldsymbol{\theta}_t, b_t(\boldsymbol{x}) = b|I(\boldsymbol{x})) \propto \log p(I|b_t(\boldsymbol{x}) = b, \boldsymbol{\theta}_t) + \log p(b_t(\boldsymbol{x}) = b|\boldsymbol{\theta}_t) + \log p(\boldsymbol{\theta}_t), \tag{2.185}$$

and that $p(I(\boldsymbol{x})|b_t(\boldsymbol{x}) = b, \boldsymbol{\theta}_t) = p(I(\boldsymbol{x})|b_t(\boldsymbol{x}) = b)$. Here, $p(I(\boldsymbol{x})|b_t(\boldsymbol{x}) = 0)$ and $p(I(\boldsymbol{x})|b_t(\boldsymbol{x}) = 1)$ can be evaluated based on the histograms of the voxel values observed outside of R_t and inside of R_t in training images, respectively. $p(b_t(\boldsymbol{x}_k) = b|\boldsymbol{\theta})$ in (2.185) can be approximated by using $f_t(\boldsymbol{\theta})$. Let $p(b_t(\boldsymbol{x}_k) = 1)$ and $p(b_t(\boldsymbol{x}_k) = 0)$ be abbreviated by p_{in} and p_{out}, respectively. Then, $p_{\text{in}} + p_{\text{out}} = 1$ holds. For the evaluation of the values of the probabilities, p_{in} and p_{out}, an inverse of a logit function [128], whose definition is described below, is often introduced. A logit function of p_{in} is defined as follows:

$$g_{\text{in}} = \text{logit}(p_{\text{in}}) \equiv \log\left(\frac{p_{\text{in}}}{p_{\text{out}}}\right). \tag{2.186}$$

The inverse of the logit function is given as follows:

$$p_{\text{in}} = \text{logit}^{-1}(g_{\text{in}}) = \exp(g_{\text{in}})/Z, \tag{2.187}$$

$$p_{\text{out}} = \text{logit}^{-1}(g_{\text{out}}) = 1/Z, \tag{2.188}$$

where

$$Z = 1 + \exp(g_{\text{in}}). \tag{2.189}$$

It should be noted that $0 < p_{\text{in}}, p_{\text{out}} < 1$ and $p_{\text{in}} + p_{\text{out}} = 1$ hold and that $p_{\text{in}} = p_{\text{out}} = 1/2$ when $g_{\text{in}} = 0$. Now the values of p_{in} and p_{out} can be evaluated by using the level-set function, $f_t(\boldsymbol{\theta})$, as

$$p_{\text{in}} = \text{logit}^{-1}\left(-f_t(\boldsymbol{x}|\boldsymbol{\theta})\right). \tag{2.190}$$

At the boundary of the region where $f_t(\boldsymbol{x}|\boldsymbol{\theta}) = 0$, $p_{\text{in}} = p_{\text{out}} = 1/2$ holds and p_{in} monotonically increases as f_t increases.

The MAP estimates shown in (2.182) can be obtained by an EM algorithm as follows:

1. Set the initial value of the parameters: $\boldsymbol{\theta}_t = \boldsymbol{\theta}_t^{\text{OLD}}$
2. E-STEP: Compute $w_k(b) = p(b_t(\boldsymbol{x}_k) = b|I, \boldsymbol{\theta}_t^{\text{OLD}})$ as follows:

$$w_k(b) \propto p(I|b_t(\boldsymbol{x}_k) = b)p(b_t(\boldsymbol{x}_k) = b|\boldsymbol{\theta}^{\text{OLD}}). \tag{2.191}$$

3. M-STEP: Solve the following problem and obtain $\boldsymbol{\theta}^{\mathrm{NEW}}$:

$$\boldsymbol{\theta}_t^{\mathrm{NEW}} = \arg\max_{\boldsymbol{\theta}_t} \sum_k \sum_{b=0}^{1} w_k(b) \log p(\boldsymbol{\theta}_t, b_t(\boldsymbol{x}_k) = b|I). \tag{2.192}$$

4. $\boldsymbol{\theta}^{\mathrm{OLD}} \leftarrow \boldsymbol{\theta}^{\mathrm{NEW}}$ and back to E-STEP if it is not converged.

The target function to be maximized shown in (2.192) can be rewritten as follows:

$$\sum_k \sum_b w_k(b) \log p(\boldsymbol{\theta}_t, b_t(\boldsymbol{x}_k) = b|I) = \sum_k \sum_b \{w_k(b) \log p(b_t(\boldsymbol{x}_k) = b|\boldsymbol{\theta})\} + \log p(\boldsymbol{\theta}_t). \tag{2.193}$$

The second term in the right side of (2.193), $p(\boldsymbol{\theta}_t)$, is the prior probability distribution of $\boldsymbol{\theta}_t$ and is given as (2.181).

In the process of the estimation of $\boldsymbol{\theta}$, no candidate points of the boundaries of target organs are explicitly extracted from given images; hence the distances between the region boundary represented by the model and the image points extracted as the candidates of the region boundary are not explicitly measured at all. The fit between the region represented by the implicit function and the given images is evaluated based on the differences in voxel values between the interior and the exterior of each target organ. This methodology is quite different from that employed for SSMs with explicit representation of region boundaries, which will be described below.

SSM with NURBS Surface Representation

SSMs with explicit representation of the surfaces of target organs are also widely employed for the segmentation of given medical images. The surfaces are explicitly represented with some parameters, and the prior probabilities of those values are represented by the SSMs. For image segmentation, the values of those parameters are estimated so that the resultant surfaces fit to the boundaries of target organs in given images. Different from the segmentation methods with implicit organ representation, the likelihoods of the parameter values are defined based on the distances between the model surfaces and the boundary candidate points extracted from the given images. In the followings, SSMs that represent the surfaces with NURBS surfaces are described.

NURBS surfaces are determined by a set of control points and a set of basis functions. Given the set of basis functions, one can vary the shape of the NURBS surfaces by changing the locations of the control points: The surfaces are parametrized by the coordinates of the control points. It is not difficult to compute the normal direction at each point on the surface, and this helps the computation of the distances between the surface and the boundary candidates extracted from given images along the normal directions.

NURBS represent surfaces thus:

$$\boldsymbol{x}(s,t) = \frac{\sum_{u=0}^{N_s-1}\sum_{v=0}^{N_t-1} w_{uv} b_u^n(s) b_v^n(t) \boldsymbol{P}_{uv}}{\sum_{u=0}^{N_s-1}\sum_{v=0}^{N_t-1} w_{uv} b_u^n(s) b_v^n(t)}, \tag{2.194}$$

where $b_*^n(s)$ denote B-spline basis functions of order n that are periodic over the range $0 \leq s \leq L$, w_{uv} denote weights of control points, and the three-vectors, $\boldsymbol{P}_{uv}$, denote the coordinates of the control points. Let

$$B_{uv}(s,t) = \frac{w_{uv} b_u^n(s) b_v^n(t)}{\sum_u \sum_v w_{uv} b_u^n(s) b_v^n(t)}. \tag{2.195}$$

Then, the Eq. (2.194) can be rewritten as follows:

$$\boldsymbol{x}(s,t) = \mathscr{B}(s,t)\boldsymbol{\theta}, \tag{2.196}$$

where $\mathscr{B}(s,t)$ is a $3 \times 3N_sN_t$ matrix such that

$$\mathscr{B}(s,t) = \begin{bmatrix} B_{00} & & & B_{01} & & & B_{02} & & & \ldots & B_{N_sN_t} & & \\ & B_{00} & & & B_{01} & & & B_{02} & & \ldots & & B_{N_sN_t} & \\ & & B_{00} & & & B_{01} & & & B_{02} & \ldots & & & B_{N_sN_t} \end{bmatrix}. \tag{2.197}$$

Here, $B_{uv} = B_{uv}(s,t)$ are the products of the B-spline basis functions as shown in (2.195), and $\boldsymbol{\theta}$ denotes the coordinates of all control points: $\boldsymbol{\theta} = [\boldsymbol{P}_{00}^T, \boldsymbol{P}_{01}^T, \ldots, \boldsymbol{P}_{N_sN_t}^T]^T$. The matrix $\mathscr{B}(s,t)$ can be determined in advance, and the shape of the surfaces can be varied by changing the values of the components of the vector $\boldsymbol{\theta}$. Figure 2.21 An example of a surface (kidney) represented by NURBS.

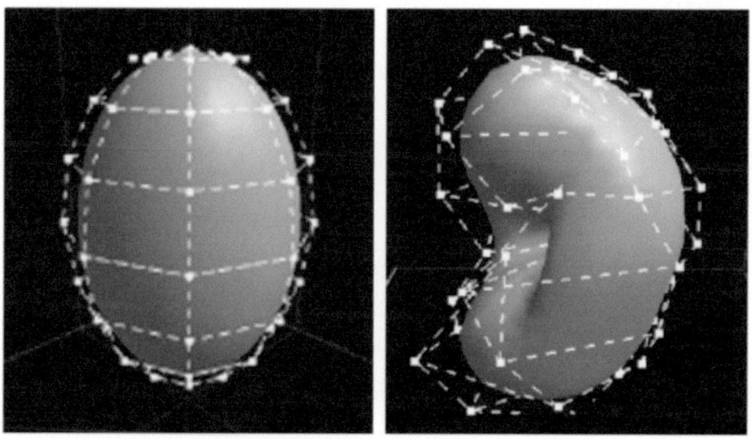

Fig. 2.21 An example of closed surface represented by NURBS appeared in [129]

When the model in (2.197) is employed, the model can be registered to given images by estimating the values of $\boldsymbol{\theta}$ by means of a MAP estimation or Bayesian estimation.

A NURBS-based SSM represents a prior probability distribution of $\boldsymbol{\theta}$, which is constructed from a set of training images. Assume the boundaries of target organs in the training images are manually extracted by some experts. Letting the boundary surface in the i-th image be denoted by S^i $(i = 1, 2, \ldots, M)$ and letting $\boldsymbol{\theta}^i$ denote the parameter of NURBS describing S^i, referring to the set of the parameters, $\{\boldsymbol{\theta}^i | i = 1, 2, \ldots, M\}$, allows estimation of the prior probability distribution, $p(\boldsymbol{\theta})$. Once the prior, $p(\boldsymbol{\theta})$, is determined, the simultaneous prior probability distribution of the points, $\boldsymbol{x}(s,t)$, of the NURBS surfaces can be determined.

For statistically registering the NURBS surfaces, statistical models of the residuals between the NURBS surfaces and the corresponding candidates of organ boundaries detected by, e.g., edge detectors, are also needed. Letting $\boldsymbol{x}_j = \boldsymbol{x}(s_j, t_j)$ $(j = 1, 2, \ldots, L)$ be a point of a NURBS surface and letting $\tilde{\boldsymbol{x}}_j$ denote a detected candidate point corresponding to $\boldsymbol{x}_j$, the conditional probability distributions of the residuals, $p(\tilde{\boldsymbol{x}}_j - \boldsymbol{x}_j | \boldsymbol{x}_j) = p(\tilde{\boldsymbol{x}}_j | \boldsymbol{x}_j)$, supply the likelihood distributions when $\tilde{\boldsymbol{x}}_j$ for each j is given. Using $p(\boldsymbol{\theta})$ and $p(\tilde{\boldsymbol{x}} | \boldsymbol{x})$ allows computation of a posterior probability of $\boldsymbol{\theta}$ as follows:

$$p(\boldsymbol{\theta} | \tilde{\boldsymbol{x}}_1, \tilde{\boldsymbol{x}}_2, \ldots, \tilde{\boldsymbol{x}}_L) \propto p(\boldsymbol{\theta}) \prod_{j=1}^{L} p(\tilde{\boldsymbol{x}}_j | \boldsymbol{x}_j) = p(\boldsymbol{\theta}) \prod_{j=1}^{L} p(\tilde{\boldsymbol{x}}_j | \mathscr{B}(s_j, t_j)\boldsymbol{\theta}), \tag{2.198}$$

where L denotes the number of the samples and it is assumed that the residuals are determined independently. The MAP estimates of $\boldsymbol{\theta}$ can be obtained by maximizing $p(\boldsymbol{\theta} | \tilde{\boldsymbol{x}}_1, \tilde{\boldsymbol{x}}_2, \ldots, \tilde{\boldsymbol{x}}_L)$ in (2.198).

For example, a Gaussian distribution can be employed for representing $p(\boldsymbol{\theta})$: The mean and the covariance matrix, $\bar{\boldsymbol{\theta}}$ and Σ, can be estimated from the set of the training parameters, $\{\boldsymbol{\theta}^i | i = 1, 2, \cdots, M\}$:

$$p(\boldsymbol{\theta}) = \frac{1}{Z} \exp\left\{ -\frac{1}{2} (\boldsymbol{\theta} - \bar{\boldsymbol{\theta}})^T \Sigma^{-1} (\boldsymbol{\theta} - \bar{\boldsymbol{\theta}}) \right\}, \tag{2.199}$$

where Z is a constant for the normalization, $\bar{\boldsymbol{\theta}}$ is a $3N_sN_t$ vector, and Σ is a $3N_sN_t \times 3N_sN_t$ matrix. The likelihood can also be modeled by a Gaussian. For example:

$$p(\tilde{\boldsymbol{x}}_j | \boldsymbol{x}_j) = \frac{1}{Z} \exp\left\{ -\frac{1}{2} \frac{\eta_i^2}{\sigma^2}, \right\} \tag{2.200}$$

where η_i denotes the distance between $\tilde{\boldsymbol{x}}_j$ and $\boldsymbol{x}_j$ measured along the direction of the unit normal vector, $\boldsymbol{n}_j$, perpendicular to the NURBS surface at $\boldsymbol{x}_j$ such that $\eta_j = \boldsymbol{n}_j^T (\tilde{\boldsymbol{x}}_j - \boldsymbol{x}_j)$.Then, the cost function, $E(\boldsymbol{\theta})$, which is derived from the negative logarithm of the posterior probability distribution, is obtained:

$$E(\boldsymbol{\theta}) \propto (\theta - \bar{\theta})^T \Sigma^{-1} (\theta - \bar{\theta}) + \frac{1}{\sigma^2} \sum_{j=1}^{N} \{\boldsymbol{n}_j^T (\mathscr{B}(s_j, t_j)\boldsymbol{\theta} - \tilde{\boldsymbol{x}}_j)\}^2 . \qquad (2.201)$$

Minimizing $E(\boldsymbol{\theta})$, the estimates, $\hat{\boldsymbol{\theta}}$, that maximize the posterior probability of $\boldsymbol{\theta}$ can be obtained. $E(\boldsymbol{\theta})$ has a quadratic form of $\boldsymbol{\theta}$, and the computation of the minimizer is straightforward if $\tilde{\boldsymbol{x}}_j$ ($j = 1, 2, \ldots, N$) are fixed. Once $\boldsymbol{\theta}$ is updated to $\hat{\boldsymbol{\theta}}$, then the locations of $\boldsymbol{x}_j$ move, and the corresponding candidate points, $\tilde{\boldsymbol{x}}_j$, change. The registration methods, hence, iteratively minimize $E(\boldsymbol{\theta})$ and update $\tilde{\boldsymbol{x}}_j$ until they are converged.

As mentioned, a NURBS-based SSMs are constructed from a set of training data, which are obtained by manually fitting a NURBS surface to the boundary of a target organ in each of the training images. The number of control points should be identical among all fitted NURBS surfaces, and each of the control points of one NURBS surface should be corresponded to one of the control points of each of the other surfaces. Making this correspondence is not straightforward.

The prior distribution, $p(\boldsymbol{\theta})$, is estimated based on the control points of the training data. Different from the SSMs with implicit representation, it is guaranteed that the surfaces represented by the SSMs are single and closed if the surfaces in the training images are all single and closed. It is *not* guaranteed, though, that the surfaces represented by the NURBS surfaces are simple: The surfaces would have self-intersections even if all of the training surfaces are simple and have no self-intersections. It is difficult to find the global maximum of the posterior probability, and the algorithm described above can find only the local maximum. The estimated parameters, $\hat{\boldsymbol{\theta}}$, hence vary, depending on the initial values of the parameters, $\boldsymbol{\theta}$.

SSM with PDM (Subspace Representation)

PDMs represent surfaces of target organs using sets of points distributed on them. Letting the coordinates of the j-th point on a surface be denoted by $\boldsymbol{x}_j \in \mathbb{R}^3$ ($j = 1, 2, \ldots, N$), a PDM represents a surface with a $3N$-vector, $\boldsymbol{x} = [\boldsymbol{x}_1^T \boldsymbol{x}_2^T \ldots \boldsymbol{x}_L^T]^T$, and a corresponding SSM represents the prior probability distribution of $\boldsymbol{x}$. Combining the prior probability distribution with the likelihood distribution of $\boldsymbol{x}$ allows computation of the posterior probability distribution of $\boldsymbol{x}$ and registration of the models to given images by estimating the coordinates of each point, $\boldsymbol{x}_i$, by maximizing the posterior probability.

A prior probability distribution of $\boldsymbol{x}$ is estimated from a set of training data. Again letting S^i ($i = 1, 2, \ldots, M$) denote surfaces manually extracted by experts from the i-th image, letting $\boldsymbol{x}_j^i$ ($j = 1, 2, \ldots, N$) denote a set of N corresponding points generated on S^i, and letting $3N$-vectors $\boldsymbol{x}^i = [(\boldsymbol{x}_1^i)^T (\boldsymbol{x}_2^i)^T \ldots (\boldsymbol{x}_N^i)^T]^T$ ($i = 1, 2, \ldots, M$) denote a set of the generated corresponding points, the prior distribution of $\boldsymbol{x}$ can be estimated from the set, $\{\boldsymbol{x}^i | i = 1, 2, \ldots, M\}$. Letting the prior be denoted by

$p(\boldsymbol{x})$, many PDM-based SSMs including the ASMs represent $p(\boldsymbol{x})$ by degenerated Gaussian functions that constrain represented surfaces to subspaces. Letting the sample mean of the training data be denoted by $\bar{\boldsymbol{x}} = (\sum_{i=1}^{N} \boldsymbol{x}^i)/M$ and letting the empirical covariance matrix of them be denoted by Σ_{emp} such that

$$\Sigma_{\text{emp}} = \frac{1}{M} \sum_{i=1}^{N} (\boldsymbol{x}^i - \bar{\boldsymbol{x}})^T (\boldsymbol{x}^i - \bar{\boldsymbol{x}}), \tag{2.202}$$

a subspace for the representation is spanned by the eigenvectors of the empirical covariance matrix. Letting the eigenvalues of the covariance matrix, Σ_{emp}, be denoted by $\lambda_1, \lambda_2, \ldots, \lambda_{3L}$, where they are in decreasing order, $\lambda_1 \geq \lambda_2 \geq \ldots$, and letting the corresponding eigenvectors be denoted by $\boldsymbol{u}_1, \boldsymbol{u}_2, \ldots, \boldsymbol{u}_{3L}$, using the eigenvectors and the mean vector, the surfaces can be linearly represented as

$$\boldsymbol{x} = \sum_{k=1}^{T} \theta_k \boldsymbol{u}_k + \bar{\boldsymbol{x}}, \tag{2.203}$$

where T denotes the number of the eigenvectors used for the representation and θ_k denotes a weight for each eigenvector. Equation (2.203) can be rewritten as follows:

$$\boldsymbol{x} = U\boldsymbol{\theta} + \bar{\boldsymbol{x}}, \tag{2.204}$$

where a $3N \times T$ matrix, U, is composed of the eigenvectors as follows:

$$U = \left[\boldsymbol{u}_1 \middle| \boldsymbol{u}_2 \middle| \ldots \middle| \boldsymbol{u}_T \right] \tag{2.205}$$

and a T-vector, $\boldsymbol{\theta}$, denotes the shape parameters that controls the shapes, where $\boldsymbol{\theta} = [\theta_1, \theta_2, \ldots \theta_T]^T$. The model can be registered by estimating the values of the shape parameters, $\boldsymbol{\theta}$, in (2.204).

Assuming that $\boldsymbol{x}$ obeys a Gaussian and is constrained to the subspace shown in (2.204):

$$p(\boldsymbol{x}) = \mathcal{N}(\boldsymbol{x}|\mu, \Sigma_{\text{sub}}). \tag{2.206}$$

Then, the mean of the Gaussian satisfies $\bar{\boldsymbol{x}}$, $\mu = \bar{\boldsymbol{x}}$, and the covariance matrix, Σ_{sub}, in (2.206) is obtained as follows:

$$\Sigma_{\text{sub}} = U \Sigma_{\text{eigen}} U^T, \tag{2.207}$$

where $\Sigma_{\text{eigen}} = \text{diag}\,(\lambda_1, \lambda_2, \ldots, \lambda_T)$. It is often assumed that the shape parameters $\boldsymbol{\theta}$ in (2.204) obey a Gaussian distribution:

$$p(\boldsymbol{\theta}) = \mathcal{N}(\boldsymbol{\theta}|\boldsymbol{0}, I), \tag{2.208}$$

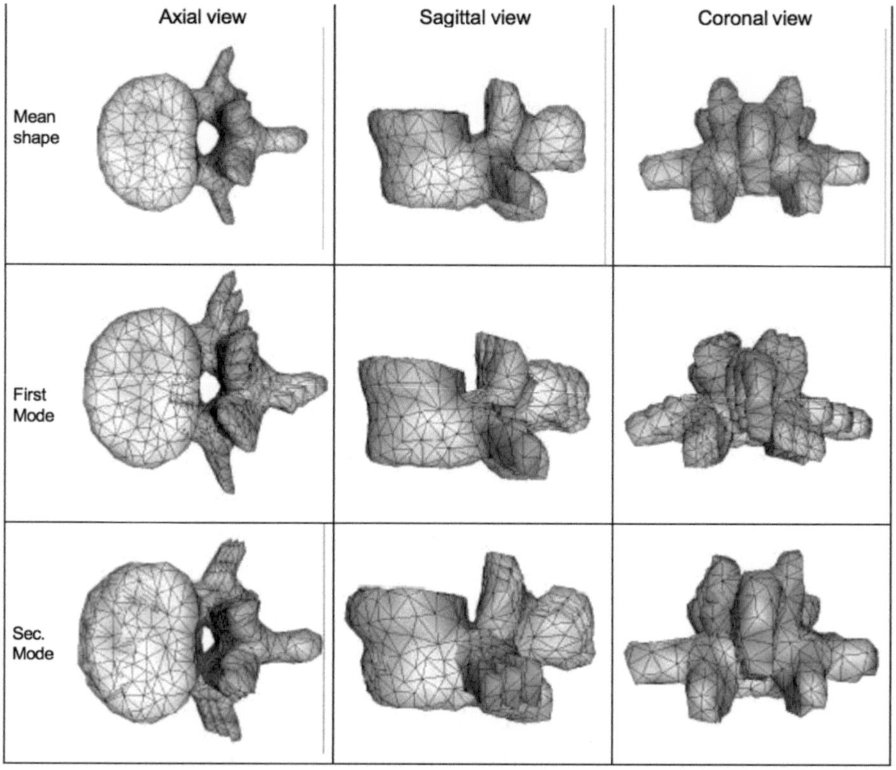

Fig. 2.22 Visualization of the shape model appeared in [130]. *Top row*: Axial, sagittal, and coronal views of the mean shape, a vertebra. *Second row*: First eigenmode of the generated model visualized by an overlay of the mean shape and two deformed shapes according to the first eigenvector. *Third row*: Mean shape and deformed shapes according to the second mode

where I denotes a unit matrix. Figure 2.22 shows an example of an SSM of a lumbar vertebra represented by a PDM that appeared in [130]. The PDM-based SSMs are registered to given images by estimating the values of the shape parameters, $\boldsymbol{\theta}$, so that every point on the surface locates on the boundary of a target organ. In the registration process, candidate points of the boundaries are first extracted from given images, and then a corresponding point, $\tilde{\boldsymbol{x}}_j$, is selected for each of a point of the PDM, $\boldsymbol{x}_j$. The likelihood of $\boldsymbol{x}$ is computed based on the distances between the corresponding points, $\tilde{\boldsymbol{x}}_j$ ($j = 1, 2, \ldots, L$), and the model points, $\boldsymbol{x}_j$. Assuming that the residuals between $\boldsymbol{x}_j$ and the corresponding candidate points $\tilde{\boldsymbol{x}}_j$ obey an isotropic Gaussian such that

$$p(\tilde{\boldsymbol{x}}_j|\boldsymbol{x}_j) = \frac{1}{(2\pi)^{3/2}\sigma^3} \exp\left\{-\frac{1}{2}\frac{\|\tilde{\boldsymbol{x}}_j - \boldsymbol{x}_j\|^2}{\sigma^2}\right\}, \tag{2.209}$$

then the posterior probability distribution, $p(\boldsymbol{x}|\tilde{\boldsymbol{x}}_1, \tilde{\boldsymbol{x}}_2, \ldots, \tilde{\boldsymbol{x}}_N)$ can be computed as

$$p(\boldsymbol{x}|\tilde{\boldsymbol{x}}_1, \tilde{\boldsymbol{x}}_2, \cdots, \tilde{\boldsymbol{x}}_N) \propto p(\boldsymbol{x}) \prod_{j=1}^{N} p(\tilde{\boldsymbol{x}}_j|\boldsymbol{x}_j). \tag{2.210}$$

Substituting (2.206) and (2.209)–(2.210) and computing the negative logarithm of the resultant posterior distribution result in the following quadratic cost function, $E(\boldsymbol{\theta})$:

$$\begin{aligned} E(\boldsymbol{\theta}) &= -\log p(\boldsymbol{x}|\tilde{\boldsymbol{x}}_1, \tilde{\boldsymbol{x}}_2, \ldots, \tilde{\boldsymbol{x}}_N) = -\log p(\boldsymbol{x}) - \sum \log p(\tilde{\boldsymbol{x}}_j|\boldsymbol{x}_j) \\ &\propto (U\boldsymbol{\theta} - \bar{\boldsymbol{x}})^T \Sigma_{\mathrm{sub}}^{-1} (U\boldsymbol{\theta} - \bar{\boldsymbol{x}}) + \frac{1}{\sigma^2} \|\tilde{\boldsymbol{x}} - U\boldsymbol{\theta} - \bar{\boldsymbol{x}}\|^2, \end{aligned} \tag{2.211}$$

where $\tilde{\boldsymbol{x}} = [\tilde{\boldsymbol{x}}_1^T, \tilde{\boldsymbol{x}}_2^T, \ldots, \tilde{\boldsymbol{x}}_N^T]^T$. The MAP estimate that maximizes the posterior probability in (2.210) can be obtained by minimizing the cost function, $E(\boldsymbol{\theta})$, shown in (2.211). If the candidate points, $\tilde{\boldsymbol{x}}$, are fixed, then the optimal parameter, $\hat{\boldsymbol{\theta}}$, that minimizes $E(\boldsymbol{\theta})$ can be computed analytically. Once the values of the shape parameters are updated to $\tilde{\boldsymbol{\theta}}$, then the surface represented by the PDM varies and the corresponding points, $\tilde{\boldsymbol{x}}_j$, should also be updated. The final estimates of the shape parameters, $\boldsymbol{\theta}$, are obtained by iteratively updating the shape parameters and the corresponding points until they converge.

The statistical models shown above are constructed from the corresponding points generated on the training surfaces, S^j ($j = 1, 2, \ldots, M$), and it is not easy to generate these corresponding points. It should be noted that many methods for generating the corresponding points on surfaces have been proposed and that different SSMs can be constructed from an identical set of training data if different methods for the corresponding generation are employed.

SSM with Point Distribution Model (MRF Representation)

MRF [54, 131] can be employed for representing PDM-based SSMs [66, 70, 132, 133]. With MRF, multivariable probability distributions are represented with products of single-variable and two-variable probabilities. Let $\boldsymbol{x}_j$ ($j = 1, 2, \ldots, N$) denote the coordinates of the N points that represent a target surface and let $I_j = I_j(\boldsymbol{x})$ denote an appearance observed in a local small region around $\boldsymbol{x}_j$ in a given image. The following equation represents an MRF model of the SSMs:

$$p(\boldsymbol{x}_1, \ldots, \boldsymbol{x}_N, I_1, \ldots, I_N) = \prod_{j=1}^{N} p(\boldsymbol{x}_j) p(I_j|\boldsymbol{x}_j) \prod_{(s,t)\in\mathscr{C}} p(\boldsymbol{x}_s, \boldsymbol{x}_t), \tag{2.212}$$

where $\mathscr{C}$ is a set of pairs of indexes: $(s,t) \in [1,2,\ldots,N]^2$. In (2.212), $p(\boldsymbol{x}_j)$ represents the prior probability distribution of $\boldsymbol{x}_j$, $p(I_j|\boldsymbol{x}_j)$ represents the likelihood of $\boldsymbol{x}_j$ when a target image is given, and $p(\boldsymbol{x}_s, \boldsymbol{x}_t)$ represents the simultaneous probability distribution of two of the points. MRF models can be represented by using undirected graphical models, in which a node represents one of the variables and an edge connecting two nodes represents the conditional dependencies of the two variables. The set of the pairs, $\mathscr{C}$, corresponds to a set of the edges in the graphical model.

$p(\boldsymbol{x}_j)$, $p(I_j|\boldsymbol{x}_j)$, and $p(\boldsymbol{x}_s, \boldsymbol{x}_t)$ can be estimated using a set of training data, $\{\boldsymbol{x}_j^i|i = 1,2,\ldots,N, j = 1,2,\ldots,M\}$, where $\boldsymbol{x}_j^i$ denotes the j-th point on the surface of a target organ in the i-th training image. The prior distribution, $p(\boldsymbol{x}_j)$, can be estimated directly from the set, $\{\boldsymbol{x}_j^i|i = 1,2,\ldots,N\}$. For example, employing a Gaussian distribution to represent $p(\boldsymbol{x}_j)$, then the mean and the covariance matrix are estimated from the set. The simultaneous distribution, $p(\boldsymbol{x}_s, \boldsymbol{x}_t)$, can also be estimated from the training set. Letting $\boldsymbol{x}_{st}^i = [(\boldsymbol{x}_s^i)^T, (\boldsymbol{x}_t^i)^T]^T$ allows estimation of $p(\boldsymbol{x}_s, \boldsymbol{x}_t)$ from the set, $\{\boldsymbol{x}_{st}^i|i = 1,2,\ldots,N\}$. Estimating $p(I_j|\boldsymbol{x}_j)$ requires a training set of local appearances in addition to the training set of the points. For example, letting an L^3-vector, I_j^i, denote an appearance observed in a $L \times L \times L$ local cube in $I^i(\boldsymbol{x})$ whose center is located at $\boldsymbol{x}_j^i$, the conditional probability of the appearance, $p(I_j|\boldsymbol{x}_j)$, can be estimated from a set of the appearances, $\{I_j^i|i = 1,2,\ldots,N\}$. The prior probability and the likelihood do not contain shape information and can be computed without using any information of other points. In (2.212), the statistical variety of the shape is represented by the simultaneous probability distributions, $\prod p(\boldsymbol{x}_s, \boldsymbol{x}_t)$.

Using the model shown in (2.212) allows segmentation of target organs in given images by estimating the marginal posterior probability distribution of each point, $p(\boldsymbol{x}_j|I)$, as follows: first, the posterior probability is temporally evaluated without using the simultaneous probability distributions, as $p'(\boldsymbol{x}_j) \propto p(\boldsymbol{x}_j)p(I_j|\boldsymbol{x}_j)$. Employing this temporal estimate as the initial state, the posterior probability of each point is then estimated based on the model shown in (2.212). The shape model that is represented by the simultaneous probabilities is now explicitly used. For this estimation, several techniques developed for inferring on undirected graphical models can be employed, e.g., belief propagation or MCMC.

Here, it should be remembered that multivariable Gaussian functions can be represented by products of single-variable and two-variable Gaussian functions. The statistical model shown in (2.206), hence, can be written as follows:

$$p(\boldsymbol{x}) = \mathcal{N}(\boldsymbol{x}|\mu, \Sigma_{\text{sub}}) = \prod_j \mathcal{N}(\boldsymbol{x}_j|\boldsymbol{\mu}_j, \Sigma_j) \prod_{(s,t)\in\mathscr{C}} \mathcal{N}(\boldsymbol{x}_{st}|\boldsymbol{\mu}_{st}, \Sigma_{st}), \tag{2.213}$$

where $\mathscr{C} = \{(s,t)|(\Sigma_{\text{sub}}^{-1})_{st} \neq 0\}$ and $(\Sigma_{\text{sub}}^{-1})_{st}$ denotes the (s,t) component of a matrix Σ_{sub}^{-1}. An inverse matrix of a covariance matrix is called a *precision matrix*, and its zero components correspond to the pairs of variables that are conditionally dependent. When a multivariable Gaussian distribution is employed for the prior probability distribution of the points of a PDM, the structure of the corresponding undirected graphical model, or the set of the edges, $\mathscr{C}$ in (2.212), is determined by the precision matrix.

Multi-atlas

The probabilistic atlases and the SSMs are constructed from training images, in which target organs are manually labeled. Here, a multi-atlas-based method that uses training images more directly for the segmentation of given images is described [134–137, 137, 138, 138–141].

A multi-atlas-based method segments target organs from given images as follows: First, each of the training images is nonrigidly registered to a given target image. Let $I^i(\boldsymbol{x})$ $(j = 1, 2, \ldots, M)$ denote the training images and R_t^i denote the region of a target organ in $I^i(\boldsymbol{x})$ and $b^i(\boldsymbol{x})$ denote binary label images where

$$b^i(\boldsymbol{x}) = \begin{cases} 1, & \text{if } \boldsymbol{x} \in R_t^i, \\ 0, & \text{otherwise.} \end{cases} \tag{2.214}$$

and letting the given target image be denoted by $I_{\text{tgt}}(\boldsymbol{x})$, the nonrigid registration deforms $I^i(\boldsymbol{x})$ so that it is aligned to the given image. Letting a two-vector, $\boldsymbol{u}^i(\boldsymbol{x})$, denote the image coordinates obtained by the registration where $I^i(\boldsymbol{u}^i(\boldsymbol{x}))$ and $I_{\text{tgt}}(\boldsymbol{x})$ are aligned, the deformed image is represented by $I^i(\boldsymbol{u}^i(\boldsymbol{x}))$, and $b^i(\boldsymbol{u}(\boldsymbol{x}))$ represents the deformed binary labeled image. Each of the deformed binary labeled images, $b^i(\boldsymbol{u}(\boldsymbol{x}))$, is an estimate of the region of a target organ in $I_{\text{tgt}}(\boldsymbol{x})$.

The multi-atlas-based method, then, computes the absolute difference between $I^i(\boldsymbol{u}^i(\boldsymbol{x}))$ and $I_{\text{tgt}}(\boldsymbol{x})$: $D^i(\boldsymbol{x}) = |I^i(\boldsymbol{u}(\boldsymbol{x})) - I_{\text{tgt}}(\boldsymbol{x})|$, and a weight, $\lambda(\boldsymbol{x})$, that is a decreasing function of $D^i(\boldsymbol{x})$, such that

$$\lambda^i(\boldsymbol{x}) = \frac{1}{D^i(\boldsymbol{x}) * g(\boldsymbol{x}|\sigma^2) + \epsilon}, \tag{2.215}$$

where $g(\boldsymbol{x}|\sigma^2) = \mathscr{N}(\boldsymbol{x}|0, \Sigma^2)$ is a Gaussian filter of which variance is σ^2 and $\epsilon > 0$. The regions indicated by the deformed binary labeled images would be more reliable and would have larger values of $\lambda^i(\boldsymbol{x})$ if the difference, $D^i(\boldsymbol{x})$, were smaller.

Finally, the method computes a weighted average of the deformed labeled images, $b^i(\boldsymbol{u}^i(\boldsymbol{x}))$ $(i = 1, 2, \ldots, M)$, as follows and classifies voxels as the inside of the organ if the average is larger than 1/2:

$$S(\boldsymbol{x}) = \frac{1}{Z(\boldsymbol{x})} \sum_{i=1}^{M} \lambda^i(\boldsymbol{x}) b^i(\boldsymbol{u}^i(\boldsymbol{x})), \tag{2.216}$$

where $Z(\boldsymbol{x}) = \sum_{i=1}^{M} \lambda^i(\boldsymbol{x})$.

The algorithm described here can be interpreted as a MAP estimation of the region, as will be described below. It should be remembered that the single-atlas-based method also determines the region of a target organ by means of MAP estimation as shown in (2.172). The difference between the multi-atlas-based method and the single-atlas method is the model used for computing the posterior probability distribution.

For the interpretation of the multi-atlas-based method, let us introduce a latent variable, $\boldsymbol{\theta}$, that represents the shape of the target organ in an image. For example, the shape of the organ in $b^i(\boldsymbol{x})$ is explicitly indicated as $b^i(\boldsymbol{x}|\boldsymbol{\theta}^i)$. As described in (2.175), a single probabilistic atlas can be estimated by averaging all of the binary labeled images, $b^i(\boldsymbol{x})$. That computation of the average is nothing but the marginalization of the binary labeled images, $b(\boldsymbol{x}|\boldsymbol{\theta})$, over the latent shape parameters as follows:

$$p(\boldsymbol{x} \in R_t) = \int p(\boldsymbol{\theta})p(\boldsymbol{x} \in R_t|\boldsymbol{\theta})d\boldsymbol{\theta} = \int p(\boldsymbol{\theta})b(\boldsymbol{x}|\boldsymbol{\theta}) \simeq \frac{1}{M}\sum_{i=1}^{M} b^i(\boldsymbol{x}|\boldsymbol{\theta}^i). \quad (2.217)$$

In the single-atlas-based method, the posterior probability distribution of each voxel being included in the target region is evaluated by multiplying the prior in (2.217) by the likelihood which is defined based only on the voxel value; no information on the organ shape is used for computing the likelihood. In the multi-atlas-based method, the differences between the shapes are more explicitly considered in the computation.

For a simple example, let again $b_t(\boldsymbol{x})$ denote a binary image, where $b_t(\boldsymbol{x}) = 1$ if $\boldsymbol{x} \in R_t$ and otherwise $b_t(\boldsymbol{x}) = 0$. Using the latent shape parameter, the posterior probability distribution can be represented thus:

$$p(b_t(\boldsymbol{x})|I_{\text{tgt}}) = \int p(\boldsymbol{\theta}|I_{\text{tgt}})p(b_t(\boldsymbol{x})|I_{\text{tgt}}, \boldsymbol{\theta})d\boldsymbol{\theta}. \quad (2.218)$$

Because $p(b_t(\boldsymbol{x})|I_{\text{tgt}}, \boldsymbol{\theta}) \propto p(I_{\text{tgt}}|b_t(\boldsymbol{x}), \boldsymbol{\theta})p(b_t(\boldsymbol{x})|\boldsymbol{\theta})$, $p(\boldsymbol{\theta}|I_{\text{tgt}})$, $p(b_t(\boldsymbol{x})|\boldsymbol{\theta})$, and $p(I_{\text{tgt}}|b_t(\boldsymbol{x}), \boldsymbol{\theta})$ need to be estimated for computing the posterior probability in (2.218). The multi-atlas-based method estimates $p(\boldsymbol{\theta}|I_{\text{tgt}})$ and $p(b_t(\boldsymbol{x})|\boldsymbol{\theta})$ by the nonrigid registration between $I^i(\boldsymbol{x})$ and I_{tgt}.

The nonrigid registration computes the transformation $\boldsymbol{u}(\boldsymbol{x})$ so that the two input images align. Once the transformation is obtained, the deformed binary labeled image, $b^i(\boldsymbol{u}(\boldsymbol{x}))$, should be an estimate of the region of the target organ in the target image, $I_{\text{tgt}}(\boldsymbol{x})$. Let the true and unknown shape parameters of the organ region in the given image be denoted by $\boldsymbol{\theta}_{\text{tgt}}$ and let $\boldsymbol{\theta}^i$ and $\hat{\boldsymbol{\theta}}^i_{\text{tgt}}$ denote the shape parameters representing the organs in $b^i(\boldsymbol{x})$ and in $b^i(\boldsymbol{u}(\boldsymbol{x}))$, respectively, $\boldsymbol{\theta}^i$ $(i = 1, 2, \ldots, M)$ obey $p(\boldsymbol{\theta})$, and $\hat{\boldsymbol{\theta}}^i_{\text{tgt}}$, which are the estimates of $\boldsymbol{\theta}_{\text{tgt}}$, obey the posterior probability distribution, $p(\boldsymbol{\theta}|I_{\text{tgt}})$. In addition, from the definition of $\hat{\boldsymbol{\theta}}^i_{\text{tgt}}$, the following equation can result:

$$p(b_t(\boldsymbol{x}) = 1|\hat{\boldsymbol{\theta}}^i_{\text{tgt}}) = b^i(\boldsymbol{u}(\boldsymbol{x})) = \begin{cases} 1, & \text{if } \boldsymbol{x} \in R_t, \\ 0, & \text{otherwise.} \end{cases} \quad (2.219)$$

The last factor to be estimated is $p(I_{\mathrm{tgt}}(\boldsymbol{x})|b_t(\boldsymbol{x}), \boldsymbol{\theta})$. As defined above, the label, $b_t(\boldsymbol{x})$, is uniquely determined by the shape parameter, $\boldsymbol{\theta}$, and hence $p(I_{\mathrm{tgt}}(\boldsymbol{x})|b_t(\boldsymbol{x}), \boldsymbol{\theta}) = p(I_{\mathrm{tgt}}(\boldsymbol{x})|\boldsymbol{\theta})$. When $\boldsymbol{\theta} = \hat{\boldsymbol{\theta}}^i_{\mathrm{tgt}}$, it is natural to assume that the probability of the voxel value $I(\boldsymbol{x})$ being equal to $I_{\mathrm{tgt}}(\boldsymbol{x})$ decreases as the absolute difference, $D(\boldsymbol{x}) = |I^i(\boldsymbol{u}(\boldsymbol{x})) - I_{\mathrm{tgt}}(\boldsymbol{x})|$, increases because $I^i(\boldsymbol{u}(\boldsymbol{x})) = I^i(\boldsymbol{u}(\boldsymbol{x})|\hat{\boldsymbol{\theta}}^i_{\mathrm{tgt}})$. Following (2.215), assume that the following equation holds:

$$p(I_{\mathrm{tgt}}(\boldsymbol{x})|b_t(\boldsymbol{x}), \hat{\boldsymbol{\theta}}^i_{\mathrm{tgt}}) \propto \lambda^i(\boldsymbol{x}) = \frac{1}{D^i(\boldsymbol{x}) * g(\boldsymbol{x}|\sigma_1^2) + \epsilon}. \tag{2.220}$$

From the discussion above, the following equation results:

$$p(b_t(\boldsymbol{x}) = 1|I_{\mathrm{tgt}}(\boldsymbol{x})) = \int p(\boldsymbol{\theta}|I_{\mathrm{tgt}}(\boldsymbol{x}))p(b_t(\boldsymbol{x})|\boldsymbol{\theta})p(I_{\mathrm{tgt}}|b_t(\boldsymbol{x}), \boldsymbol{\theta})d\boldsymbol{\theta} \simeq \frac{1}{Z}\sum_{i=1}^{M} \lambda(\boldsymbol{x})b^i(\boldsymbol{u}(\boldsymbol{x})), \tag{2.221}$$

where Z is a normalization term. As shown in (2.216), $S(\boldsymbol{x})$ represents the posterior probability distribution, and MAP estimates can be obtained by binarizing $S^i(\boldsymbol{x})$ with a threshold $1/2$.

2.3.6 CA-Based Segmentation

Organ segmentation is a central topic in the field of medical image analysis, and a number of segmentation algorithms have been presented [142–145]. This section focuses on segmentation algorithms of organs/tissues in a human torso based on a CA model.

Since shape features of an organ play an important role in segmentation, researchers have attempted to incorporate them into their segmentation frameworks. Deformable model-based segmentation is a typical example. Pioneering works using a deformable model were reported in the early 1970s [146–148] followed by several epoch making works, Snakes [149], ASM [1], and active appearance model (AAM) [150], the first of which used a local shape feature, or curvature-based feature, to make extracted boundaries smooth and the latter two of which employed global shape or appearance features, or CA, to make extracted shapes more accurate anatomically. An alternative CA-based approach is atlas-based segmentation, which was initially developed for brain segmentation of magnetic resonance imaging (MRI) images and was then imported into organ segmentation of a human torso from CT, positron-emission tomography (PET), and MRI images.

This section describes CA-based segmentation algorithms for organs/tissues in a human torso starting with probabilistic atlas-based segmentation with an example of organ segmentation of a human trunk CT volume, followed by ASM, which employs

PDM, level-set-based segmentation with a shape prior, and ensemble learning-based segmentation with a shape prior.

2.3.6.1 Probabilistic Atlas-Based Segmentation

A probabilistic atlas represents the existence probability of an organ at each voxel and was originally explored in brain segmentation followed by organ segmentation of a human torso. Details of a probabilistic atlas can be found in Sect. 2.3.5.1. This section focuses on segmentation algorithms of organs in a human torso based on a probabilistic atlas. A typical probabilistic atlas-based segmentation algorithm is MAP estimation of an organ [151], in which a prior probability is defined by a probabilistic atlas. Given a feature vector $\boldsymbol{x} \in \mathbb{R}^d$, posterior probability is represented by the following equation using Bayes' theorem:

$$p(n|\boldsymbol{x}) = \frac{p(\boldsymbol{x}|n)p(n)}{p(\boldsymbol{x})}, \tag{2.222}$$

where $p(\boldsymbol{x}|n)$ shows the likelihood of a vector $\boldsymbol{x}$ of organ n, which is widely assumed to be a mixture of Gaussian distributions as follows:

$$p(\boldsymbol{x}|n) = \sum_{m=1}^{M} \frac{1}{(2\pi)^{L/2}|\Sigma_{n,m}|^{1/2}} \exp\left(-\frac{1}{2}(\boldsymbol{x} - \boldsymbol{\mu}_{n,m})^T \Sigma_{n,m}^{-1}(\boldsymbol{x} - \boldsymbol{\mu}_{n,m})\right), \tag{2.223}$$

where $\boldsymbol{\mu}_{n,m}$ and $\Sigma_{n,m}$ are an m-th average vector and an m-th covariance matrix of an organ n in an unseen image which can be estimated by an EM algorithm or variational Bayes [33]. An atlas-guided EM algorithm is an optional choice to achieve low computational cost and high accuracy [124]. Parzen window estimation is an alternative choice to define a vector x of organ n (refer details of nonparametric probability density function estimation to [34]). A probability (density) function is defined by the following equation:

$$p(\boldsymbol{x}) = \sum_{n=1}^{N} p(\boldsymbol{x}|n)p(n) \tag{2.224}$$

where N is equal to number of organs to be segmented plus one that corresponds to the background.

As mentioned earlier in this section, prior probability is given by a probabilistic atlas which makes the segmentation results more accurate especially from the point of view of anatomy. Figure 2.23a, b are an axial section of an original abdominal CT volume dataset (a) with prior probability of liver, or a probabilistic atlas, in which whiter color represents higher probability of existence of liver (b). Parts (c) and (d) illustrate the likelihood of liver and that of background, in which numbers of Gaussian of the mixture distributions are 2 for liver and 3 for background. Parts (e)

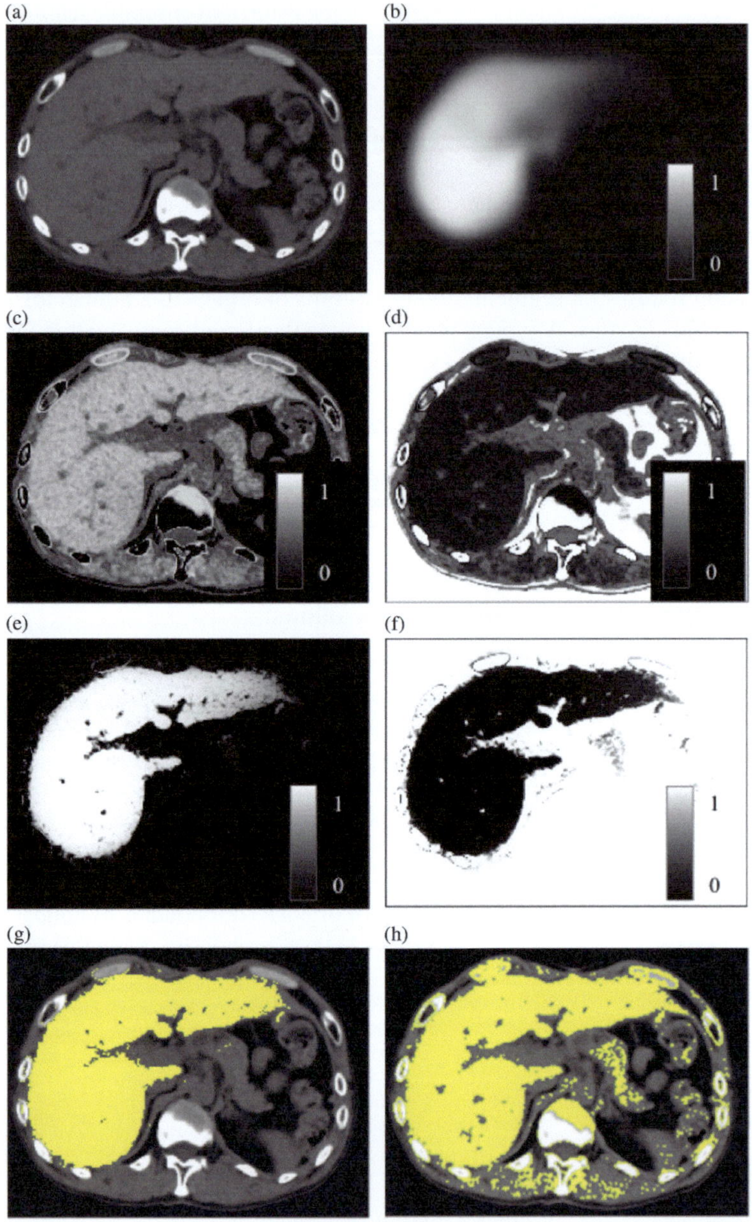

Fig. 2.23 Example of maximum a posterior-based liver segmentation using a probabilistic atlas of liver. (**a**) Original CT image. (**b**) Probabilistic atlas of liver. (**c**) Likelihood of liver. (**d**) Likelihood of background. (**e**) Posterior probability of liver. (**f**) Posterior probability of background. (**g**) Segmentation result with a probabilistic atlas. (**h**) Segmentation result without a probabilistic atlas

and (f) present posterior probability of liver and that of background, respectively, and parts (g) and (h) show segmentation results with and without the probabilistic atlas of liver, respectively. The likelihood parts (c) and (d) tell us that the liver is algorithmically enhanced but parts of surrounding tissues and organs, such as muscle and pancreas, are also enhanced, which might lead to false positives in segmentation. In contrast, the probabilistic atlas of part (b) appropriately restricts the existence area of liver, resulting in better segmentation results of part (g) compared with the results of part (h) (performed without the probabilistic atlas). The JIs are 0.851 for part (g) and 0.536 for (h), respectively.

Extensions of a probabilistic atlas of organs in a human torso can be found in several papers. For example, a multi-organ probabilistic atlas can be easily derived by normalizing probabilities of multiple organs at each voxel under constraints that the summation of probabilities of whole organs/tissues is equal to one. Figure 2.24 presents a result of twelve-organ segmentation using a multi-organ probabilistic atlas from a noncontrast CT volume [124, 152], in which MAP-based segmentation using a multi-organ probabilistic atlas was followed by a multiple level-set method with interaction mechanism between neighboring level-set functions. In [153], probabilistic atlases of multiple organs were constructed and used in graph-cuts-based fine segmentation, in which an energy term is defined using probabilities of multiple organs.

A modification of a probabilistic atlas was reported in [151], where MRF was adopted to improve prior probability. In the iterative scheme of multi-organ segmentation, a prior probability of each voxel of interest was updated at each step so as to reduce unnatural non-smooth boundaries, holes, and over-extracted regions by referring to the regions of the organs neighboring to each target organ segmented at the previous step.

Alternative important improvement for organ segmentation was a patient-specific probabilistic atlas [154–156]. Since a conventional atlas accounts for a whole distribution of existence probability, it is effective in describing shapes around a mean shape but not for atypical shapes located in marginal areas of the distribution. To improve the segmentation performance for an organ with atypical shape, patient-specific atlases were constructed using an SSM [154], multi-atlas [155], and sparse modeling-based [156] approaches. It was proven that a patient-specific probabilistic atlas was effective in segmenting an organ with a typical shape to a statistically significant degree [156].

In general, a probabilistic atlas is constructed using healthy organs or mostly healthy organs, such as an organ with small lesions. Consequently this might not account for an organ with large pathological lesions, which frequently change the appearance as well as the shape of an organ radically, resulting in low posterior probability, even though a probabilistic atlas indicates high existence probability of a target organ. In such cases, the probability of pathological lesions might be helpful to recover the low posterior probability of a voxel in a lesion and segment an organ with large lesions. An algorithm was proposed [156] to construct a probabilistic atlas of pathological lesions and a probabilistic atlas of a liver simultaneously using sparse modeling with lesion bases. The result of applying the algorithm to livers

(a) (b)

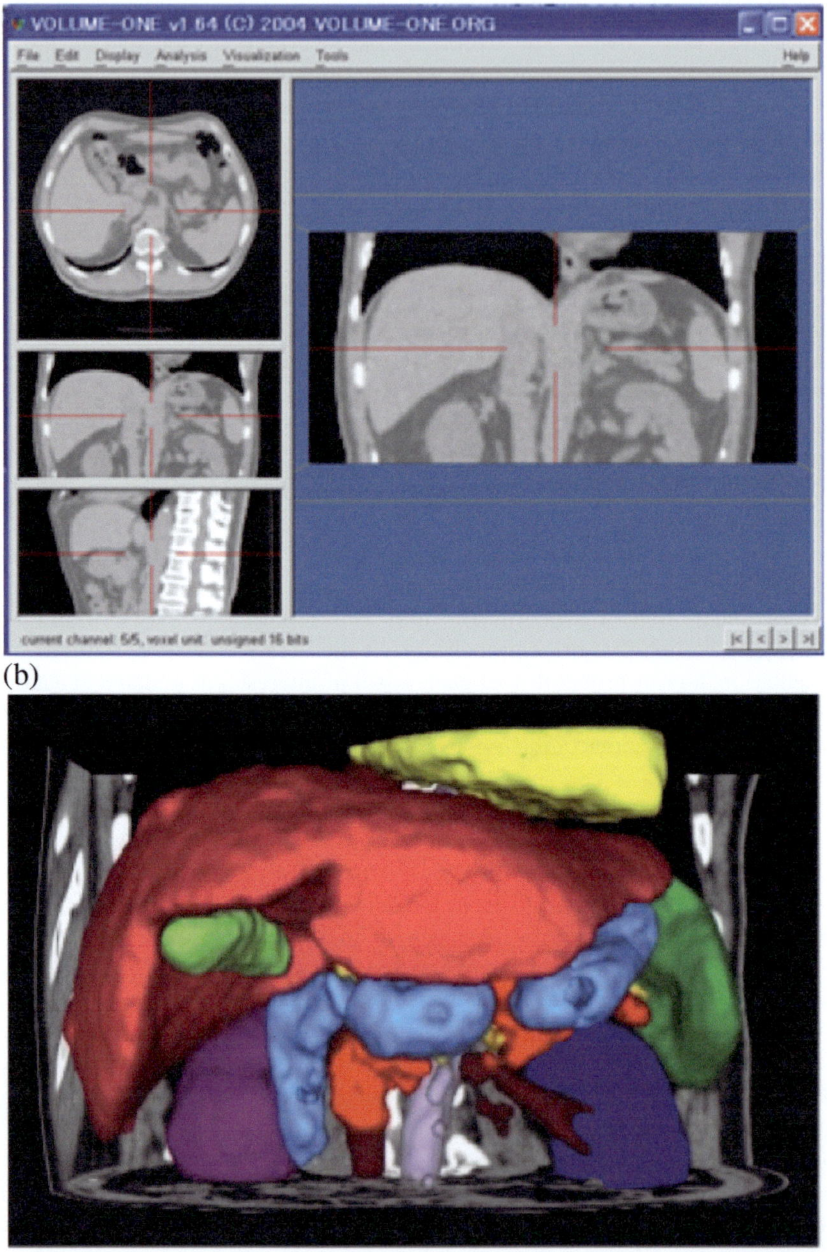

Fig. 2.24 MAP-based segmentation using a multi-organ probabilistic atlas was followed by a multiple level-set method. (**a**) An input noncontrast upper abdominal CT volume. (**b**) Multi-organ segmentation result. The twelve target organs are the esophagus, heart, stomach, liver, gallbladder, pancreas, spleen, left and right kidneys, inferior vena cava, aorta, and splenic vein between the liver and spleen

with large pathological lesions in CT volumes proved its statistical effectiveness comparing with a conventional probabilistic atlas-based algorithm.

2.3.6.2 Active Shape Model

The ASM [1] is one of the most successful deformable models to segment a target object, or an organ/tissue, which iterates computation of displacement vectors based on edge information in an image and update of pose and shape parameters of a model according to the PDM explained in Sect. 2.3.2. Details of the segmentation process will be explained below.

An ASM requires initial location of a model which might be given by a user or an automated or semiautomated process. Once an initial boundary is given, ASM repetitively finds new suggested locations for model points and updates pose and shape parameters of a model. Typically the suggested locations for model points are searched along profiles normal to the model boundary through each model point using a statistical model for the grey-level profile about a point. Parameters of shape and pose, consisting of location, scale, and orientation, are updated to best fit the new found points according to the following equation:

$$\boldsymbol{x} = T_{X_t,Y_t,s,\theta}(\boldsymbol{\mu} + V\boldsymbol{\alpha}), \tag{2.225}$$

where $T_{X_t,Y_t,s,\theta}$ is a function of translation (X_t, Y_t), scaling s, and rotation θ around an origin. The parameters $\boldsymbol{\mu}$, V, and $\boldsymbol{\alpha}$ are an average shape vector, a matrix of eigenshape vectors, and a shape vector that corresponds to a principal component vector. All these parameters are updated so that mean squared error between suggested locations and model points is minimized. In the actual update process, pose parameters are adjusted efficiently using a least-squares approach followed by an update of shape parameters based on the updated pose parameters. In practice, weighted adjustment for the update process and a multi-resolution scheme are employed to achieve higher performance.

A major limitation of the ASM is caused by a local minimum problem because it searches the best fit between edges in an image and a model using an iterative scheme starting from an initial location. If the initial shape is far from the true boundary of an organ, the searching process may fail. Another limitation is that a point distribution model does not take into account gray-level statistical variation in an organ across patients. The AAM was developed to overcome this limitation, in which a prior model is trained using not only shape but also grayscale values, and used for segmentation. Details can be found in [150].

ASM is not a solitary example of a segmentation algorithm that utilizes a PDM. A number of algorithms in which a PDM plays a major role in segmentation can be found in [142–145].

2.3.6.3 Level Set with CA

This section describes a combination of level-set-based segmentation and an SSM of an organ, or a statistical model of a signed distance function, called a level-set distribution model (LSDM) in Sect. 2.3.2.

An important finding was made by Leventon [157], who introduced an LSDM into a level-set-based segmentation algorithm consisting of following steps: Letting a statistical shape variation of an organ be modeled by a principal component vector $\boldsymbol{\alpha}$ under the assumption that the distribution is a Gaussian distribution:

$$p(\boldsymbol{\alpha}) = \frac{1}{(2\pi)^{k/2}|\Sigma_k|^{1/2}} \exp\left(-\frac{1}{2}\boldsymbol{\alpha}^T \Sigma_k^{-1} \boldsymbol{\alpha}\right). \tag{2.226}$$

the segmentation algorithm finds a set of shape parameter vector $\boldsymbol{\alpha}$ and pose parameter vector $\boldsymbol{p}$ using the following equation:

$$\langle \boldsymbol{\alpha}_{\mathrm{MAP}}, \boldsymbol{p}_{\mathrm{MAP}} \rangle = \arg\max_{\boldsymbol{\alpha},\boldsymbol{p}} P(\boldsymbol{\alpha}, \boldsymbol{p}|u, \nabla I), \tag{2.227}$$

where $P(\boldsymbol{\alpha}, \boldsymbol{p}|u, \nabla I)$ is a posterior probability of parameters of shape and pose given a boundary u and gradient image ∇I. In practice, the posterior probability is transformed using the formula for Bayes' theorem and several assumptions on probabilistic distributions of parameters. The maximization is performed using a gradient ascent algorithm. Subsequently the estimated shape and pose parameters are incorporated into the following update equation of a geodesic active contour [158]:

$$\frac{\partial u}{\partial t} = g(c + \kappa)\,|u| + \nabla u \cdot \nabla g, \tag{2.228}$$

where g is a stopping function based on image gradient, c is a constant value, and κ is the curvature of a boundary u. The updated shape of the boundary u at time $t + 1$ can be computed from $u(t)$ by

$$u(t+1) = u(t) + \lambda_1 \left(g(c + \kappa)|\nabla u(t)| + \nabla u(t) \cdot \nabla g\right) + \lambda_2 \left(u^*(t) - u(t)\right), \tag{2.229}$$

where $u^*(t)$ is the estimated boundary with parameter $\boldsymbol{\alpha}_{\mathrm{MAP}}$, $\boldsymbol{p}_{\mathrm{MAP}}$ and two parameters λ_1, λ_2 balance the influence of the gradient-curvature model and the shape model. The above process is repeated until convergence of deformation.

Figure 2.25 shows the segmentation results of corpora callosa from MR brain images. The algorithm was tested on unknown sections that were not used for the training. It is found from parts (a), (b), and (c) that the MAP estimator of shape and pose guides the model to the true boundary. Part (d) presents the result without the shape model, which failed in segmentation. In addition to this example, Leventon et al. showed results of applying the algorithm to 2D slices of MR images of femur

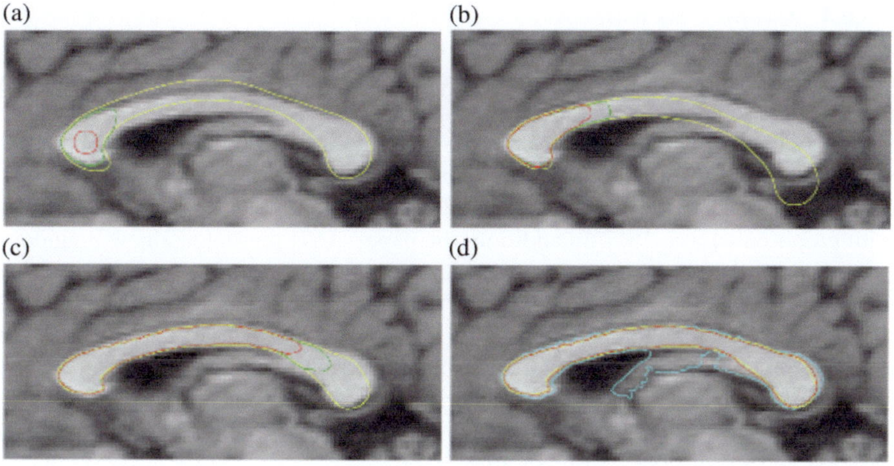

Fig. 2.25 Shape-based segmentation that combines level-set method with a statistical shape model. The *red curve* is the zero level set of the evolving surface. The *green curve* is the next step in the curve evolution. The *yellow curve* is the MAP estimate of the position and shape of the curve. The *cyan contour* is the standard evolution without the shape influence

and corpus callosum as well as 3D CT volumes to segment vertebrae, all of which showed successful segmentation results.

An alternative approach was presented in the paper [125], in which LSDMs for multiple objects with neighbor constraints were employed and an energy functional including Mahalanobis distances computed in eigenshape spaces of the models was minimized. The proposed method was employed to extract left and right ventricles from a 2D cardiac MR image as well as eight subcortical structures in an MRI brain image.

In the paper [126], an implicit representation of the shape was combined with the Chan Vese region-based energy functionals [159]. After the training phase of a statistical shape model, the segmentation phase was carried out, in which the region-based functional is minimized. In practice, shape parameters for the eigenshapes and pose parameters to handle pose variations are iteratively updated to generate a new level set that determines the segmenting curve implicitly. The image statistics inside and outside the curve are used to compute the update function for the next iteration, and the iterative scheme is continued until convergence is reached for segmentation. The experimental results of left ventricle segmentation on cardiac MRI and prostate segmentation on pelvic MRI acquired with an endorectal coil proved that the proposed segmentation algorithm was computationally efficient and robust to noise.

2.3.6.4 Graph-Cuts with CA

A recent noteworthy development in image segmentation is the graph- cuts approach [160–164], in which the global minimum for a sub-modular function of a binary label problem is guaranteed. This section introduces techniques that incorporate a shape prior into a graph-cuts approach in order to achieve accurate segmentation. So far, general shape constraints, such an ellipse and a star shape, were combined with graph-cuts [165, 166]. In addition, the paper [167] presented a new boundary term that can take into account more general shape constraints or a compact shape constraint in a graph-cuts segmentation. Although the above priors are general and can be used in segmentation of a variety of objects, it might not be applicable to more complicated shape objects. To give a specific shape constraint on graph-cuts, a user-defined rigid template was incorporated into a graph-cuts-based segmentation [168]. To deal with statistical variations in the shape of an organ, the papers [169, 170] derived a shape prior from an SSM.

A limitation of the above approaches is that a single shape might not account for a shape in a given test image. A shape-based energy computed by the Parzen window method was presented in [153], which is a population statistic-based method, but the combinatorial problem of multiple shape information remains unsolved. The paper [171] extended the algorithm of [169] to multi-shape graph-cuts that can take into account multiplicity of shapes prior in the segmentation framework by using fusion move with the quadratic pseudo-Boolean optimization (QPBO) algorithm. In the remainder of this section, a multi-shape graph-cuts algorithm [163] is presented. A new pairwise energy, $S_{p,q}$, for multiple shape priors was proposed and linearly combined with the conventional graph-cuts energy:

$$S_{p,q} = \min\left(\sqrt{\frac{1}{2}(1-\cos\theta_{A_p})}, \sqrt{\frac{1}{2}(1-\cos\theta_{A_q})}\right), \tag{2.230}$$

where θ_{A_p} is the angle between a vector connecting voxels p and q and a gradient vector of a signed distance ϕ_{A_p} from the boundary of a shape prior A_p. The method selects five shapes from an eigenshape space of an SSM which are similar to a region by MAP-based segmentation. The selected shapes are forwarded to multi-shape graph-cuts to perform segmentation with multiple shape priors. Unlike a conventional single-shape graph-cuts, multi-shape graph- cuts use multiple labels, each of which corresponds to a shape prior in graph-cuts-based fine segmentation. In addition, negative labels are used for background to differentiate between background labels and object labels. The fusion move with the QPBO algorithm finds a combinatorial optimal solution of labels or shapes.

Figure 2.26 shows lung segmentation results from a CT volume and shape priors. Although a single shape graph-cuts with shape 1 of part (e) led to inaccurate lung segmentation (red arrow in part (c)), the remaining shape priors of parts (f) to (i) corrected the segmentation as presented in part (d). Details of the algorithm as well as the experimental results using 97 CT volumes can be found in [171].

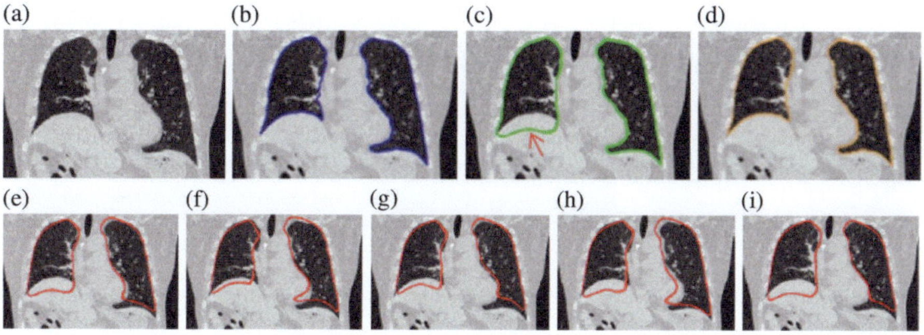

Fig. 2.26 Influence of shape priors on segmentation results. (**a**) original CT, (**b**) true boundary, (**c**) segmentation result by a single- shape graph-cuts with single-shape prior of part (**e**), and (**d**) segmentation result by the multi-shape graph-cuts with five shape priors of figures (**e**) to (**i**)

2.3.6.5 Ensemble Learning with CA

Ensemble learning, such as AdaBoost [172] and random forest [173], has been prominent in recent segmentation research. Sophisticated segmentation examples by ensemble learning have been presented so far. A possible limitation is, however, that the segmentation result tends to be unnatural in terms of shape because the segmentation is performed in dependently voxel-by-voxel. A number of methods have been proposed to make the segmentation results natural in shape. The combination of MRF with AdaBoost [174], Spatial Boost [175], and Spatial AdaBoost [174] are typical examples. In [176, 177], a wide range of spatial information was employed for segmentation of subcortical structures in brain images and shape-based retrieval in cluttered images. However, in the context of medical image analysis, it is important to deal with organ- or tissue-specific features in the boosting- based segmentation.

A boosting algorithm that can take into account a target object's specific shape prior was presented in [178], in which a new shape loss function evaluating directions of normal vectors of a target surface was proposed and minimized together with a conventional error loss function in the training process. The trained ensemble classifier was applied to extract lung fissures that are the thin pleural-covered potential spaces separating the lung lobes. Figure 2.27 shows an original CT image and the difference in outputs before binarization between a classifier trained by AdaBoost and that by Shapeboost. It is found from the figure that the shape of the algorithmically more enhanced area by Shapeboost is more similar to true fissures than those using Adaboost. The segmentation results of graph-cuts based on the enhanced results is satisfactory as shown in part (e).

Another alternative approach incorporates a novel shape loss function that evaluates distance between an extracted shape and a subspace of SSM of an organ and minimizes a total loss function including not only a conventional error loss but also the proposed shape loss [179]. The method was successfully applied to spleen

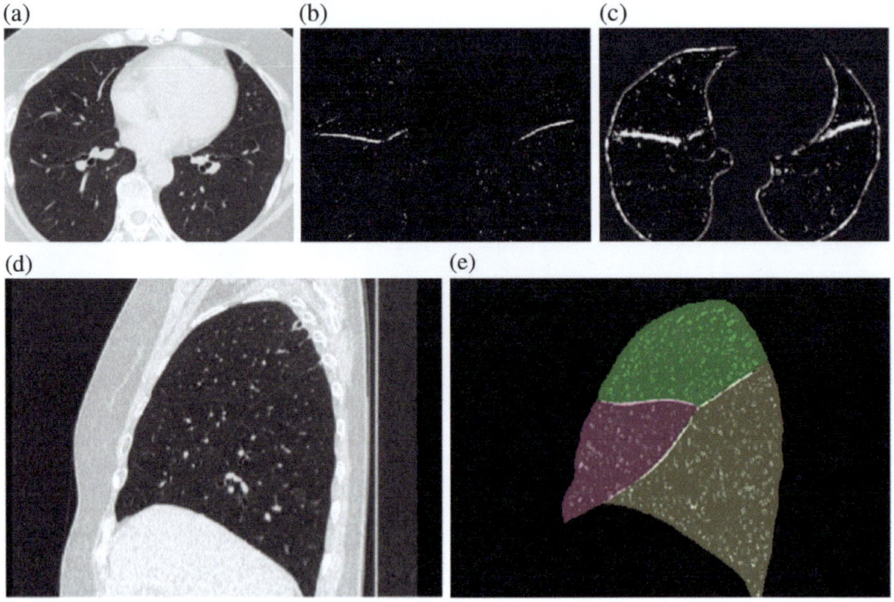

Fig. 2.27 Outputs of ensemble classifiers trained by boosting algorithms without and with the proposed shape loss. (**a**) axial section of an original CT volume, (**b**) algorithmically more enhanced lung fissures by the proposed Shapeboost and (**c**) by AdaBoost. (**d**) sagittal section of an original CT volume and (**e**) algorithmically enhanced fissures by the proposed Shapeboost overlaid with the lung lobe segmentation result using a graph-cuts algorithm with an energy term derived from the algorithmically enhanced fissures [178]

segmentation and validated using 80 CT volumes. Details on the topic of spleen segmentation, in which the ensemble learning is applied, can be found in Sect. 3.9.3.

2.3.7 *Multiple Organs, Anomaly, and Lesions*

In this section, two important tasks are described: multiple organ segmentation and anomaly recognition. The methods described in the previous section segmented one target organ by using its single model. Such methods would be insufficient for the solutions because they cannot handle the two problems. First, any information on the dependencies between the locations and shapes of the multiple organs cannot be used if each organ is segmented independently. For example, some of the resultant regions of the organs can overlap each other and be inconsistent with human anatomy. Second, the models described in the previous sections cannot represent some anatomical anomalies, such as variations in the number of some anatomical entities. In the following section, some approaches to these issues are introduced.

2.3.7.1 Multiple Organs

A CA model (SSM) for multiple organs can be constructed by simply extending an SSM for a single organ, such as a PDM [69] or an LSDM [180]. For example, a collection of SSMs, each of which corresponds to an organ and is constructed individually, might be a solution for an SSM for multiple organs. This simple solution, however, suffers from an overlap problem between neighboring organs. The relationship between neighboring organs should be incorporated to reduce the overlap and can be modeled by connecting feature vectors, each of which corresponds to an organ, followed by statistical analysis of the concatenated vectors [181]. Other modeling approaches include logarithm of the odds (LogOdds) [128], label space [182], and isometric log-ratio mapping [183], which were originally designed for multiple objects (organs) and thus are unaffected by overlap of neighboring organs. This section describes an SSM for multiple organs based on PDM or LSDM followed by LogOdds and label space-based models. Note that all of the models mentioned above are SSMs in which all shape features are simultaneously exploited without any weight, or priority, in the modeling process. In contrast, if features highly correlated with target organs are available, a conditional SSM [184–188] will work, which will be also explained in this section.

1. SSM for multiple organs based on a point distribution model and an LSDM
   Let $\boldsymbol{x}$ be a feature vector of M organs, which consists of feature vectors $\boldsymbol{x}_1, \boldsymbol{x}_2, \ldots, \boldsymbol{x}_M$, each of which corresponds to each organ. Specifically, two of the vectors consist of point coordinates of multiple organs when based on a PDM and a signed distance function (SDF) when based on an LSDM. An SSM for multiple organs is a statistical model of a probability distribution of $\boldsymbol{x}$ or $p(\boldsymbol{x})$. Since the original feature vectors are defined in a very high-dimensional space and might be contaminated by noise, statistical analysis is applied to reduce dimension and noise. Let $\boldsymbol{v}$ be a shape feature vector whose length is L and element is $\boldsymbol{v}_1, \boldsymbol{v}_2, \ldots, \boldsymbol{v}_L$, which are shape features extracted by statistical analysis of a training data set of $\boldsymbol{x}$, e.g., principal component scores (PCS) extracted by principal component analysis (PCA). Consequently, the probability density function $p(\boldsymbol{x})$ is approximated by $p(\boldsymbol{v})$. The following equation is a probability density function if the distribution of v can be assumed to be Gaussian:

$$p(\boldsymbol{v}) = \frac{1}{(2\pi)^{L/2}|\Sigma|^{1/2}} \exp\left(-\frac{(\boldsymbol{v}-\boldsymbol{\mu})^T \Sigma^{-1}(\boldsymbol{v}-\boldsymbol{\mu})}{2}\right) \tag{2.231}$$

   where $\boldsymbol{\mu}$ is an average vector and Σ is a covariance matrix of shape feature vector $\boldsymbol{v}$.

   Variation in location and shape of multiple organs is an accumulation of individual differences of an organ, resulting in a larger variation in pose and shape. This makes model construction difficult and results in low performance. One possible solution to handle such large variation is to separate variations in

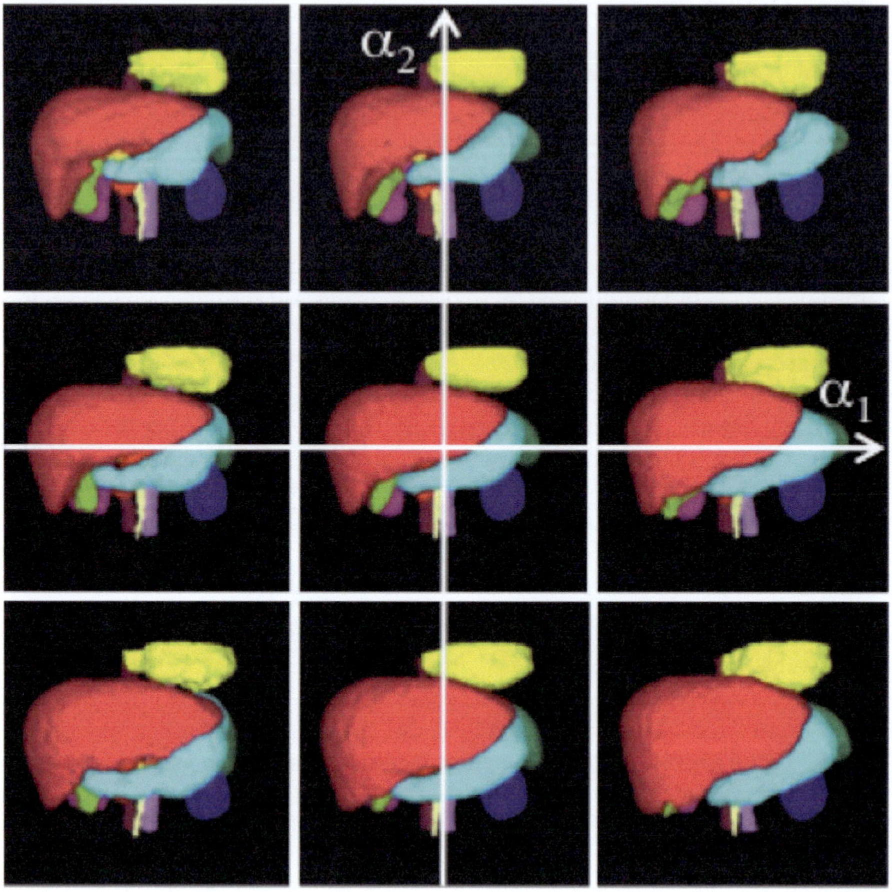

Fig. 2.28 An SSM of 14 organs: heart, esophagus, stomach, liver, gallbladder, pancreas, spleen, left and right kidneys, inferior vena cava, aorta, splenic vein, portal vein, superior mesenteric vein

pose from those in shape in a statistical modeling process. One group [181] proposed an SSM, or an LSDM, for multiple organs, in which Procrustes alignment was applied to a training label dataset of each organ individually to separate pose, or rotation and translation parameters, from shape. The extracted parameter sets of rotation and translation as well as shape were analyzed by PCA to build a rotation model, a translation model, and a shape model for each organ. The final step was to integrate all models of multiple organs by concatenating PCS vectors of organs into a vector followed by PCA of the concatenated vectors. Figure 2.28 shows fourteen organs generated from nine different parameter sets in a subspace of the proposed SSM, in which the horizontal axis corresponds to the first PCS and the vertical axis corresponds to the second PCS. In terms of generalization, specificity, and overlap between neighboring organs, the performance of the proposed algorithm was proven to be superior to that of a multi-organ model without separation of pose and shape. The SSM for multiple organs will be incorporated into a multi-organ segmentation algorithm [124] to boost segmentation performance.

2. LogOdds and Label space
LogOdds is an example of a class of functions that map the space of discrete distributions to Euclidean space and is employed for multi-organ shape representation [128]. The multinomial LogOdds function logit $\mathbb{P}_M \rightarrow \mathbb{R}^M$ and the generalized logistic operations are used to bridge between the manifold of signed distance maps (SDMs) and the linear space of LogOdds:

$$[\text{logit}(\text{p})]_i \equiv \log\left(\frac{p_i}{p_M}\right), \tag{2.232}$$

where $p_i(\in [0,1])$ is the i-th probability and $p_M(\in [0,1])$ is the last entry of a discrete distribution $p(\in \mathbb{P}_M)$, where $\mathbb{P}_M$ is an open probability simplex, or the space of discrete distributions for M labels which correspond to background label and $M-1$ object (organs) labels.
The inverse of the logit function is the generalized logistic function:

$$[\mathscr{P}(t)]_i \equiv \begin{cases} \frac{e^{t_i}}{1+\sum_{j=1}^{M} e^{t_j}}, & \text{for } i \in \{1, 2, \ldots, M-1\}, \\ \frac{1}{1+\sum_{j=1}^{M}} e^{t_j}, & \text{if } i = M, \end{cases} \tag{2.233}$$

Note that logit $\{\text{logit}(p)|p \in \mathbb{P}_M\}$ is the $(M-1)$-dimensional space of LogOdds induced from $\mathbb{P}_M$ and is equivalent to an $M-1$-dimensional real vector space that provides closed operations for addition and scalar multiplication, which is not the case for an SDM-based model or LSDM. Pohl et al. [128] showed that the LogOdds variant was superior to the SDM model in an experiment segmenting 20 subject brains into subcortical structures.

Label space [182] is a multi-organ shape representation that maps M organ labels to the vertices of a regular simplex, in which a scalar label value is changed to a vertex coordinate position in an M dimensional space. As demonstrated in Fig. 2.29, the regular simplex is a hyperdimensional analogue of an equilateral

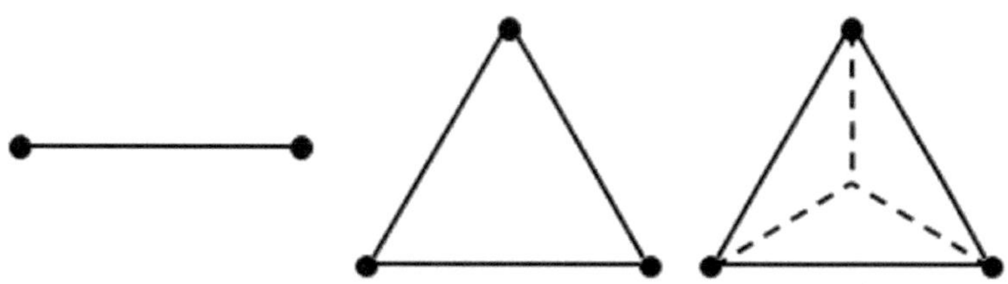

Fig. 2.29 The first three label space configurations. A unit interval for two labels (organ and background), a triangle for three labels (organ 1, organ 2, and background), and a tetrahedron for four labels (organ 1, organ 2, organ 3, and background)

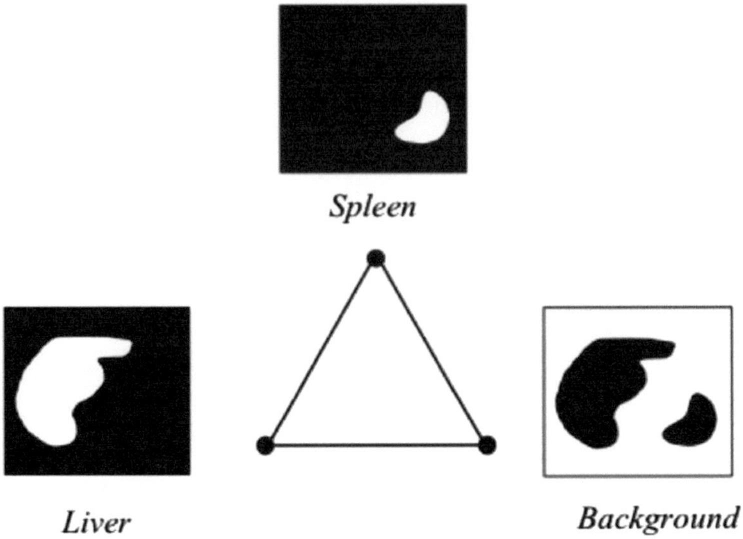

Fig. 2.30 An example of organ assignment to each vertices of a triangle

triangle with $M + 1$ vertices capable of being represented in M dimensions. Figure 2.30 is an example of an assignment of two organs and background to vertices of a triangle. Lying in a linear vector space, this space has several desirable properties: all labels are equally separated in space, addition and scalar multiplication are natural, label uncertainty is expressed as a weighted combination of label vertices, and interpolation is unbiased toward any label including the background. Malcolm et al. demonstrated that algebraic operations may be done directly [182]. Label uncertainty is expressed naturally as a mixture of labels, interpolation is unbiased toward any label or the background, and registration may be performed directly.

An alternative probabilistic multi-organ shape representation is the isometric Log-Ratio, which forms a vector space, isometric, and thus isomorphic to the probability simplex, and results in a nonsingular covariance [183]. These authors claimed that these properties did not exist together in any previously offered probabilistic CA work. They demonstrated how the lack of some of these properties degraded the results, e.g., statistical analysis using linear PCA.

The main advantages of the above multi-organ modelings are that they do not suffer from overlap between neighboring organs. However, they sometimes result in unnatural shapes, in particular, boundaries contacting neighboring tissues or organs.

3. Conditional SSMfor multiple organs

When features highly correlating with a feature vector $\boldsymbol{x}$ of a target organ are available, it is effective to model the distribution of $\boldsymbol{x}$ using a conditional SSM [184–188], for example, a conditional SSM of gallbladder given a liver. Let organ

1 and organ 2 be a target organ and a conditional organ, respectively. Feature vectors of both organs are denoted by $\boldsymbol{x}_{1i}$ and $\boldsymbol{x}_{2i}$ $(i = 1, 2, \ldots, N)$, where N is the number of training labels. Average vectors are given by $\boldsymbol{\mu}_1$ and $\boldsymbol{\mu}_2$, and covariance matrices Σ_{kl} $(k, l = 1, 2)$ are computed using the following equation:

$$\Sigma_{kl} = \frac{1}{N-1} \sum_{i=1}^{N} (\boldsymbol{x}_{ki} - \boldsymbol{\mu}_k)(\boldsymbol{x}_{li} - \boldsymbol{\mu}_l)^T \tag{2.234}$$

Average $\boldsymbol{m}$ and covariance matrix $\boldsymbol{K}$ of conditional distribution of $\boldsymbol{x}_1$ given $\boldsymbol{x}_2$, $p(\boldsymbol{x}_1|\boldsymbol{x}_2)$ are given below:

$$\boldsymbol{m} = \boldsymbol{\mu}_1 + \Sigma_{12}\Sigma_{12}^{-1}(\boldsymbol{x}_2 - \boldsymbol{\mu}_1), \tag{2.235}$$

$$\boldsymbol{K} = \Sigma_{11} - \Sigma_{12}\Sigma_{22}^{-1}\Sigma_{21}. \tag{2.236}$$

A conditional SSM can be derived by statistical analysis of $p(\boldsymbol{x}_1|\boldsymbol{x}_2)$, for example, eigenvalue decomposition of $\boldsymbol{K}$. Variance and covariance of a conditional probability distribution of $\boldsymbol{x}_1$ given $\boldsymbol{x}_2$, or $p(\boldsymbol{x}_1|\boldsymbol{x}_2)$, are smaller than those of a probability distribution of $\boldsymbol{x}_1$ or $p(\boldsymbol{x}_1)$. Therefore, performance of a conditional SSM of organ 1 given organ 2 is better than performance of an SSM of organ 1 without any condition. Figure 2.31 shows probabilistic atlases of a gallbladder generated by a non-conditional SSM, or a model of $p(\boldsymbol{x}_1)$, and a conditional SSM of a gallbladder given a liver or a model of $p(\boldsymbol{x}_1|\boldsymbol{x}_2)$. As shown in this figure, an area with high probability of existence in panel (a) merged with the liver, but did not do so in panel (b), which means that the conditional SSM appropriately restricted the area of the gallbladder with high probability.

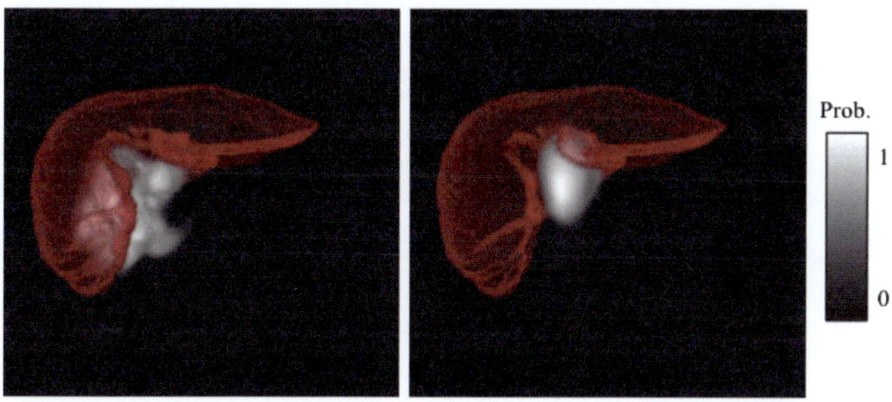

Fig. 2.31 Probabilistic atlases generated by two SSMs of a gallbladder. The atlases were computed from labels reconstructed by normal random numbers generated in subspaces of SSMs or (**a**) nonconditional SSM and (**b**) conditional SSM given a liver

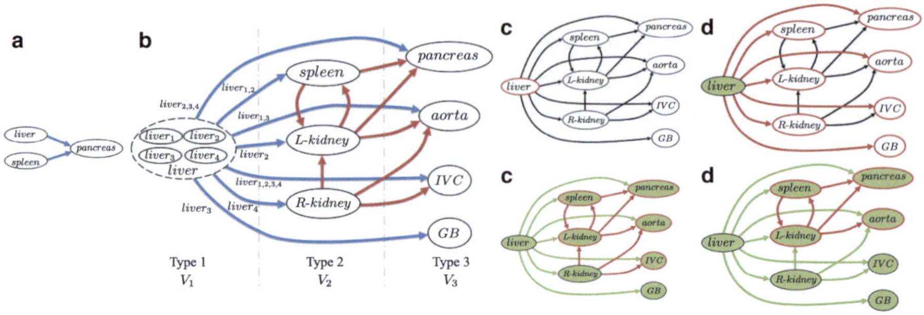

Fig. 2.32 Organ correlation graph (OCG) and OCG-based multi-organ segmentation. (**a**, **b**) OCG. *Blue* and *red edges* indicate the directed edges from a node in Types 1 and 2, respectively. (**c**)–(**f**) Sequential segmentation steps based on OCG. A *red border* indicates the nodes to be segmented. *Green color* indicates segmented nodes

The selection of a condition organ given a target organ is an important task. Okada et al. [188] proposed an organ correlation graph (OCG) which encodes the spatial correlations among organs inherent in human anatomy as presented in Fig. 2.32. Panels (a) and (b) show OCGs, and panels (c)–(f) present sequential segmentation steps based on the OCG of (b) (see [189] for details of segmentation steps). Panel (b) indicates an OCG of eight abdominal organs (liver, spleen, left and right kidney, pancreas, gallbladder, aorta, and inferior vena cava). This was constructed automatically based on shape predictability by partial least squares regression (PLSR) [189] under the constraints on the three types of organs in panel (b). A set of PDM-based conditional SSMs were constructed from 86 CT datasets obtained with four imaging conditions. These were used in the OCG-based segmentation, resulting in high segmentation accuracy of eight organs.

Finally, we discuss the relationship between an SSM of $p(\boldsymbol{x}_1, \boldsymbol{x}_2)$ and a conditional SSM of $p(\boldsymbol{x}_1|\boldsymbol{x}_2)$. The following is an equation that connects $p(\boldsymbol{x}_1, \boldsymbol{x}_2)$ and $p(\boldsymbol{x}_1|\boldsymbol{x}_2)$. $p(\boldsymbol{x}_1, \boldsymbol{x}_2) = p(\boldsymbol{x}_1|\boldsymbol{x}_2)p(\boldsymbol{x}_2)$ where $p(\boldsymbol{x}_2)$ is a marginal distribution of $\boldsymbol{x}_2$. As indicated by this equation, $p(\boldsymbol{x}_1, \boldsymbol{x}_2)$ is composed of not only $p(\boldsymbol{x}_1|\boldsymbol{x}_2)$ but also $p(\boldsymbol{x}_2)$, which means that, in principle, an SSM of $p(\boldsymbol{x}_1, \boldsymbol{x}_2)$ is able to describe shapes that can be represented by a conditional SSM of $p(\boldsymbol{x}_1|\boldsymbol{x}_2)$ and an SSM of $p(\boldsymbol{x}_2)$. It is, however, more difficult to construct precisely and use effectively an SSM of $p(\boldsymbol{x}_1, \boldsymbol{x}_2)$ than an SSM of $p(\boldsymbol{x}_1|\boldsymbol{x}_2)$, because of the large variance of distribution in a subspace. It should be noted that variances of $p(\boldsymbol{x}_1, \boldsymbol{x}_2)$, $p(\boldsymbol{x}_1)$, and $p(\boldsymbol{x}_1|\boldsymbol{x}_2)$ are decreased in the following order:

$$\mathrm{Var}(\boldsymbol{x}_1, \boldsymbol{x}_2) \geq \mathrm{Var}(\boldsymbol{x}_1) \geq \mathrm{Var}(\boldsymbol{x}_1|\boldsymbol{x}_2) \tag{2.237}$$

where $\mathrm{Var}(\underline{x})$ means variances of $\boldsymbol{x}$. The higher the correlation between $\boldsymbol{x}_1$ and $\boldsymbol{x}_2$, the smaller the variance of $p(\boldsymbol{x}_1|\boldsymbol{x}_2)$, which means that a conditional SSM will be more effective for image analysis, such as segmentation. When using a conditional SSM, attention must be given to the reliability of the condition, which might be contaminated by noise and measurement error, in particular when the conditional features are measured by an automated process. Tomoshige et al. [190] proposed a relaxation scheme of condition with an error model in measurement, which is applicable to multiple organ modeling and segmentation.

2.3.7.2 Anatomical Anomaly

Definition

Many parts of the human body have a large variations in their morphology, such as dimensions, shapes, or topologies of structures. Additionally, some anatomical entities may or may not be present, or the number of entities may vary. In the conventional anatomy, such variations are sometimes categorized into normal variants and congenital abnormalities [191–193]. The former are variations that are generally asymptomatic, even though they may predispose to pathological conditions [191]. In other words, these variations are considered to be safely ignored by clinicians. Conversely, congenital abnormalities are variations that are usually symptomatic and thus are clinically significant. However, in some instances, there are no clear-cut distinctions between normal variants and congenital abnormalities [193].

In the context of CA, it is useful to consider another categorization of interindividual variations, that is, continuous morphological varieties and discrete anatomical anomalies (Fig. 2.33). For example, the different shapes of the liver (without any topological change) are categorized as a continuous morphological variety, whereas the variety of the number of the vertebrae is categorized as an anatomical anomaly. This categorization is important in practice because of their different difficulties in statistical modeling.

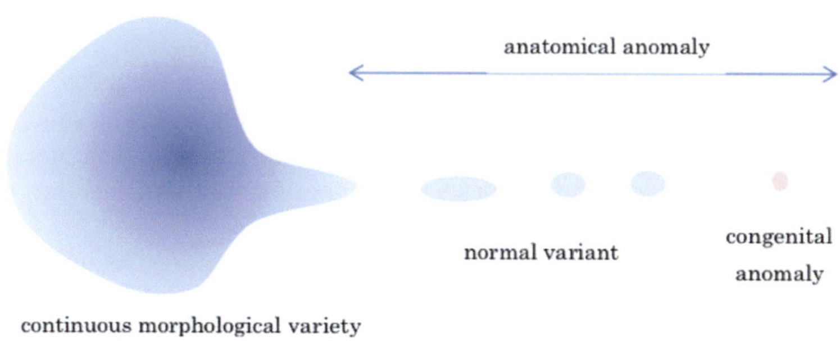

Fig. 2.33 Continuous morphological variety and anatomical anomalies

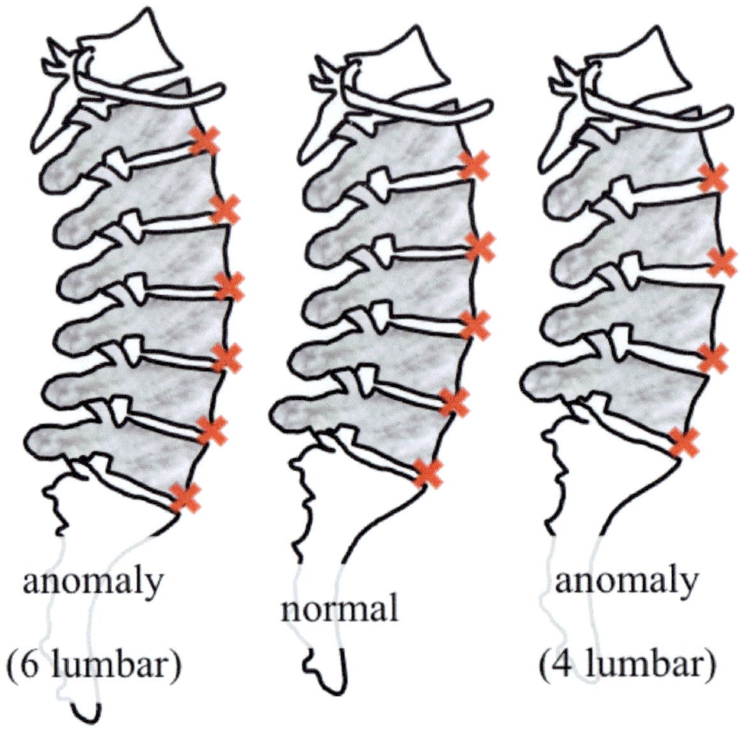

Fig. 2.34 Alterations in the number of lumbar vertebrae

Examples of Normal Variants

In this section, several examples of frequent normal variants are presented.

Alteration of the number of vertebrae or ribs is one of the most frequent anomalies. Usually, human beings have seven cervical (C1–C7), 12 thoracic (T1–T12), and five lumbar (L1–L5) vertebrae, as well as 12 pairs of ribs. However, alterations in the total number of vertebrae is as frequent as 8.2%, according to a survey [194] (Fig. 2.34). Additionally, a transitional vertebra, which has an intermediate shape between a lumbar and a sacral vertebra, is sometimes found. These variants may cause difficulties in the identification and labeling of each vertebra.

There are also variations in position and number of structures, including:

- Ureteropelvic duplication
- Abberrant right hepatic artery
- Pancreas divisum
- Cavum septum pellucidum

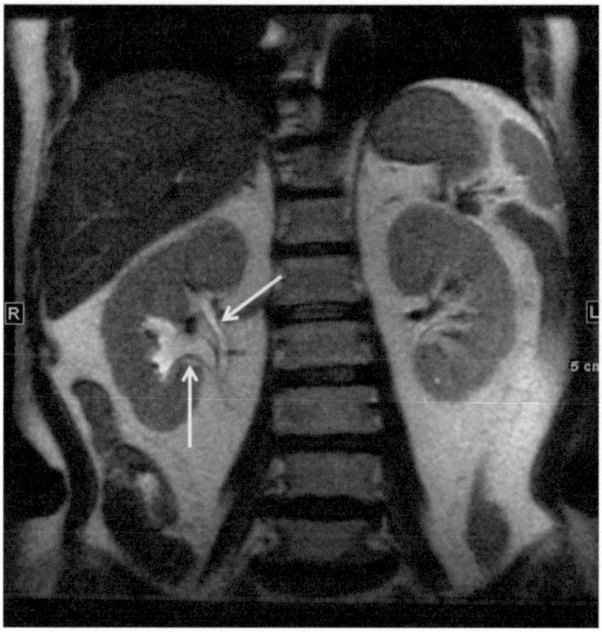

Fig. 2.35 Ureteropelvic duplication (T1-weighted MR image, coronal cross- section)

The occurrence rates of these four variants are reported as 0.5–1% [195], 11–26.5% [196], 8% [197], and 3% [198], respectively (Figs. 2.35, 2.36, 2.37, and 2.38). Because of these high prevalences, these variants should be considered in modeling of the corresponding organs. However, it is difficult to represent both normal and variant subjects with a single model, because these variants accompany alterations in topologies of the organs (i.e., the ureter, the arterial system, the pancreatic duct, and the brain ventricles).

The branching patterns of vessels may also vary widely. The larger arteries and veins may have normal variants with topology changes. Note that such normal variants sometimes become a problem in surgical operations; thus it may have some clinical impact.

Statistical Modeling of Normal Variants

Although many statistical modeling methods for continuous variations have been introduced, few methods for discrete normal variants have been reported. Among them, Mori et al. [199] reported a graph-based method to identify the branching pattern of the abdominal arteries. Another work by Hanaoka et al. proposed a method to detect anomalies of the number of vertebrae [70].

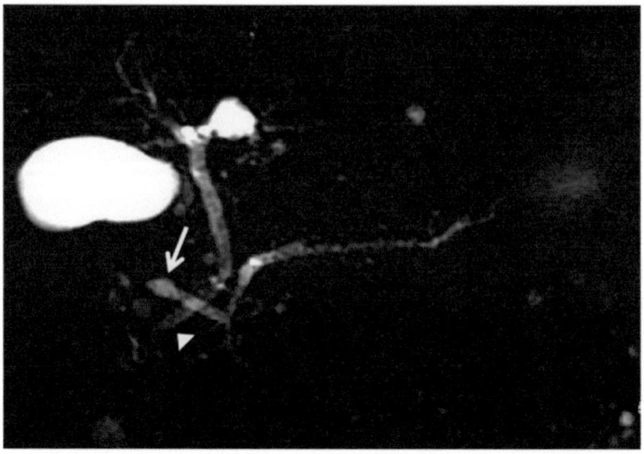

Fig. 2.36 Pancreas divisum. The main pancreatic duct continues to the papilla duodeni minor (*arrow*), not to the papilla duodeni major (*arrowhead*) (MR cholangiopancreatography, frontal view)

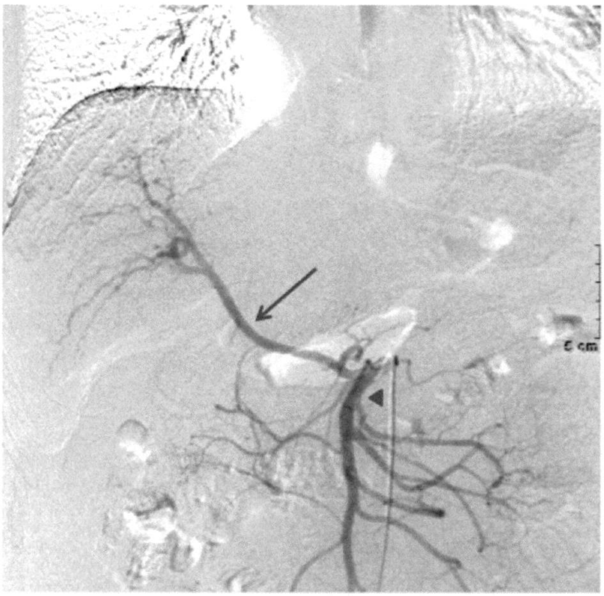

Fig. 2.37 Aberrant right hepatic artery. The right hepatic artery (*arrow*) arises from the superior mesenteric artery (*arrowhead*) instead of the celiac artery (Conventional angiography, frontal view)

However, as described above, it is difficult to represent both normal and variant subjects with a single model where these variants change the topology of the target organ. The application of a PDM to such a situation is an example. To build a PDM, point-to-point correspondence must be established. However, the points that

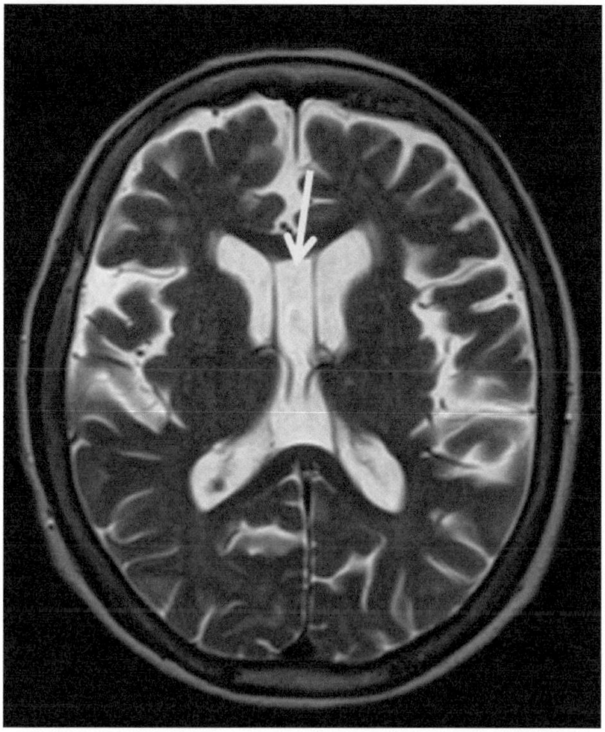

Fig. 2.38 A large cavum septum pellucidum (T2-weighted MR image, axial cross-section)

are defined on an abnormal extra structure have no corresponding points in normal subjects.

One possible solution is to build two models for normal and variant subjects. However, collecting a sufficiently large number of variant subjects is usually not practical. Otherwise, level-set methods [158] can be used to build a statistical shape model with topology changes. Another promising approach is multi-atlas-based methods [200]. Because variant and normal subjects can be combined to form a multi-atlas in multi-atlas segmentation methods, it can handle variants with high frequency without explicitly considering the existence of an abnormality.

Up to now, no general method has been introduced to build an SSM that can represent both continuous (e.g., the position, dimensions, or pose) and discrete (e.g., the branching pattern, number of objects, or topology) properties of the shape variety. Future development of new modeling methodology is needed.

References

1. Cootes TF, Taylor CJ, Cooper DH, Graham J (1995) Active shape models-their training and application. Comput Vis Image Underst 61(1):38–59
2. Cootes TF, Taylor CJ (1993) Active shape model search using local grey-level models: a quantitative evaluation. In: BMVC, Surrey, vol 93, pp 639–648. Citeseer
3. Pizer SM, Fletcher PT, Joshi S, Thall A, Chen JZ, Fridman Y, Fritsch DS, Gash AG, Glotzer JM, Jiroutek MR et al (2003) Deformable m-reps for 3d medical image segmentation. Int J Comput Vis 55(2–3):85–106
4. Styner M, Oguz I, Xu S, Brechbühler C, Pantazis D, Levitt JJ, Shenton ME, Gerig G (2006) Framework for the statistical shape analysis of brain structures using SPHARM-PDM. Insight J (1071):242
5. Wu FTH, Ng-Thow-Hing V, Singh K, Agur AM, McKee NH (2007) Computational representation of the aponeuroses as NURBS surfaces in 3D musculoskeletal models. Comput Methods Programs Biomed 88(2):112–122
6. Chen SY, Guan Q (2011) Parametric shape representation by a deformable Nonuniform Rational Basis Spline model for cardiac functional measurements. IEEE Trans Biomed Eng 58(3):480–487
7. Yushkevich P, Fletcher PT, Joshi S, Thall A, Pizer SM (2003) Continuous medial representations for geometric object modeling in 2D and 3D. Image Vis Comput 21(1):17–27
8. Besl PJ, McKay ND (1992) A method for registration of 3-D shapes. IEEE Trans Pattern Anal Mach Intell 14(2):239–256
9. Rangarajan A, Chui H, Bookstein FL (1997) The softassign procrustes matching algorithm. In: Information processing in medical imaging. Springer, Berlin/New York, pp 29–42
10. Subsol G, Thirion J-P, Ayache N (1998) A scheme for automatically building three-dimensional morphometric anatomical atlases: application to a skull atlas. Med Image Anal 2(1):37–60
11. Fleute M, Lavallée S, Julliard R (1999) Incorporating a statistically based shape model into a system for computer-assisted anterior cruciate ligament surgery. Med Image Anal 3(3):209–222
12. Kaus MR, von Berg J, Weese J, Niessen W, Pekar V (2004) Automated segmentation of the left ventricle in cardiac MRI. Med Image Anal 8(3):245–254
13. Shang Y, Dossel O (2004) Statistical 3D shape-model guided segmentation of cardiac images. In: Computers in Cardiology, Cambridge. IEEE, pp 553–556
14. Shen D, Herskovits EH, Davatzikos C (2001) An adaptive-focus statistical shape model for segmentation and shape modeling of 3-D brain structures. IEEE Trans Med Imaging 20(4):257–270
15. Zhao F, Zhang H, Wahle A, Thomas MT, Stolpen AH, Scholz TD, Sonka M (2009) Congenital aortic disease: 4D magnetic resonance segmentation and quantitative analysis. Med Image Anal 13(3):483–493
16. Fleute M, Lavallée S, Desbat L (2002) Integrated approach for matching statistical shape models with intra-operative 2D and 3D data. In: Medical image computing and computer-assisted intervention–MICCAI 2002, Tokyo, vol 2489. Springer, pp 364–372
17. Dam EB, Fletcher PT, Pizer SM (2008) Automatic shape model building based on principal geodesic analysis bootstrapping. Med Image Anal 12(2):136–151
18. Frangi AF, Niessen WJ, Rueckert D, Schnabel JA (2001) Automatic 3D ASM construction via atlas-based landmarking and volumetric elastic registration. In: Information processing in medical imaging, Davis. Springer, pp 78–91
19. Rueckert D, Frangi AF, Schnabel JA (2003) Automatic construction of 3-D statistical deformation models of the brain using nonrigid registration. IEEE Trans Med Imaging 22(8):1014–1025
20. Kelemen A, Székely G, Gerig G (1999) Elastic model-based segmentation of 3-D neuroradiological data sets. IEEE Trans Med Imaging 18(10):828–839

21. Brett AD, Taylor CJ (2000) Construction of 3D shape models of femoral articular cartilage using harmonic maps. In: Medical image computing and computer-assisted intervention–MICCAI 2000, Pittsburgh. Springer, pp 1205–1214
22. Thompson PM, Toga AW (1997) Detection, visualization and animation of abnormal anatomic structure with a deformable probabilistic brain atlas based on random vector field transformations. Med Image Anal 1(4):271–294
23. Lamecker H, Lange T, Seebass M (2004) Segmentation of the liver using a 3D statistical shape model. Konrad-Zuse-Zentrum für Informationstechnik, pp 1–25
24. Praun E, Sweldens W, Schröder P (2001) Consistent mesh parameterizations. In: Proceedings of the 28th annual conference on computer graphics and interactive techniques, Los Angeles. ACM, pp 179–184
25. Meier v, Fisher E (2002) Parameter space warping: shape-based correspondence between morphologically different objects. IEEE Trans Med Imaging 21(1):31–47
26. Wang Y, Chiang M-C, Thompson PM (2005) Automated surface matching using mutual information applied to Riemann surface structures. In: Medical image computing and computer-assisted intervention–MICCAI 2005, Palm Springs, vol 3750. Springer, pp 666–674
27. Kotcheff ACW, Taylor CJ (1998) Automatic construction of eigenshape models by direct optimization. Med Image Anal 2(4):303–314
28. Davies RH (2002) Learning shape: optimal models for analysing shape variability. Ph.D. thesis, University of Manchester
29. Thodberg HH (2003) Minimum description length shape and appearance models. In: Information processing in medical imaging, Ambleside. Springer, pp 51–62
30. Cootes TF, Taylor CJ (1994) Using grey-level models to improve active shape model search. In: Proceedings of the 12th IAPR international conference on pattern recognition, Jerusalem. Vol. 1-conference A: computer vision & image processing, vol 1. IEEE, pp 63–67
31. Ando S (2000) Consistent gradient operators. IEEE Trans Pattern Anal Mach Intell 22(3):252–265
32. Friedman J, Hastie T, Tibshirani R (2008) Sparse inverse covariance estimation with the graphical lasso. Biostatistics 9(3):432–441
33. Bishop CM et al (2006) Pattern recognition and machine learning, vol 4. Springer, New York
34. Duda RO, Hart PE, Stork DG (2012) Pattern classification. John Wiley & Sons, New York
35. Fukunaga K (2013) Introduction to statistical pattern recognition. Academic press, Boston
36. Marr D (1982) Vision: a computational investigation into the human representation and processing of visual information. W.H. Freeman and Company, San Francisco
37. Quinlan JR (1986) Induction of decision trees. Mach Learn 1(1):81–106
38. Breiman L, Friedman JH, Olshen RA, Stone CJ (1984) Classification and regression trees. Wadsworth Brooks, Monterey
39. Cristianini N, Shawe-Taylor J (2000) An introduction to support vector machines and other kernel-based learning methods. Cambridge university press, Cambridge
40. Shawe-Taylor J, Cristianini N (2004) Kernel methods for pattern analysis. Cambridge university press, Cambridge
41. Rosenblatt F (1958) The perceptron: a probabilistic model for information storage and organization in the brain. Psychol Rev 65(6):386
42. Minsky M, Papert S (1969) Perceptrons. MIT Press, Cambridge, MA
43. Fukushima K (1988) Neocognitron: a hierarchical neural network capable of visual pattern recognition. Neural Netw 1(2):119–130
44. LeCun Y, Boser B, Denker JS, Henderson D, Howard RE, Hubbard W, Jackel LD (1989) Backpropagation applied to handwritten zip code recognition. Neural Comput 1(4):541–551
45. Rumelhart DE, Hinton GE, Williams RJ (1986) Learning representations by back-propagating correlations. Nature 333:533–536
46. Rumelhart DE, Hinton GE, Williams RJ (1988) Learning representations by back-propagating errors. Cogn Model 5:3

47. Olshausen BA et al (1996) Emergence of simple-cell receptive field properties by learning a sparse code for natural images. Nature 381(6583):607–609
48. Hinton G, Osindero S, Teh Y-W (2006) A fast learning algorithm for deep belief nets. Neural Comput 18(7):1527–1554
49. LeCun Y, Bengio Y (1998) Convolutional networks for images, speech, and time series. In: The handbook of brain theory and neural networks. MIT Press, Cambridge, pp 255–258
50. Poggio T, Torre V (1984) Ill-posed problems and regularization analysis in early vision. Massachusetts Institute of Technology, Cambridge
51. Poggio T, Torre V, Koch C (1989) Computational vision and regularization theory. Image Underst 3(1–18):111
52. Cootes TF, Taylor CJ (1992) Active shape models – 'smart snakes'. In: BMVC92, Leeds. Springer, pp 266–275
53. Marroquin J, Mitter S, Poggio T (1987) Probabilistic solution of ill-posed problems in computational vision. J Am Stat Assoc 82(397):76–89
54. Geman S, Geman D (1984) Stochastic relaxation, Gibbs distributions, and the Bayesian restoration of images. IEEE Trans Pattern Anal Mach Intell 6(6):721–741
55. Marr D, Hildreth E (1980) Theory of edge detection. Proc R Soc Lond Ser B Biol Sci 207(1167):187–217
56. Haralick RM (1984) Digital step edges from zero crossing of second directional derivatives. IEEE Trans Pattern Anal Mach Intell 6(1):58–68
57. Canny J (1986) A computational approach to edge detection. IEEE Trans Pattern Anal Mach Intell 8(6):679–698
58. Sethian JA (1999) Level set methods and fast marching methods: evolving interfaces in computational geometry, fluid mechanics, computer vision, and materials science, vol 3. Cambridge university press, Cambridge
59. Sethian JA (1996) A fast marching level set method for monotonically advancing fronts. Proc Natl Acad Sci 93(4):1591–1595
60. Piegl LA (1993) Fundamental developments of computer-aided geometric modeling. Academic Press, London/San Diego
61. Cootes TF, Hill A, Taylor CJ, Haslam J (1994) Use of active shape models for locating structures in medical images. Image Vis Comput 12(6):355–365
62. Böhm W, Farin G, Kahmann J (1984) A survey of curve and surface methods in CAGD. Comput Aided Geom Des 1(1):1–60
63. Delingette H, Hebert M, Ikeuchi K (1992) Shape representation and image segmentation using deformable surfaces. Image Vis Comput 10(3):132–144
64. Piegl L (1991) On Nonuniform Rational Basis Spline: a survey. IEEE Comput Graph Appl 11(1):55–71
65. Zhang S, Zhan Y, Dewan M, Huang J, Metaxas DN, Zhou XS (2011) Deformable segmentation via sparse shape representation. In: Medical image computing and computer-assisted intervention–MICCAI 2011, Athens. Springer, pp 451–458
66. Hontani H, Watanabe W (2010) Point-based non-rigid surface registration with accuracy estimation. In: 2010 IEEE conference on computer vision and pattern recognition (CVPR), San Francisco. IEEE, pp 446–452
67. Frangi AF, Rueckert D, Schnabel JA, Niessen WJ (2002) Automatic construction of multiple-object three-dimensional statistical shape models: application to cardiac modeling. IEEE Trans Med Imaging 21(9):1151–1166
68. Tobon-Gomez C, Butakoff C, Aguade S, Sukno F, Moragas G, Frangi AF (2008) Automatic construction of 3d-ASM intensity models by simulating image acquisition: application to myocardial gated SPECT studies. IEEE Trans Med Imaging 27(11):1655–1667
69. Heimann T, Meinzer H-P (2009) Statistical shape models for 3D medical image segmentation: a review. Med Image Anal 13(4):543–563
70. Hanaoka S, Masutani Y, Nemoto M, Nomura Y, Yoshikawa T, Hayashi N, Yoshioka N, Ohtomo K (2011) Probabilistic modeling of landmark distances and structure for anomaly-proof landmark detection. In: Proceedings of the third international workshop on

mathematical foundations of computational anatomy-geometrical and statistical methods for modelling biological shape variability, Toronto, pp 159–169
71. Lowe DG (1999) Object recognition from local scale-invariant features. In: The proceedings of the seventh IEEE international conference on computer vision, Kerkyra, vol 2. IEEE, pp 1150–1157
72. Lowe DG (2004) Distinctive image features from scale-invariant keypoints. Int J Comput Vis 60(2):91–110
73. Lindeberg T (1993) Scale-space theory in computer vision. Springer, Berlin/New York
74. Dalal N, Triggs B (2005) Histograms of oriented gradients for human detection. In: IEEE computer society conference on computer vision and pattern recognition, CVPR 2005, San Diego, vol 1. IEEE, pp 886–893
75. Bay H, Tuytelaars T, Van Gool L (2006) Surf: speeded up robust features. In: Computer vision–ECCV 2006, Graz. Springer, pp 404–417
76. Tomasi C, Manduchi R (1998) Bilateral filtering for gray and color images. In: Sixth international conference on computer vision, Bombay. IEEE, pp 839–846
77. Gerig G, Kubler O, Kikinis R, Jolesz FA (1992) Nonlinear anisotropic filtering of MRI data. IEEE Trans Med Imaging 11(2):221–232
78. Sato Y, Nakajima S, Shiraga N, Atsumi H, Yoshida S, Koller T, Gerig G, Kikinis R (1998) Three-dimensional multi-scale line filter for segmentation and visualization of curvilinear structures in medical images. Med Image Anal 2(2):143–168
79. Grenander U, Miller MI (1998) Computational anatomy: an emerging discipline. Q Appl Math LVI(4):617–694
80. Miller MI, Trouve A, Younes L (2002) On the metrics and Euler-Lagrange equations of computational anatomy. Ann Rev Biomed Eng 4:375–405
81. Trouvé A (1998) Diffeomorphisms groups and pattern matching in image analysis. Int J Comput Vis 28(3):213–221
82. Miller MI, Trouve A, Younes L (2006) Geodesic shooting for computational anatomy. J Math Imaging Vis 24(2):209–228
83. Younes L (2010) Shapes and diffeomorphisms. Applied mathematical science, vol 171. Springer, Heidelberg
84. Pennec X, Arsigny V (2013) Exponential barycenters of the canonical Cartan connection and invariant means on lie groups. In: Matrix information geometry. Springer, Berlin/Heidelberg, pp 123–166
85. Holm DD, Trouve A, Younes L (2009) The Euler-Poincare theory of metamorphosis. Q Appl Math 67:661–685
86. Durrleman S (2010) Statistical models of currents for measuring the variability of anatomical curves, surfaces and their evolution. Ph.D. thesis, Universite de Nice-Sophia Antipolis
87. Beg MF, Miller MI, Trouve A, Younes L (2005) Computing large deformation metric mappings via geodesic flow of diffeomorphisms. Int J Comput Vis 61:139–157
88. Trouve A, Younes L (2005) Metamorphoses through lie group action. Found Comput Math 5(2):173–198
89. Holm DD, Schmah T, Stoica C (2009) Geometric mechanics and symmetry. Oxford University Press, New York
90. Arsigny V, Commowick O, Pennec X, Ayache N (2006) A log-Euclidean framework for statistics on diffeomorphisms. In: MICCAI, Copenhagen, vol 9, pp 924–931
91. Vercauteren T, Pennec X, Perchant A, Ayache N (2008) Symmetric log-domain diffeomorphic registration: a demons-based approach. Lect Notes Comput Sci 5241:754–761
92. Mansi T, Pennec X, Sermesant M, Delingette H, Ayache N (2010) Logdemons revisited: consistent regularisation and incompressibility constraint for soft tissue tracking in medical images. In: MICCAI, Beijing, vol 13, pp 652–659
93. Lorenzi M, Pennec X (2012) Geodesics, parallel transport and one-parameter subgroups for diffeomorphic image registration. Int J Comput Vis 12:1–17
94. Bruveris M, Holm DD (2015) Geometry of image registration: the diffeomorphism group and momentum maps. In: Geometry, mechanics, and dynamics. Springer, New York, pp 19–56

95. Holm DD, Bruveris M, Gay-Balmaz F, Ratiu TS (2011) The momentum map representation of images. J Nonlinear Sci 21(1):115–150
96. Younes L, Arrate F, Miller MI (2009) Evolutions equations in computational anatomy. NeuroImage 45(1, Supplement 1):S40–S50
97. Vialard FX, Risser L, Rueckert D, Cotter CJ (2012) Diffeomorphic 3d image registration via geodesic shooting using an efficient adjoint calculation. Int J Comput Vis 97:229–241
98. do Carmo M (1992) Reimannian geometry. Birkhauser, Boston/Basel/Berlin
99. Harms P (2010) Sobolev metrics on shape space of surfaces. Ph.D. thesis, University of Wien
100. Khesin BA, Wendt R (2009) The geometry of infinite dimensional Lie groups. Volume 51 of Ergebnisse der mathematic und ihrer Grenzgebiete. 3. Folge/a series of modern surveys in mathematics. Springer, Berlin/London
101. Hernandez M, Bossa M, Olmos S (2009) Registration of anatomical images using paths of diffeomorphisms parameterized with stationary vector flows. Int J Comput Vis 85:291–306
102. Mansi T, Voigt I, Leonardi B, Pennec X, Durrleman S, Sermesant M, Delingette H, Taylor A, Boudjemline Y, Pongiglione G, Ayache N (2011) A statistical model for quantification and prediction of cardiac remodelling: application to tetralogy of fallot. IEEE Trans Med Images 30(9):1605–1616
103. Postnikov MM (2001) Geometry VI: Riemannian geometry. Encyclopedia of mathematical science. Springer, Berlin/London
104. Vaillant M, Miller MI, Trouve A, Younes L (2004) Statistics on diffeomorphisms via tangent space representations. NeuroImage 23(S1):S161–S169
105. Vialard FX, Risser L, Holm DD, Rueckert D (2011) Diffeomorphic atlas estimation using Karcher mean and geodesic shooting on volumetric images. In: Medical image understanding and analysis. IEEE, Piscataway
106. Fletcher PT, Lu C, Pizer M, Joshi S (2004) Principal geodesic analysis for the study of nonlinear statistics of shape. IEEE Trans Med Imaging 23(8):995–1005
107. Arsigny V (2006) Processing data in lie groups: an algebraic approach. Ph.D. thesis, École politechnique, Palaiseau
108. Younes L, Qiu A, Winslow R, Miller M (2008) Transport of relational structures in groups of diffeomorphisms. J Math Imaging Vis 32(1):41–56
109. Khan A, Beg M (2008) Representation of time-varying shapes in the large deformation diffeomorphic framework. In: 5th IEEE international symposium on biomedical imaging ISBI, Paris, pp 1521–1524
110. Trouvé A, Vialard FX (2012) Shape splines and stochastic shape evolutions: a second order point of view. Quart Appl Math 70:219–251
111. Younes L (2007) Jacobi fields in groups of diffeomorphisms and applications. Q Appl Math 65:113–134
112. Durrleman S, Pennec X, Trouve A, Braga J, Gerig G, Ayache N (2012) Toward a comprehensive framework for the spatiotemporal statistical analysis of longitudinal shape data. IJCV 103(1):22–59
113. Singh N, Fletcher PT, Preston JS, Ha L, King R, Marron JS, Wiener M, Joshi S (2010) Multivariate statistical analysis of deformation momenta relating anatomical shape to neuropsychological measures. In: MICCAI, Beijing, pp 529–537
114. Bossa M, Hernandez M, Olmos S (2007) Contributions to 3d diffeomorphic atlas estimation: application to brain images. In: MICCAI, Brisbane, vol 10, pp 667–674
115. Durrleman S, Fillard P, Pennec X, Trouve A, Ayache N (2011) Registration, atlas estimation and variability analysis of white matter fiber bundles modeled as currents. NeuroImage 55(3):1073–1090
116. Vialard FX, Risser L, Rueckert D, Holm DD (2012) Diffeomorphic atlas estimation using geodesic shooting on volumetric images. Ann BMVA 5:1–12
117. Durrleman S, Pennec X, Trouvé A, Gerig G, Ayache N (2009) Spatiotemporal atlas estimation for developmental delay detection in longitudinal datasets. In: MICCAI, London, vol 12, pp 297–304

118. Durrleman S, Pennec X, Trouve A, Ayache N, Braga J (2012) Comparison of the endocranial ontogenies between chimpanzees and bonobos via temporal regression and spatiotemporal registration. J Hum Evol 62(1):74–88
119. Gu M, Nielsen MA, Dowling MR, Doherty AC (2006) Quantum computation as geometry. Science 24(311(5764)):1133–1135
120. Antoine Maintz JB, Viergever MA (1998) A survey of medical image registration. Med Image Anal 2(1):1–36
121. Mazziotta JC, Toga AW, Evans A, Fox P, Lancaster J (1995) A probabilistic atlas of the human brain: theory and rationale for its development the international consortium for brain mapping (ICBM). Neuroimage 2(2PA):89–101
122. Thompson PM, Schwartz C, Toga AW (1996) High-resolution random mesh algorithms for creating a probabilistic 3D surface atlas of the human brain. NeuroImage 3(1):19–34
123. Thompson PM, Woods RP, Mega MS, Toga AW (2000) Mathematical/computational challenges in creating deformable and probabilistic atlases of the human brain. Hum Brain Mapp 9(2):81–92
124. Shimizu A, Ohno R, Ikegami T, Kobatake H, Nawano S, Smutek D (2007) Segmentation of multiple organs in non-contrast 3D abdominal CT images. Int J Comput Assist Radiol Surg 2(3-4):135–142
125. Yang J, Staib LH, Duncan JS (2004) Neighbor-constrained segmentation with level set based 3-D deformable models. IEEE Trans Med Imaging 23(8):940–948
126. Tsai A, Yezzi A Jr, Wells W, Tempany C, Tucker D, Fan A, Grimson WE, Willsky A (2003) A shape-based approach to the segmentation of medical imagery using level sets. IEEE Trans Med Imaging 22(2):137–154
127. Furukawa D, Shimizu A, Kobatake H (2007) Automatic liver segmentation method based on maximum a posterior probability estimation and level set method. In: 3D segmentation in the clinic: a grand challenge. CSIRO, Collingwood, pp 117–124
128. Pohl KM, Fisher J, Bouix S, Shenton M, McCarley RW, Grimson WEL, Kikinis R, Wells WM (2007) Using the logarithm of odds to define a vector space on probabilistic atlases. Med Image Anal 11(5):465–477
129. Tsagaan B, Shimizu A, Kobatake H, Miyakawa K (2002) An automated segmentation method of kidney using statistical information. In: Medical image computing and computer-assisted intervention–MICCAI 2002, Tokyo, vol 2488. Springer, pp 556–563
130. Lorenz C, Krahnstöver N (2000) Generation of point-based 3D statistical shape models for anatomical objects. Comput Vis Image Underst 77(2):175–191
131. Koller D, Friedman N (2009) Probabilistic graphical models: principles and techniques. MIT, Cambridge
132. Sawada Y, Hontani H (2012) A study on graphical model structure for representing statistical shape model of point distribution model. In: Medical image computing and computer-assisted intervention–MICCAI 2012, Nice. Springer, pp 470–477
133. Hontani H, Tsunekawa Y, Sawada Y (2013) Accurate and robust registration of nonrigid surface using hierarchical statistical shape model. In: IEEE conference on computer vision and pattern recognition (CVPR), Portland. IEEE, pp 2977–2984
134. Bai W, Shi W, O'Regan DP, Tong T, Wang H, Jamil-Copley S, Peters NS, Rueckert D (2013) A probabilistic patch-based label fusion model for multi-atlas segmentation with registration refinement: application to cardiac MR images. IEEE Trans Med Imaging 32(7):1302–1315
135. Cabezas M, Oliver A, Lladó X, Freixenet J, Cuadra MB (2011) A review of atlas-based segmentation for magnetic resonance brain images. Comput Methods Programs Biomed 104(3):e158–e177
136. Zhang D, Wu G, Jia H, Shen D (2011) Confidence-guided sequential label fusion for multi-atlas based segmentation. In: Medical image computing and computer-assisted intervention–MICCAI 2011, Toronto. Springer, pp 643–650
137. Isgum I, Staring M, Rutten A, Prokop M, Viergever MA, van Ginneken B (2009) Multi-atlas-based segmentation with local decision fusion–application to cardiac and aortic segmentation in CT scans. IEEE Trans Med Imaging 28(7):1000–1010

138. Artaechevarria X, Muñoz-Barrutia A, Ortiz-de Solórzano C (2009) Combination strategies in multi-atlas image segmentation: application to brain MR data. IEEE Trans Med Imaging 28(8):1266–1277
139. Aljabar P, Heckemann RA, Hammers A, Hajnal JV, Rueckert D (2009) Multi-atlas based segmentation of brain images: atlas selection and its effect on accuracy. Neuroimage 46(3):726–738
140. Lötjönen JMP, Wolz R, Koikkalainen JR, Thurfjell L, Waldemar G, Soininen H, Rueckert D (2010) Fast and robust multi-atlas segmentation of brain magnetic resonance images. Neuroimage 49(3):2352–2365
141. van Rikxoort EM, Isgum I, Arzhaeva Y, Staring M, Klein S, Viergever MA, Pluim JPW, van Ginneken B (2010) Adaptive local multi-atlas segmentation: application to the heart and the caudate nucleus. Med Image Anal 14(1):39–49
142. Dawant BM, Zijdenbos AP (2000) Image segmentation. Handb Med Imaging 2:71–127
143. Chenyang Xu, Pham DL, Prince JL (2000) Image segmentation using deformable models. Handb Med Imaging 2:129–174
144. Bankman IN (2009) Segmentation. Academic Press, Cambridge
145. Pham DL, Xu C, Prince JL (2000) Current methods in medical image segmentation. Ann Rev Biomed Eng 2(1):315–337
146. Fischler MA, Elschlager RA (1973) The representation and matching of pictorial structures. IEEE Trans Comput 22(1):67–92
147. Widrow B (1973) The rubber-mask technique. Pattern Recognit 5(3):175–211
148. Chien YP, Fu K-S (1974) Recognition of X-ray picture patterns. IEEE Trans Syst Man Cybern SMC-4(2):145–156
149. Kass M, Witkin A, Terzopoulos D (1988) Snakes: active contour models. Int J Comput Vis 1(4):321–331
150. Cootes TF, Edwards GJ, Taylor CJ (2001) Active appearance models. IEEE Trans Pattern Anal Mach Intell 23(6):681–685
151. Park H, Bland PH, Meyer CR (2003) Construction of an abdominal probabilistic atlas and its application in segmentation. IEEE Trans Med Imaging 22(4):483–492
152. Shimizu A, Kubo M, Furukawa D, Kobatake H, Nawano S (2008) Abdomen standardization for multi-organ segmentation of CT volumes. Int J Comput Assist Radiol Surg 3:s195–s196
153. Linguraru MG, Pura JA, Pamulapati V, Summers RM (2012) Statistical 4D graphs for multi-organ abdominal segmentation from multiphase CT. Med Image Anal 16(4):904–914
154. Shimizu A, Kimoto T, Kobatake H, Nawano S, Shinozaki K (2010) Automated pancreas segmentation from three-dimensional contrast-enhanced computed tomography. Int J Comput Assist Radiol Surg 5(1):85–98
155. Wolz R, Chengwen C, Misawa K, Fujiwara M, Mori K, Rueckert D (2013) Automated abdominal multi-organ segmentation with subject-specific atlas generation. IEEE Trans Med Imaging 32(9):1723–1730
156. Umetsu S, Shimizu A, Watanabe H, Kobatake H, Nawano S (2014) An automated segmentation algorithm for CT volumes of livers with a typical shapes and large pathological lesions. IEICE Trans Inf Syst 97(4):951–963
157. Leventon ME, Grimson WEL, Faugeras O (2000) Statistical shape influence in geodesic active contours. In: Proceedings of IEEE conference on computer vision and pattern recognition, Hilton Head, vol 1. IEEE, pp 316–323
158. Caselles V, Kimmel R, Sapiro G (1997) Geodesic active contours. Int J Comput Vis 22(1):61–79
159. Chan TF, Vese LA (2001) Active contours without edges. IEEE Trans Image Process 10(2):266–277
160. Greig DM, Porteous BT, Seheult AH (1989) Exact maximum a posteriori estimation for binary images. J R Stat Soc Ser B (Methodol) 51:271–279
161. Boykov Y, Veksler O, Zabih R (1998) Markov random fields with efficient approximations. In: Proceedings of 1998 IEEE computer society conference on computer vision and pattern recognition, Santa Barbara. IEEE, pp 648–655

162. Boykov Y, Veksler O, Zabih R (2001) Fast approximate energy minimization via graph cuts. IEEE Trans Pattern Anal Mach Intell 23(11):1222–1239
163. Lempitsky V, Rother C, Roth S, Blake A (2010) Fusion moves for Markov random field optimization. IEEE Trans Pattern Anal Mach Intell 32(8):1392–1405
164. Ishikawa H (2009) Higher-order gradient descent by fusion-move graph cut. In: IEEE 12th international conference on computer vision, Kyoto. IEEE, pp 568–574
165. Funka-Lea G, Boykov Y, Florin C, Jolly M-P, Moreau-Gobard R, Ramaraj R, Rinck D (2006) Automatic heart isolation for CT coronary visualization using graph-cuts. In: 3rd IEEE international symposium on biomedical imaging: Nano to Macro, Prague. IEEE, pp 614–617
166. Slabaugh G, Unal G (2005) Graph cuts segmentation using an elliptical shape prior. In: IEEE international conference on image processing, ICIP 2005, Genoa, vol 2. IEEE, pp II–1222
167. Dasc P, Vekslera O, Zavadskyb V, Boykova Y (2008) Semiautomatic segmentation with compact shape prior. Image Vis Comput 27(1–2):206–219
168. Freedman D, Zhang T (2005) Interactive graph cut based segmentation with shape priors. In: IEEE computer society conference on computer vision and pattern recognition, CVPR 2005, San Diego, vol 1. IEEE, pp 755–762
169. Shimizu A, Nakagomi K, Narihira T, Kobatake H, Nawano S, Shinozaki K, Ishizu K, Togashi K (2011) Automated segmentation of 3D CT images based on statistical atlas and graph cuts. In: Medical computer vision. Recognition techniques and applications in medical imaging, Nagoya. Springer, pp 214–223
170. Malcolm J, Rathi Y, Tannenbaum A (2007) Graph cut segmentation with nonlinear shape priors. In: IEEE international conference on image processing, ICIP 2007, San Antonio, vol 4. IEEE, pp IV 365
171. Nakagomi K, Shimizu A, Kobatake H, Yakami M, Fujimoto K, Togashi K (2013) Multi-shape graph cuts with neighbor prior constraints and its application to lung segmentation from a chest CT volume. Med Image Anal 17(1):62–77
172. Freund Y, Schapire RE (1997) A decision-theoretic generalization of on-line learning and an application to boosting. J Comput Syst Sci 55(1):119–139
173. Breiman L (2001) Random forests. Mach Learn 45(1):5–32
174. Nishii R, Eguchi S (2005) Supervised image classification by contextual AdaBoost based on posteriors in neighborhoods. IEEE Trans Geosci Remote Sens 43(11):2547–2554
175. Avidan S (2006) SpatialBoost: adding spatial reasoning to AdaBoost. In: Computer vision–ECCV 2006, Graz. Springer, pp 386–396
176. Morra JH, Tu Z, Apostolova LG, Green AE, Toga AW, Thompson PM (2008) Automatic subcortical segmentation using a contextual model. In: Medical image computing and computer-assisted intervention–MICCAI 2008, New York. Springer, pp 194–201
177. Amores J, Sebe N, Radeva P, Gevers T, Smeulders A (2004) Boosting contextual information in content-based image retrieval. In: Proceedings of the 6th ACM SIGMM international workshop on multimedia information retrieval, New York. ACM, pp 31–38
178. Shimizu A, Kobayashi M, Nakagomi K, Kobatake H, Yakami M, Fujimoto K, Togashi K (2012) Interlobar fissure extraction from a chest CT volume based on a new loss function for a boosting algorithm. Int J Comput Assist Radiol Surg 7(supplement 1):s322–s324
179. Shimizu A, Shindo K, Kobatake H, Nawano S, Shinozaki K (2013) Proposal of a novel boosting algorithm regularized by a statistical shape feature and its application to organ segmentation. Med Imaging Technol 31(2):121–131
180. Cremers D, Rousson M, Deriche R (2007) A review of statistical approaches to level set segmentation: integrating color, texture, motion and shape. Int J Comput Vis 72(2):195–215
181. Saito A, Nakada M, Oost E, Shimizu A, Watanabe H, Nawano S (2013) A statistical shape model for multiple organs based on synthesized-based learning. In: Abdominal imaging. Computation and clinical applications. Springer, Berlin/Heidelberg, pp 280–289
182. Malcolm J, Rathi Y, Shenton ME, Tannenbaum A (2008) Label space: a coupled multi-shape representation. In: Medical image computing and computer-assisted intervention–MICCAI 2008, New York. Springer, pp 416–424

183. Changizi N, Hamarneh G (2010) Probabilistic multi-shape representation using an isometric log-ratio mapping. In: Medical image computing and computer-assisted intervention–MICCAI 2010, Beijing. Springer, pp 563–570
184. de Bruijne M, Lund MT, Tankó LB, Pettersen PC, Nielsen M (2007) Quantitative vertebral morphometry using neighbor-conditional shape models. Med Image Anal 11(5):503–512
185. Baka N, de Bruijne M, Reiber JHC, Niessen W, Lelieveldt BPF (2010) Confidence of model based shape reconstruction from sparse data. In: IEEE international symposium on biomedical imaging: from Nano to Macro, Rotterdam. IEEE, pp 1077–1080
186. Blanc R, Seiler C, Székely G, Nolte L-P, Reyes M (2012) Statistical model based shape prediction from a combination of direct observations and various surrogates: application to orthopaedic research. Med Image Anal 16(6):1156–1166
187. Syrkina E, Blanc R, Székely G (2011) Propagating uncertainties in statistical model based shape prediction. In: SPIE medical imaging, vol 7962. International Society for Optics and Photonics, Bellingham
188. Okada T, Linguraru MG, Hori M, Summers RM, Tomiyama N, Sato Y (2013) Abdominal multi-organ CT segmentation using organ correlation graph and prediction-based shape and location priors. In: Medical image computing and computer-assisted intervention–MICCAI 2013, Nagoya. Springer, pp 275–282
189. Rao A, Aljabar P, Rueckert D (2008) Hierarchical statistical shape analysis and prediction of sub-cortical brain structures. Med Image Anal 12(1):55–68
190. Tomoshige S, Oost E, Shimizu A, Watanabe H, Nawano S (2014) A conditional statistical shape model with integrated error estimation of the conditions; application to liver segmentation in non-contrast CT images. Med Image Anal 18(1):130–143
191. De Filippo M, Calabrese M, Quinto S, Rastelli A, Bertellini A, Martora R, Sverzellati N, Corradi D, Vitale M, Crialesi G, Sarli L, Roncoroni L, Garlaschi G, Zompatori M (2008) Congenital anomalies and variations of the bile and pancreatic ducts: magnetic resonance cholangiopancreatography findings, epidemiology and clinical significance. La Radiol Medica 113(6):841–859
192. Yu J, Turner MA, Fulcher AS, Halvorsen RA (2006) Congenital anomalies and normal variants of the pancreaticobiliary tract and the pancreas in adults: part 1, biliary tract. Am J Roentgenol 187(6):1536–1543
193. Guebert GM, Yochum TR, Rowe LJ (1987) Congenital anomalies and normal skeletal variants. In: Essentials of skeletal radiology. Williams & Wilkins, Baltimore
194. Carrino JA, Campbell PD Jr, Lin DC, Morrison WB, Schweitzer ME, Flanders AE, Eng J, Vaccaro AR (2011) Effect of spinal segment variants on numbering vertebral levels at lumbar MR imaging. Radiology 259(1):196–202
195. Chao C-T (2013) Ureteropelvic duplication as focus of recurrent infection. QJM 106(5):471–472
196. Rammohan A, Palaniappan R, Pitchaimuthu A, Rajendran K, Perumal SK, Balaraman K, Ramasamy R, Sathyanesan J, Govindan M (2014) Implications of the presence of an aberrant right hepatic artery in patients undergoing pancreaticoduodenectomy. World J Gastrointest Surg 6(1):9
197. DiMagno MJ, Wamsteker E-J (2011) Pancreas divisum. Curr Gastroenterol Rep 13(2):150–156
198. Nopoulos P, Swayze V, Flaum M, Ehrhardt JC, Yuh WTC, Andreasen NC (1997) Cavum septi pellucidi in normals and patients with schizophrenia as detected by magnetic resonance imaging. Biol Psychiatry 41(11):1102–1108
199. Mori K, Oda M, Egusa T, Jiang Z, Kitasaka T, Fujiwara M, Misawa K (2010) Automated nomenclature of upper abdominal arteries for displaying anatomical names on virtual laparoscopic images. In: Medical imaging and augmented reality. Springer, Berlin, pp 353–362
200. Heckemann RA, Hajnal JV, Aljabar P, Rueckert D, Hammers A (2006) Automatic anatomical brain MRI segmentation combining label propagation and decision fusion. NeuroImage 33(1):115–126

Chapter 3
Understanding Medical Images Based on Computational Anatomy Models

Shouhei Hanaoka, Naoki Kamiya, Yoshinobu Sato, Kensaku Mori, Hiroshi Fukuda, Yasuyuki Taki, Kazunori Sato, Kai Wu, Yoshitaka Masutani, Takeshi Hara, Chisako Muramatsu, Akinobu Shimizu, Mikio Matsuhiro, Yoshiki Kawata, Noboru Niki, Daisuke Fukuoka, Tomoko Matsubara, Hidenobu Suzuki, Ryo Haraguchi, Toshizo Katsuda, and Takayuki Kitasaka

S. Hanaoka
The University of Tokyo, 113-8654, Tokyo, Japan
e-mail: hanaoka-tky@umin.ac.jp

N. Kamiya
Aichi Prefectural University, 480-1198, Aichi, Japan
e-mail: n-kamiya@ist.aichi-pu.ac.jp

Y. Sato
Nara Institute of Science and Technology, 630-0192, Nara, Japan
e-mail: yoshi@is.naist.jp

K. Mori
Nagoya University, 464-8601, Nagoya, Japan
e-mail: kensaku@is.nagoya-u.ac.jp

H. Fukuda
Division of Radiology, Tohoku Medical and Pharmaceutical University, 981-8558, Miyagi, Japan
e-mail: hirofuku@tohoku-mpu.ac.jp

Y. Taki • K. Sato
Department of Nuclear Medicine and Radiology, Institute of Development, Aging and Cancer, Tohoku University, 980-8575, Miyagi, Japan
e-mail: yasuyuki.taki.c7@tohoku.ac.jp; kns@tohoku.ac.jp

K. Wu
Department of Biomedical Engineering, School of Material Science and Engineering, South China University of Technology, 510006, Guangdong, China
e-mail: kaiwu@scut.edu.cn

Y. Masutani
Hiroshima City University, 731-3194, Hiroshima, Japan
e-mail: masutani@hiroshima-cu.ac.jp

T. Hara • C. Muramatsu • D. Fukuoka
Gifu University, 501-1194, Gifu, Japan
e-mail: hara@info.gifu-u.ac.jp; chisa@fjt.info.gifu-u.ac.jp; dfukuoka@gifu-u.ac.jp

H. Kobatake, Y. Masutani (eds.), *Computational Anatomy Based on Whole Body Imaging*, DOI 10.1007/978-4-431-55976-4_3

Abstract This chapter presents examples of medical image understanding algorithms using computational anatomy models explained in Chap. 2. After the introductory in Sect. 3.1, Sect. 3.2 shows segmentation algorithms for vertebrae, ribs, and hip joints. Segmentation algorithms for skeletal muscle and detection algorithms for lymph nodes are explained in Sects. 3.3 and 3.4, respectively. Section 3.5 deals with algorithms for understanding organs/tissues in the head and neck regions and starts with computational neuroanatomy, followed by analysis and segmentation algorithms for white matter, brain CT, oral regions, fundus oculi, and retinal optical coherence tomography (OCT). Algorithms useful in the thorax, specifically for the lungs, tracheobronchial tree, vessels, and interlobar fissures from a thoracic CT volume, are presented in Sect. 3.6. Section 3.7 provides algorithms for breast ultrasound imaging, i.e., mammography and breast MRI. Cardiac imaging algorithms in an echocardiographic image sequence and MR images as well as coronary arteries in a CT volume are explained in Sect. 3.8. Section 3.9 deals with segmentation algorithms of abdominal organs, including the liver, pancreas, spleen, kidneys, gastrointestinal tract, and abdominal blood vessels, followed by anatomical labeling of segmented vessels.

Keywords Segmentation • Head • Neck • Thoracic • Abdomen • CT • MR • X-ray • Fundus image • OCT

A. Shimizu (✉)
Tokyo University of Agriculture and Technology, 184-8588, Tokyo, Japan
e-mail: simiz@cc.tuat.ac.jp

M. Matsuhiro • Y. Kawata • N. Niki • H. Suzuki
The University of Tokushima, 770-8501, Tokushima, Japan
e-mail: mike@tokushima-u.ac.jp; kawata@tokushima-u.ac.jp; niki@tokushima-u.ac.jp; hidenobu-s@tokushima-u.ac.jp

T. Matsubara
Nagoya Bunri University, 492-8520, Aichi, Japan
e-mail: matsubara.tomoko@nagoya-bunri.ac.jp

R. Haraguchi
Graduate School of Applied Informatics, University of Hyogo, 650-0047, Hyogo, Japan
e-mail: haraguch@ai.u-hyogo.ac.jp

T. Katsuda
Butsuryo College of Osaka, 593-8328, Osaka, Japan
e-mail: katsuda@butsuryo.ac.jp

T. Kitasaka
Aichi Institute of Technology, 470-0392, Aichi, Japan
e-mail: kitasaka@aitech.ac.jp

3.1 Introduction

There has been tremendous progress in medical imaging of the human body as represented by multi-detector computed tomography (CT), allowing higher scanning speeds and higher spatial resolution than ever before. A human torso is often scanned within a few seconds with 0.5 mm spatial resolution, which results in between several hundred and more than one thousand image sections. Although such CT volume includes rich information for imaging diagnosis, it forces doctors to interpret a large number of axial images. For the purpose of assisting physicians, it is important to develop automated image understanding algorithms of human anatomy.

This chapter presents examples of segmentation algorithms for organs/tissues in the human body, mainly in the trunk. As is well known, by avoiding unnatural segmentation results, a computational anatomy (CA) model that learns statistical variations of an organ/ tissue is very helpful for segmentation. It achieves higher segmentation accuracy than one without a CA model. Chapter 3 focuses on algorithms of medical imaging based on CA models presented in Chap. 2. Since the CT scanner is one of the most common imaging devices in the clinic, this chapter mainly concerns CT data. It should be kept in mind, however, that multimodality medical imaging devices such as PET/CT and PET/MRI are being used more and more. This chapter also describes segmentation algorithms for these. The outline of this chapter is as follows.

Section 3.2 discusses segmentation algorithms of the axial skeleton, i.e., the vertebrae, ribs, followed by hip joints, all of which are the main concern in skeleton segmentation. After reviewing automated segmentation algorithms, a landmark detection framework based on a statistical model for whole spinal anatomical landmarks is introduced and discussed. Subsequently, segmentation and reconstruction algorithms of the hip joint are presented. A skeletal muscle segmentation algorithm from CT volume data is discussed in Sect. 3.3. In Sect. 3.4, segmentation algorithms of lymph nodes in medical images are reviewed followed by a lymph node segmentation algorithm from abdominal CT images. Section 3.5 deals with topics in the brain, head, and neck, and it starts with computational neuroanatomy that is a discipline focused on analyzing and modeling the anatomy of individual brains and the structural variability across a population. In section 3.5.1, voxel-based morphometry (VBM) and deformation-based morphometry (DBM) are introduced. Section 3.5.2 presents algorithms for understanding white matter from diffusion-weighted MR images (DWI) with some clinical applications. A brain CT understanding algorithm based on a normal brain CT model is presented for detection of intracranial hemorrhage in Sect. 3.5.3. Several oral segmentation algorithms using statistical models and algorithms for image understanding of fundus oculi from fundus images and retinal OCT images of retinal layers are also described with their applications to computer-aided diagnosis in Sects. 3.5.4 and 3.5.5. Section 3.6 focuses on thoracic organs, including the tracheobronchial tree, lungs, vessels, and fissures in a thoracic CT volume in the first half of this section. Several segmentation algorithms of these organs followed by an anatomical labeling process are discussed. Since breast ultrasound imaging, mammography, and breast MRI are used in the diagnosis and follow-up of breast cancer, computational model-based segmentation algorithms are presented in Sect. 3.7. In Sect. 3.8, image understanding algorithms of the heart in an echocardiographic image sequence and MR images, and coronary arteries in a CT volume, are provided. Section 3.9 presents CA model-based segmentation algorithms of the abdominal organs, in which a brief survey of segmentation algorithms for each organ is given followed by details of a state-of-the-art algorithm. Multi-organ segmentation schemes based on computational models are also described and discussed in this section. Segmentation algorithms of the abdominal aorta and abdominal vessels are presented followed by anatomical labeling of the segmented vessels.

3.2 Bone

Shouhei Hanaoka

Vertebrae and Ribs Segmenting the vertebrae and ribs presents a special challenge. Usually human beings have seven cervical (C1–C7), 12 thoracic (T1–T12), and five lumbar (L1–L5) vertebrae, as well as 12 pairs of ribs. Therefore, an ideal spine segmentation algorithm needs to detect all vertebrae, identify the number (anatomical name) of each vertebra, and, finally, segment them. The terms "detection," "identification," and "segmentation" were used first by Klinder [161], who developed the first fully automatic segmentation method for all 24 vertebrae from CT volume data.

Some detection, identification, and segmentation methods are listed in Table 3.1. In the rest of this section, these three processes are separately discussed, followed by a brief review of segmentation methods for ribs.

Detection of Vertebrae The detection phase may include preprocessing to locate anchor structures such as the pelvis [47] or the spinal canal [321]. For example, Yao et al. [321] used a watershed algorithm to segment the spinal column. Another method is the generalized Hough transform (GHT) to find the spinal canal [235] or the pelvis [47].

Several strategies can be considered for detecting vertebrae. They can be detected directly by GHT [161], template matching [116], feature extraction, or machine learning-based methods [124, 327].

Another strategy is to detect the intervertebral discs. For instance, Kim et al. [154] and Hanaoka et al. [112] applied a ray-casting search algorithm, where intervertebral discs are searched as low-density structures sandwiched between two high-density endplates (the upper and lower surfaces of the vertebral body). Kim also used a fence-like deformable model to segment neighboring vertebrae precisely. Kelm et al. [193] used a marginal space learning method in which positions, rotations, and scales of the discs are hierarchically estimated. First, candidates of target object positions are estimated, followed by estimation of position–rotation, and, finally, of position–rotation–scale. They expanded this marginal space learning

Table 3.1 List of bone detection, identification, and segmentation papers

	Authors	Objects of interest		Modality	Pathology	Detection target	Detection method	Success ratio (%)	Identification method	Success ratio (%)	Segmentation method	Mean distance error (mm)	
[161]	Klinder et al. (2009)	Spine	C,T,L	CT	(+) incl. scoliosis	Vertebrae	GHT	92	Appearance model registration	95	Surface mesh	1.12	
[154]	Kim et al. (2009)	Spine	lower T, L	CT	*	Discs	Ray search		*		3D deformable fence	*	
[47]	Buerger et al. (2013)	Spine	T, L	MR (T1WI Dixon)	*	Pelvis, lungs	GHT, mesh model fitting	100	Counting from sacrum	100	Surface mesh	1.69	
[268]	Štern et al. (2011)	Spine	T, L (v. body only)	CT MR (T2WI)	(+)	*			*		Parametric shape model	1.17 (CT) 1.85 (MR)	(Landmark to landmark distances)
[52]	Carballido-Gamio et al. (2004)	Spine	L (v. body only)	MR (T1WI)	*	*			*		Normalized cuts	6.225†	(6.64 voxel of mean error * 0.9375 mm/ voxel)

(continued)

Table 3.1 (continued)

	Authors	Objects of interest		Modality	Pathology	Detection target	Detection method	Success ratio (%)	Identification method	Success ratio (%)	Segmentation method	Mean distance error (mm)	
[124]	Huang et al. (2009)	Spine	C, T, L (v. body only)	MR (T2WI, 2D)	*	Vertebrae	Wavelet features + modified AdaBoost	97.78	*		Normalized cuts	*	
[138]	Kadoury et al. (2013)	Spine	T, L	CT MR(T1WI)	*	Vertebrae	Artificated deformable model and high-order MRF (**)		*		The same method as (**)	2.2 (T), 2.8 (L), 2.9 (T), 3.0 (L)	
[308]	Whitmarsh et al. (2013)	Spine	L2, L3	CT	*	*			*		Multi-atlas	0.3	
[126]	Ibragimov et al. (2013)	Spine	L	CT	*	*			*		Landmark detection and atlas-based registration	0.76	
[327]	Zhan et al. (2012)	Spine	C, T, L	MR (scout scan)	(+) incl. scoliosis	Vertebrae and discs	Wavelet features + AdaBoost cascade	97.7	*		(−)		
[193]	Kelm et al. (2013)	Spine	T, L(CT) C,T,L (MR)	CT MR (T2WI)	(+) (−)		iterated marginal space learning	98.04 (CT) 98.64 (MR)	*		Graph cuts	*	

[184]	Ma et al. (2010)	Spine	T	CT	*	Vertebrae	Marginal space learning	*	Mean shape fitting	73 ∼ 91	Surface mesh	0.95	
[116]	Hayashi et al. (2011)	Spine	T2–12, L (v. body only)	CT	*	Vertebrae	Template matching		By rib detection		(−)	3.6–5.5	(Mean of Hausdorff distances evaluated on cross sections)
[208]	Naegel (2007)	Spine	T, L†	CT	*	Vertebrae	Morphology filtering	*	By detecting the lowest ribs	100	Watershed algorithm	*	
[112]	Hanaoka et al. (2011)	Spine	C3–7, T, L	CT	(−)	Vertebrae	Ray search	95	Counting from sacrum	60	Graph cuts with Rieman-nian metrics	1.11 (C), 1.43 (T), 1.11 (L)	
[194]	Mirzaalian et al. (2013)	Spine	C3–7†, T, L	CT			Iterated marginal space learning	*	*		Surface mesh	1.37	(Symmetric point to mesh)
[321]	Yao et al. (2006)	Spine	T, L	CT	(+) incl. metas-tases	Spinal canal, discs	Watershed algorithm, intensity profile, etc.	97.2			Four-part deformable model fitting	*	

(continued)

Table 3.1 (continued)

	Authors	Objects of interest		Modality	Pathology	Detection target	Detection method	Success ratio (%)	Identification method	Success ratio (%)	Segmentation method	Mean distance error (mm)	
[160]	Klinder et al. (2007)	Ribs		CT	*	Ribs	Ray search	94.4	ICP registration of rib centerlines	94.1	Surface mesh	0.36	
[267]	Staal et al. (2007)	Ribs		CT	*	Ribs	Ridge detection and spin-grass classifier		Heuristic algorithms	98.4	Seeded region growing	*	
[167]	Lamecker et al. (2004)	Pelvis		CT	*						Surface mesh	1.8	
[251]	Seim et al. (2008)	Pelvis		CT	*						Surface mesh + graph cuts	0.7	
[320]	Yao et al. (2003)	Pelvis		CT	*						Flexible mesh template matching	*	
[323]	Yokota et al. (2009)	Pelvis and femur		CT	(+)						Hierarchical SSM	1.2	

*No information provided in the paper
†Not clearly described; see the original paper

to detect multiple discs. The method showed a detection success rate of 98% for both CT and MR images.

Identification of Vertebrae Identification of all 24 vertebrae would seem to be a very trivial task, as long as the following conditions are met: (1) the field of view of the given image includes the whole spine, (2) the target spine has the normal number of 24 vertebrae, and (3) all 24 vertebrae are successfully detected. Unfortunately, these conditions are rarely satisfied. In fact, to the best of our knowledge, the work by Klinder et al. [161] is the only one which can identify and segment all 24 vertebrae in a CT image. The method uses appearance model registration to distinguish each vertebra. The success rate in their identification method was 95%. This method successfully handled images with various field-of-view sizes, including those encompassing only a part of the spine, and with various pathologies, e.g., scoliosis.

Ma et al. [184] introduced a method employing mean shape fitting to identify and segment thoracic vertebrae. The success rate was 73% if only one vertebra was segmented, but it increased to 91% when seven continuous vertebrae were simultaneously identified.

Another method (that of Hanaoka et al. [112]) could detect 22 vertebrae, excluding C1 and C2. The success rate of identification was no more than 60%, because their identification method was based on a "counting-from-the-bottom" strategy. However, it should be noted that one of the failures was due to an anomalous number of lumbar vertebrae (the patient had six lumbar vertebrae). According to a survey by Carrino et al. [54], no fewer than 8.2% of people have an anomalous number of vertebrae.

Hanaoka et al. also introduced a whole-body landmark detection method [113] which identifies landmarks for each of the 24 vertebrae. This particular method will be discussed later.

Segmentation of the Vertebrae After detection and identification, segmentation of each vertebra must be performed. The most popular segmentation method is surface mesh fitting [47, 161, 184, 194]. For example, Ma et al. used learning-based edge detection and a coarse-to-fine deformable surface mesh model to segment the thoracic vertebrae. Klinder et al. [161] also used a surface mesh-based statistical shape model (SSM) with an image gradient-based external energy term.

Kadoury et al. [138] recently reported a unique method in which detection/identification and segmentation are simultaneously processed. They assume that global deformations of the spinal column will manifest similar local deformations of the vertebrae because of the same type of pathological deviation. Under this assumption, they modeled scoliotic spinal columns as an articulated deformable model, embedded it into a local linear embedding (LLE) manifold, and tried to represent local shape appearances of the vertebrae as a linear combination of shapes of neighbor samples on the manifold. The segmentation problem is formulated as a high-order Markov random field (MRF) and solved as a single optimization problem.

Štern et al., among others, modeled pathological changes in vertebral body shapes by a parametrical shape model with 25 parameters. Carballido-Gamio et al. [52] and Huang et al. [124] used a normalized cut method to segment vertebral bodies. Recently Whitmarsh et al. reported a multi-atlas-based segmentation method [308] for lumbar vertebrae with a mean distance error of no more than 0.30 mm.

Segmentation of Ribs Staal et al. [267] reported an automatic rib segmentation and labeling method. After 1D ridges are extracted from the given image, line elements are constructed, classified, and then grouped as rib centerlines. These centerlines are used in the final region-growing algorithm. They reported an identification success rate of 98.4%, excluding the first ribs. Klinder et al. [160] reported another method with an identification success rate of 94.4% and a mean distance error of 0.36 mm. Note that a rib detection method sometimes serves as an identification method for thoracic vertebrae (e.g., in [116] or [208]).

Spinal Landmark Detection In this section, a landmark detection framework for whole spinal anatomical landmarks is introduced and discussed. The framework was developed by Hanaoka et al. [113].

The framework can determine the positions of over 100 landmarks concurrently, taking spatial correlations of all landmark pairs into account. The outline of the framework is illustrated in Fig. 3.1. First, a set of landmark candidate lists is generated by sensitivity-optimized single-landmark detectors. Each landmark detector will detect its target landmark and output approximately 100 candidate positions. Then, a Markov chain Monte Carlo (MCMC)-based combinatorial optimization algorithm will find the most probable combination of candidate positions through maximum a posteriori (MAP) estimation.

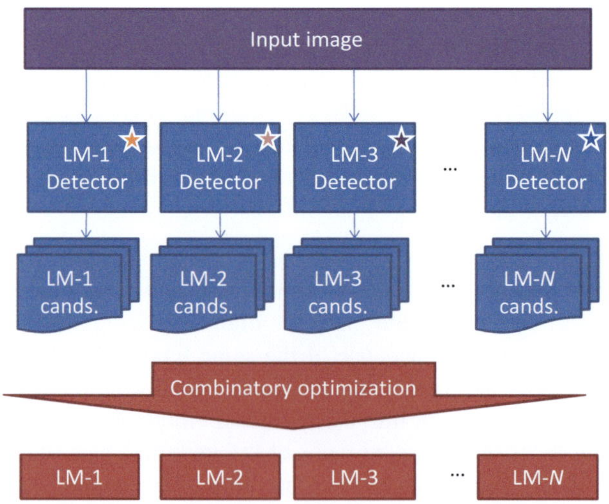

Fig. 3.1 Outline of the landmark detection framework (Cands = candidates)

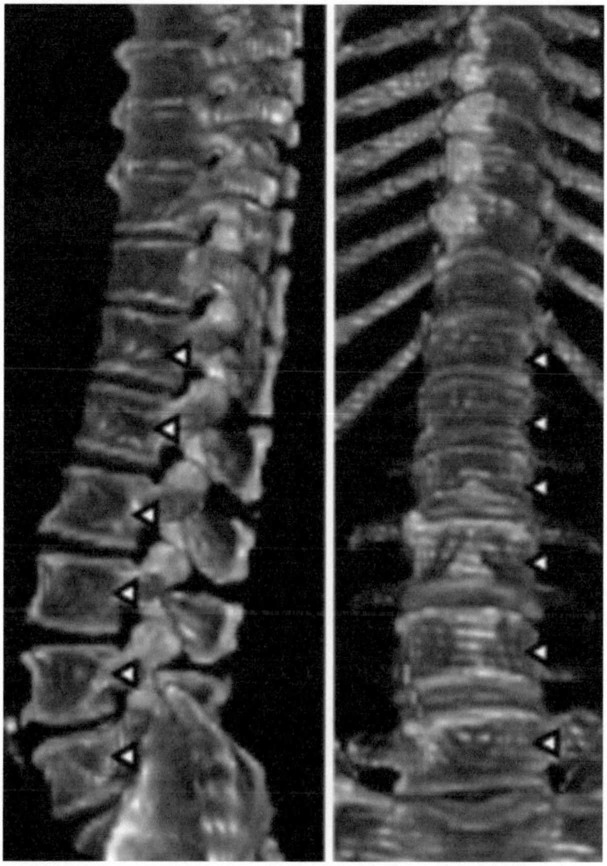

Fig. 3.2 Spine with six lumbar vertebrae

The unique feature of the framework is that it can handle subjects with segmentation anomalies of the spinal column. As mentioned, no fewer than 8.2% of people have an anomalous total number of vertebrae. Such anomalies are very problematic in both defining and detecting vertebral landmarks (Fig. 3.2). To overcome this, a series of *anomaly landmark position set converters* is introduced. One converter can convert any landmark position set in a subject with a certain type of anomaly into a virtually normalized landmark position set (Fig. 3.3). Because of this converter, the posterior probabilities can be calculated even in subjects with an anomalous number of vertebrae. The framework can determine the type of anomaly in a given unseen image by (1) applying all anomaly type hypotheses sequentially and (2) adopting the hypothesis with the largest posterior probability.

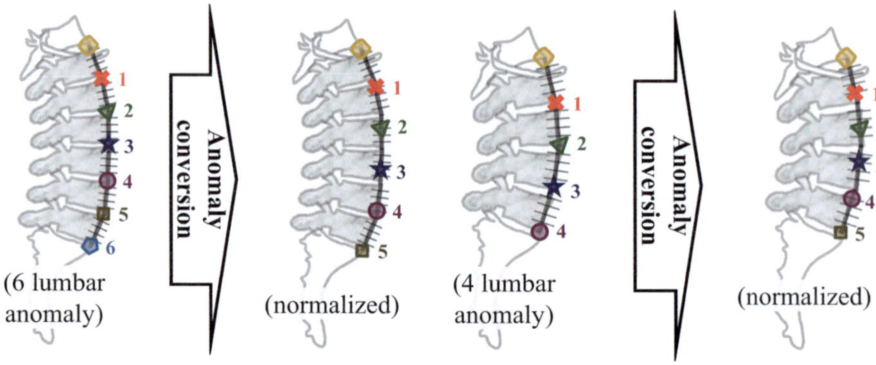

Fig. 3.3 Examples of spinal bone number anomalies (four or six lumbar vertebrae) and their anomaly conversion results

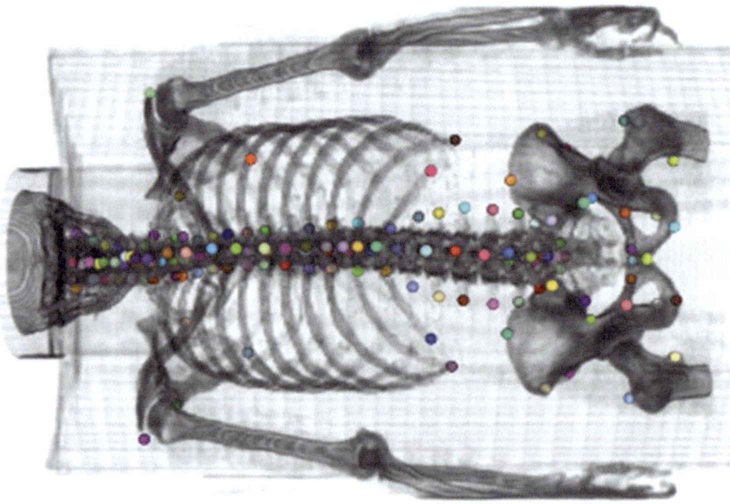

Fig. 3.4 Landmark detection result in a subject with 13 (i.e., one more than normal) thoracic vertebral bodies

Through an experiment with artificial detector outputs, the framework achieved a 97.6% success rate in anomaly type determination. The results show the potential of this framework to detect an anomalous number of vertebrae by trying several anatomical variant hypotheses sequentially (Fig.3.4).

Hip Joint The hip joint consists of the femoral head (which is a part of the femur) and the acetabulum (a part of the pelvis), with their respective articular cartilages. The acetabulum is formed by the ischium, ilium, and pubis. SSMs and related methods have been utilized for reconstruction and segmentation of the hip joint from two-dimensional (2D) and three-dimensional (3D) medical imaging data.

Main applications of early studies were on 3D shape reconstruction from sparse or incomplete data such as 3D point data acquired in the operating room [234], 2D X-ray images [320], and 3D ultrasound data [29]. These applications of the SSMs were intended to statistically interpolate and extrapolate sparse and incomplete data in order to reconstruct an approximate 3D shape when 3D CT/MR data were not available, and they are still actively studied [330]. Even when 3D CT/MR data are available, however, accurate 3D reconstruction from them is still not an easy task, and SSMs play an important role, especially for automated segmentation. Furthermore, 3D CT/MR data have more detailed information, and thus it is worth investigating more complex statistical models of the hip joint, rather than a single SSM of the pelvis or femur, to fully utilize these data. In this section, SSM-based segmentation of the hip joint from CT/MR data is reviewed.

One of the earliest works on application of SSMs to CT/MR data segmentation was done by Lamecker et al. [167]. In their study, an SSM of the single pelvis was used, and the SSM was initialized by placing the mean shape manually, which raised several issues. One issue is how the SSMs were automatically initialized, including landmark detection, for this purpose. In addition, the particular issue in the hip joint is how the consistency of the geometric relationship between the acetabulum and femoral head was maintained. Therefore, multi-structure modeling is a key problem. Especially for diseased hips, keeping the consistency becomes more difficult due to joint space narrowing as well as severe deformation. In the following paragraphs, methods for the SSM initialization are described, and then several approaches for keeping the shape consistency are discussed.

Regarding the initialization, GHT has been successfully applied by using the mean shape as the template to automatically determine the initial pose of the pelvis SSM [251]. The problem of using GHT is the trade-off between computational cost and parameter range/resolution. Another approach is to manually provide anatomical landmarks, [106, 248]. Because the landmarks are well-localized features, interoperator variability of the initial pose and shape parameters estimated from the specified landmark points is expected to be small even if the landmarks are provided manually. Therefore, more objective initialization will be possible compared with manual specification of the pose itself. Because determination of the pelvic coordinate system is clinically important, some methods [164, 324] use automatic methods specific to the pelvis to determine the coordinate system, which can be the reference frame of the SSM. Similarly, the hip joint center has been used for initialization for a small field-of-view (FOV) MR dataset [249]. As described in the previous section, initialization will be improved by using automatic landmark detection in the future.

To maintain the consistency between the acetabulum and femoral head, multi-object integrated modeling of the pelvis and femur has been investigated. The simplest integration is simply to avoid overlap of the two SSMs of the pelvis and femur during their fitting to the 3D data [139]. However, it overcomes only one of the inconsistencies (although it is still effective). A more elaborate approach is to model joint-specific motion in addition to bone-specific shapes. Both analytical and statistical approaches have been proposed for incorporating the articular motion. In the analytical approach, one composite SSM of the pelvis and femur is constructed, in which the shape and motion parameters are separately represented by assuming that the hip joint is a ball-and-socket joint, that is, three degrees of freedom rotation centered at the hip joint [140]. Therefore, variability of hip joint motion can be added by the minimum numbers of parameters without reducing the ability of shape representation by the SSM. The motion of diseased hip joints will not be a simple rotation. In the statistical approach, the combined shape of the pelvis and femur of each patient is regarded as a single shape, and one composite SSM of the pelvis and femur is constructed [323]. This composite SSM includes not only shape variations but also motion variations of the joint. Its shape and motion parameters are not separated. One drawback is that the ability of shape variability representation is reduced. Therefore, hierarchical modeling is combined so as to realize a coarse-fine fitting as shown in Fig. 3.5a. Coarse fitting is first performed using the composite pelvis and femur SSM to provide initialization for the subsequent stages, and then finer fitting is performed by using the divided SSMs to gradually increase fitting accuracy as well as to provide initialization for the next stage while keeping the consistency. Furthermore, a conditional SSM are applied for further improvement of segmentation accuracy [325]. This method was shown to be particularly effective for CT segmentation of diseased hip joints even in the presence of joint narrowing and severe deformation. Figure 3.5b shows a typical segmentation result.

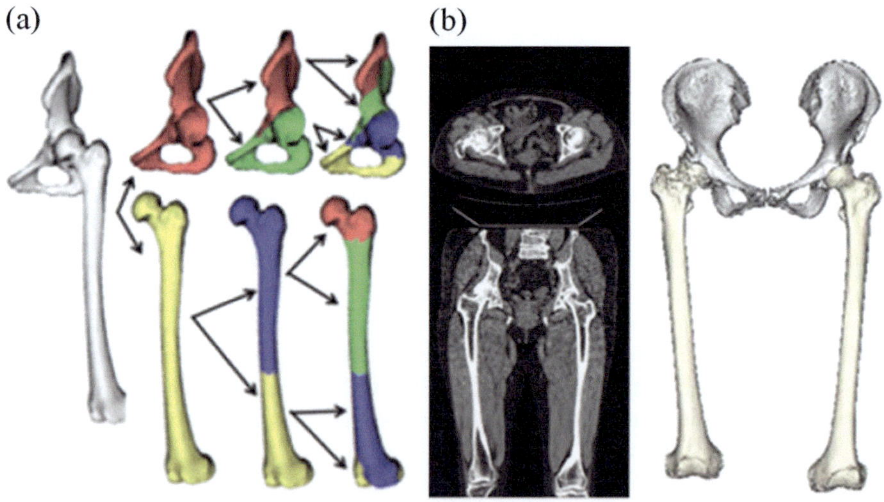

Fig. 3.5 Segmentation of the pelvis and femur from CT images using a hierarchical SSM (H-SSM) of the hip joint. (**a**) H-SSM of the hip joint. (**b**) Typical segmentation result

3.3 Skeletal Muscle

Naoki Kamiya, Yoshinobu Sato

Introduction In this section, we describe an automatic recognition method for skeletal muscle based on CA. As skeletal muscle does not undergo enhancement in a way that makes it stand out from surrounding tissues, we focus on non-contrast CT images, a challenge to computer recognition.

Skeletal muscle has distinct features on CT (see Fig. 3.6). Muscle is found throughout the body and is visible in virtually all cross-sectional images. However, the main area of interest for the physician is usually the organ observation. Therefore, automatic analysis of skeletal muscle using a computer algorithm may be useful for incidental findings or findings that may be overlooked.

To improve quality of life in a rapidly aging society, various estimation methods for skeletal muscle have been devised to assess the therapeutic effects of rehabilitation. Solutions to accurately measure skeletal muscle in image recognition based on CA have been described [95, 145, 146]. Recognition of knee extensor and flexor muscles in MR images is being explored using probabilistic methodology [6]. An SSM-based method has been used in the thigh using MRI [33]. An atlas-based method using breast MRI is employed in the pectoral muscle regions [110].

3.3.1 *Anatomical Modeling of Skeletal Muscles*

Figure 3.7 shows the correspondence of the anatomical features. These features are utilized by an anatomical modeling method. These three features details described below.

Landmarks It is important to recognize the origin and insertion points of a muscle in the bone as landmarks. A landmark is a one-dimensional datum indicating one point in three-dimensional space. Various methods exist for landmark recognition. As skeletal muscle is attached to planar configurations, if the selection of the bone is correct, some positional error is acceptable in the shape model described below.

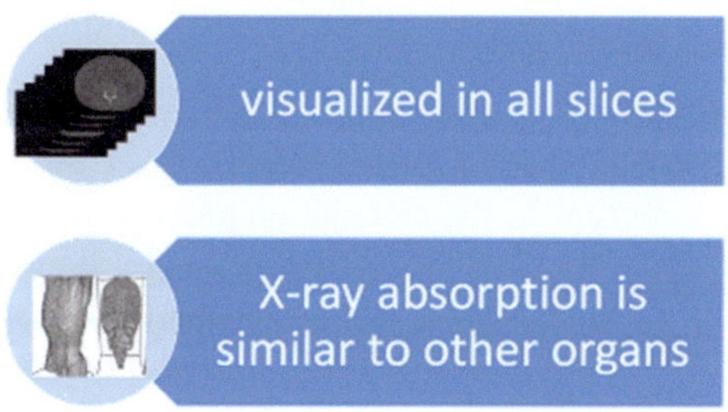

Fig. 3.6 Representative features of skeletal muscle in CT images

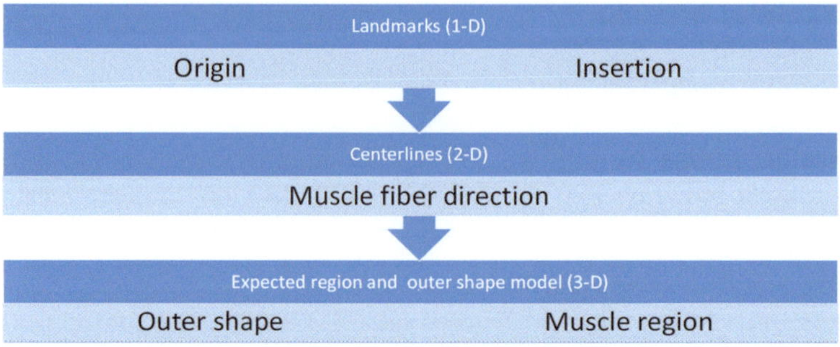

Fig. 3.7 Correspondence of the anatomical features on computer

For the recognition of landmarks, we use skeleton images classified automatically using an atlas derived from non-contrast CT images. This classified skeleton image is generated by a method considering the connectivity of the skeleton [332].

Landmarks are selected by first selecting the bone from the atlas based on the anatomical definition and then finding the most proximal point on the selected bone for the origin and the most distal point for the insertion. Classified skeletal image atlases and the virtual image unfolding-based method by this author and other studies [134, 216] describe this.

Anatomical Centerlines Anatomical centerlines indicate the direction of the long axes of the muscle fibers. Therefore, in each skeletal muscle, a pair of origin and insertion landmarks are connected by a straight line or curve. For use as an approximate running direction of the muscle fibers that supports recognition, there is no serious problem even if using a straight line or low-dimensional mathematical function.

Because of differences in height, body shape, gender, subcutaneous fat distribution, and other factors, there are large individual differences in the shape of the human body. In our experience, the geometric centerline is different from the anatomic one. Thus, in surface skeletal muscles, it is preferable to generate a centerline passing through multiple points placed as additional landmarks between the origin and insertion. We select additional landmarks using the surface of the subcutaneous fat, which is relatively easy to recognize by its low-density values.

Figure 3.8 shows anatomic centerlines generated in the identification of surface and deep muscle.

Model Generation The modeling of skeletal muscle can involve displaying the distribution in three-dimensional space or using an SSM to indicate the outline of the muscle. In the recognition process, it is necessary to select whether to use both models or to use the distribution model alone. Many muscles, such as the temporalis, have a complicated shape that requires the distribution model. Muscles with well-defined insertion points, belly, and attachment points, such as the biceps, can be analyzed with the shape model.

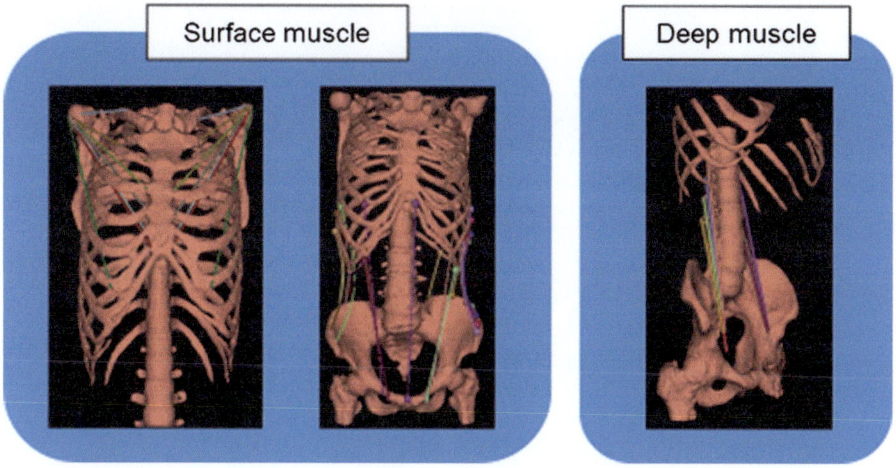

Fig. 3.8 Anatomic centerlines of the surface muscle and deep muscle

3.3.1.1 Muscle Distribution Model

During the generation of a muscle distribution model, there is a method using only obtained landmarks and a method using landmarks and centerlines. The simplest distribution model is a method for determining the muscle area using closure landmarks. Closure landmarks are used to determine the closure region to detect the initial region of the muscle. This is effective when the distance between the origin and insertion is small and the shape of the muscle is simple. However, it is not suitable when the muscles and other organs are adjacent to each other, because the separation is difficult in the recognition process. A method using landmarks and centerlines solves this problem. This is the technique for closure region, including the centerline and landmarks. The closure region depends on the target muscle, but is defined by the shape information and the statistical variance, which is obtained by using the landmarks. This can be realized by specifically fitting a surface to the centerline, placing auxiliary points at regular intervals on the centerlines, and performing surface interpolation using these auxiliary points. This approach is strongly independent of differences in body shape and is effective for recognition of the surface muscles. Figure 3.9 shows the distribution model in the chest and abdominal region.

3.3.1.2 SSM

The distribution model described above is intended to depict the locations of muscles on the basis of the centerline and landmarks. However, the muscle-specific shape is not included in the distribution model. We describe a technique for

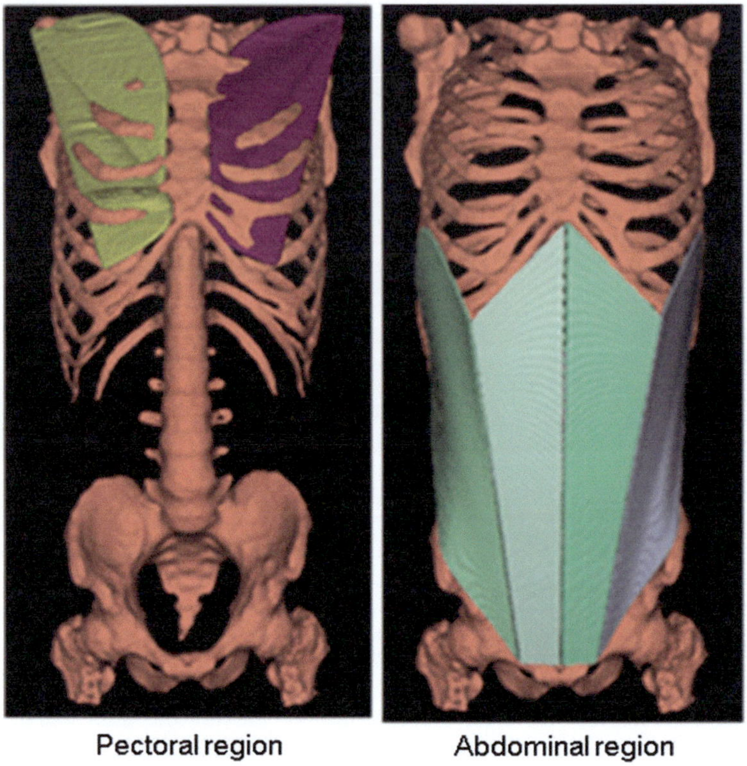

Fig. 3.9 Distribution model of the skeletal muscle

statistically modeling the outer shape of the muscle, expressed as a mathematical function. We describe the modeling of the psoas major muscle, which has a characteristic shape. SSM requires muscle area data by semiautomatic or manual extraction of the target muscle.

The psoas major muscle is spindle shaped. It is assumed that the cross section of the outer shape can be represented by a quadratic function that is symmetric to the centerline. Based on this assumption, a muscle area is extracted from the training data to determine the distance to the outer shape from each centerline. The approximate curve of the quadratic function is fitted to the distribution of the distance values along the centerlines. The curve generated can be the quadratic function whose vertex is located at the midpoint of the centerline. Here it should be noted that two parameters are present in the quadratic function, *gradient alpha* and *intercept beta*. We define the value that represents the muscle-specific shape as gradient alpha. The fitting parameter which accommodates individual differences in muscle mass is defined as intercept beta. The gradients resulting from the approximate curve fitting are saved as the shape parameters. Fitting parameters are determined from the test data in the recognition process described later.

We used SSMs to recognize the deeply situated psoas major muscle and the superficially situated rectus abdominis muscle. If the modeling can be applied to a characteristic shape, it is not limited to superficial or deep locations.

Recognition The choice of recognition method depends on the model. First, using only the distribution model. In this pattern, it is necessary in identifying the skeletal muscle by using a grayscale value inside and outside of the boundary. Second, using an SSM. The SSM-based method needs, for calculating the fitting parameters, intercept beta, which is a parameter of remaining that has not been determined in the model building process. These parameters are calculated automatically on the test data by generating the centerlines. For example, in psoas major muscle recognition, where the cross section of the midpoint position of the centerline is thickest, the maximum diameter of the muscle is determined from the test data and defines intercept beta. Finally, it fits the model function to the landmarks and performs the recognition of the muscle using the grayscale value from the CT image.

Ordinarily, each model is only intended to indicate a representative area or boundary of the muscle region. Therefore, in the recognition process, as well as in other organ recognitions, a precise extraction process is needed.

Figure 3.10 shows the recognition results of the rectus abdominis muscle and psoas major muscle using the SSM.

Recognition of Lower Limb Muscles In the above sections, recognition methods for relatively isolated single muscles were described. In the lower limb muscles, multiple muscles are densely and closely interrelated and adjacent to each other. Therefore, the above described methods may not be useful for the lower limb, and segmentation methods suitable for densely interrelated muscles need to be developed.

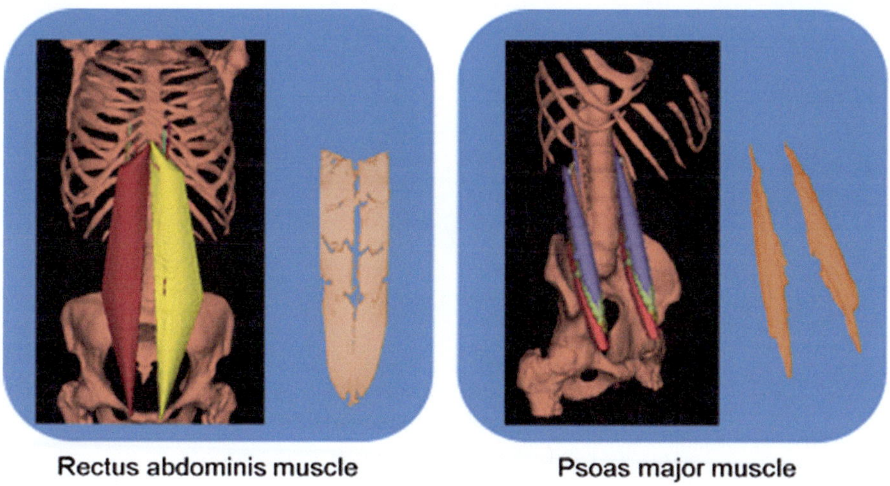

Fig. 3.10 Recognition results of the rectus abdominis muscle and psoas major muscle

Lower limb muscle segmentation was mainly motivated by biomechanics research. To overcome the limitations of conventional musculoskeletal biomechanical simulations using line segments as muscle models, precise geometries of muscles are needed to be reconstructed. Early works used manual [39] or interactive segmentation [106, 107] to reconstruct musculoskeletal anatomy models from MR data to perform patient-specific simulations. More recently, attention to lower limb muscle segmentation has been paid for diagnostic purposes, and automated segmentation methods have been investigated. Random walks [34] and multi-atlas segmentation [95, 301] data were used in order to deal with incomplete boundaries among the muscle regions. MR data were usually considered to be suitable [34, 39, 106, 107, 301], but CT data have also been also utilized [95]. Figure 3.11 shows typical results of automated segmentation of the hip and thigh muscles from CT data [95]. In this method, the bone regions (the pelvis and femur) of the segmentation target CT data were nonrigidly registered to the atlas data of the training dataset for spatial normalization, and then the muscle regions were segmented. The muscle regions were also used for spatial normalization in the next stage of the method, and some muscles were further segmented and used for further normalization.

Conclusion We described automatic recognition methods based on CA, a general technique of skeletal muscle recognition in non-contrast CT images using anatomic features. We present the recognition of landmarks, construction of anatomical centerlines, and anatomical shape model generation. The SSM describes the outer shape of the muscle, and the distribution model indicates the existence area of the muscle.

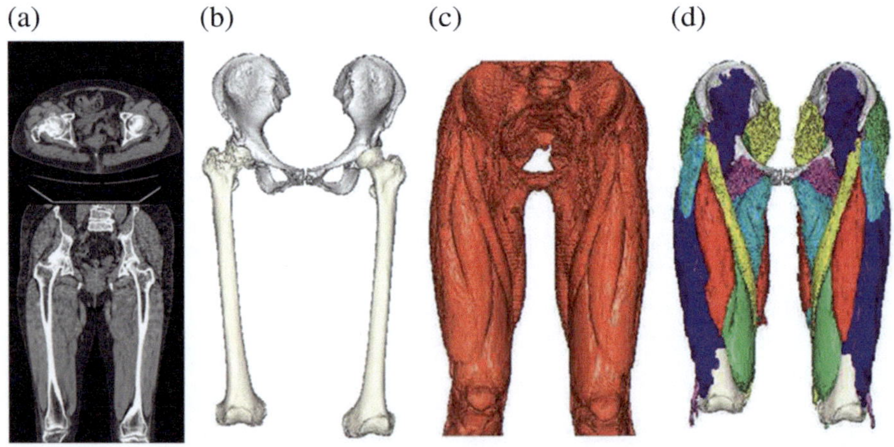

Fig. 3.11 Typical segmentation result of the bones and muscles in the pelvis and thigh from CT data. (**a**) Input CT data, (**b**) segmentation of the pelvis and femurs, (**c**) segmentation of muscle tissues, (**d**) segmentation of individual muscles

Automatic recognition of skeletal muscle based on a CA model is robust. When separation from other organs is difficult, the shape model provides a useful initial region. However, this requires careful selection of the appropriate modeling function for outer shape determination and also requires accurate extraction.

3.4 Lymph Nodes

Kensaku Mori

3.4.1 Overview of Lymph Node Segmentation on Medical Images

The lymph nodes are part of the reticuloendothelial system. They are a vital part of the immune system and are significant actors in many cancers, both hematologic and solid. Typically a human has several hundreds of lymph nodes in their body. Most of the tissues in the body, except for the brain, have specific patterns of lymphatic drainage, with lymphatic channels draining into specific lymph node chains. Lymph nodes may become enlarged because of immune responses to infection/inflammation or because of infiltration with cancer cells, although tumor-involved nodes may also be normal in size. Such lymph nodes can be observed on medical imaging including computed tomography (CT) or magnetic resonance imaging (MRI) images. In clinical imaging, it is important to detect lymph nodes whose diameters are ≥ 5 mm. Quantitative evaluation of lymph nodes is important for diagnosis and subsequent staging after surgical, medical, and/or radiation therapy.

Most enlarged lymph nodes can be identified as elliptical-shaped structures on CT images. Figure 3.12 shows examples of lymph nodes observed on axial CT slices. On contrast-enhanced CT images, these regions can be observed as foci whose intensity values are higher than surrounding structures such as fat. In the mediastinal area, the existing areas of the lymph nodes are almost fixed.

Several studies have been conducted for automated detection of nodes on chest and abdominal CT images [27, 77]. The basic framework of detection consists of (a) blob-like structure enhancement for lymph node candidate detection and (b) false-positive reduction.

Several methods have been proposed for "initial candidate" selection: (a) the Hessian-based method [212], (b) the directional difference filter-based method [77], and (c) the Haar-like feature method with machine learning augmentation [27]. The Hessian-based approach tries to detect lymph node candidates by the fact enlarged lymph nodes show a spherical or elliptical shape on CT images. The directional difference filter method uses a similar approach. The Haar-like feature-based approach directly computes features that enlarged lymph nodes have to extract lymph node candidates.

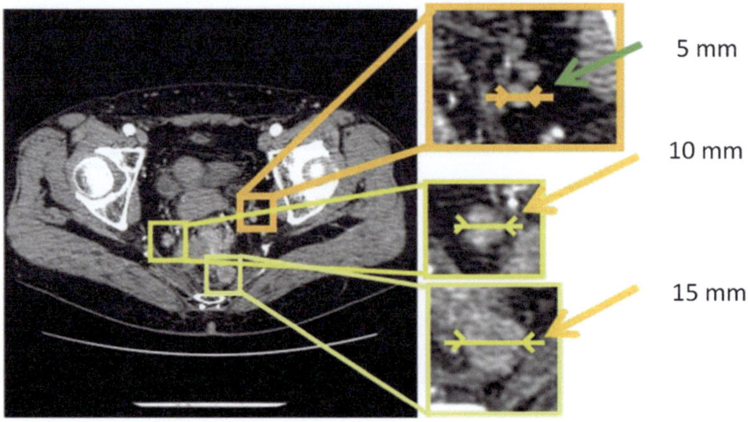

Fig. 3.12 Examples of lymph nodes depicted on axial CT slice images

A feature-based approach is generally used to avoid false positives. Several classifiers are constructed to discriminate false-positive regions from lymph node candidate regions. Most methods use a machine learning approach to classify true-positive and false-positive regions.

A lymph node atlas showing the usual distribution of nodes assists in the false-positive reduction process. Feuerstein et al. (2012) [77] used a patient-specific mediastinal atlas to reduce false positives.

3.4.2 *Overview of Lymph Node Segmentation from Abdominal CT Images*

This section introduces the lymph node extraction process on abdominal CT images. The basic flow of abdominal lymph node detection consists of four steps: (1) preprocessing, (2) blob-like structure enhancement, (3) initial candidate detection from enhanced image, and (4) false-positive reduction.

3.4.2.1 Preprocessing

In this step, a Gaussian smoothing filter of kernel size σ is applied. Because the sizes of enlarged lymph nodes vary, it is necessary to adjust the threshold target size. This size control can be achieved by changing the size of the kernel of the Gaussian smoothing filter and the size of the hypersurface fitting area explained in the next subsection.

3.4.2.2 Blob-Like Structure Enhancement

As stated in the previous sections, enlarged lymph nodes are spherical or elliptical in shape, which can be described as blob-like structures. Intensity structure analysis based on a Hessian matrix can distinguish between sheet-, line-, and blob-like shapes. As enlarged lymph nodes may have a blob structure, a Hessian-based blob structure enhancement filter is used. A blob structure enhancement filter was developed by Sato and Frangi in 1999. The basic idea of this method is that specific patterns emerge from the eigenvalues of the Hessian matrix.

The partial derivative of the intensity function f can be derived by simple numerical differentiation of f or hyper-curve fitting to the function f. After computing the Hessian matrix at (x, y, z), three eigenvalues $\lambda_1, \lambda_2, \lambda_3$ $(0 \geq \lambda_1 \geq \lambda_2 \geq \lambda_3)$ are computed. Before applying this blob-like structure enhancement filter, the Gaussian filter whose standard deviation is σ is executed. The eigenvalues obtained by applying the Gaussian filtering of σ are described as $\lambda_1^{(\sigma)}, \lambda_2^{(\sigma)}, \lambda_3^{(\sigma)}$.

Although there are several variations in blob-like structure enhancement, Nakamura et al. (2013) [212] used the following blob-like structure enhancement filter:

for a CT voxel $\mathbf{x} = (x, y, z)$, the blobness value at $\mathbf{x}$ can be expressed as

$$G_B(\boldsymbol{x}) = \begin{cases} |\lambda_3| \cdot \left(\frac{\lambda_2}{\lambda_3}\right)^{\gamma} \cdot \left(\frac{\lambda_1}{\lambda_2}\right)^{\gamma} & \lambda_1 < 0 \\ 0 & \text{otherwise} \end{cases}, \tag{3.1}$$

where λ_1, λ_2, and λ_3 are the eigenvalues of the Hessian matrix computed at the voxel $\mathbf{x}$. The Hessian matrix of the voxel located at (x, y, z) can be calculated by

$$H = \begin{pmatrix} \left.\frac{\partial^2 f}{\partial x^2}\right|_{(x,y,z)} & \left.\frac{\partial^2 f}{\partial x \partial y}\right|_{(x,y,z)} & \left.\frac{\partial^2 f}{\partial x \partial z}\right|_{(x,y,z)} \\ \left.\frac{\partial^2 f}{\partial y \partial x}\right|_{(x,y,z)} & \left.\frac{\partial^2 f}{\partial y^2}\right|_{(x,y,z)} & \left.\frac{\partial^2 f}{\partial y \partial z}\right|_{(x,y,z)} \\ \left.\frac{\partial^2 f}{\partial z \partial x}\right|_{(x,y,z)} & \left.\frac{\partial^2 f}{\partial z \partial y}\right|_{(x,y,z)} & \left.\frac{\partial^2 f}{\partial z^2}\right|_{(x,y,z)} \end{pmatrix} \tag{3.2}$$

where f means a function that approximates intensity distribution around the point $\mathbf{p}$. The function f can be expressed as

$$f(x, y, z) = ax^2 + by^2 + cz^2 + dxy + eyz + fzx + gx + hy + lz + m, \tag{3.3}$$

where $a, b, c, d, e, f, g, h, l$, and m are coefficients of the function. A set of these coefficients are denoted as $\mathbf{w}$. These coefficients can be obtained by solving the least mean square problem that minimizes a residual error e defined as

$$e = \sum_i \sum_j \sum_k (f(i + x, j + y, k + z) - I(i + x, j + y, k + z))^2. \tag{3.4}$$

Optimum **w** is obtained as the solution of

$$\frac{\partial e}{\partial \boldsymbol{w}} = 0. \tag{3.5}$$

As stated above, it is necessary to enhance lymph nodes of different sizes. If the input image is smoothed by the Gaussian smoothing filter of kernel size σ, a hypersurface fitting process is performed in the area of 3 σ. Blob structure-enhanced images of different scales are generated by computing $G_B(\mathbf{x})$ for all voxels of the input image.

3.4.2.3 Lymph Node Candidate Region Detection

After obtaining the blob structure-enhanced image, initial candidate regions of lymph nodes are extracted by simply thresholding the enhanced images. However, some lymph nodes are observed as contacting other regions like fat, muscle, and other organs on CT images. Intensity contrasts between lymph nodes and other structures are not high. False-positive identifications can occur in high-contrast regions such as border areas of organs or air-containing regions. Eq. (3.1) is easily and strongly affected by the magnitude of λ_3 [54]. Furthermore, $|\lambda_3|$ becomes high in areas containing linear/tubular structures such as blood vessels. If $|\lambda_3|$ becomes high at a certain voxel, such a voxel has a higher possibility of false positivity. Considering this situation, lymph node candidate region image **C** can be obtained as

$$\mathbf{C}(\boldsymbol{x}) = \begin{cases} 1 \; G_B(\boldsymbol{x}) \geq T_{BSE} \text{and} \; |\lambda_3| \leq T_{\lambda_3} \\ 0 \qquad\qquad\qquad\qquad\qquad \text{otherwise} \end{cases}, \tag{3.6}$$

where T_{BSE} and T_{λ_3} are the threshold values for the blobness and the contrast, respectively [212].

We obtain lymph node candidate region images for different scales (different σ).

3.4.2.4 False-Positive Region Reduction

Lymph node candidate region images contain many false-positive regions. A typical method of false-positive reduction is based on feature analysis. An example of a feature value set for false-positive region reduction is summarized in Table 3.2. A machine learning approach, such as a support vector machine, AdaBoost, or an artificial neural network can be used for this purpose. The typical procedure of false-positive reduction based on a machine learning approach consists of two steps: (a) the training step and (b) the learning step. In the training step, features of true-

Table 3.2 Feature value list for false-positive reduction

Shape features (16 features)	Ratio of V_C and SA_C, Ratio of the surface area of inscribed sphere and SA_C, Sphericity, Inscribed sphericity, Circumscribed sphericity, Ratio of major axis and minor axis, Length of major and miner axis, Cross-section area at the midpoint of the each axis of BB, Distance between the corner point of BB and the centroid of BB, Ratio of SA_C and the surface area of S_V, Overlapping volume between R_C and S_V and the ratio of that volume and the volume of S_V, Ratio of r_V and the maximum distance from back ground region
Intensity features (9 features)	Average value, Variance value, Maximum value, Minimum value, Median value, First quartile, Third quartile, Kurtosis, Skewness
Feature values of intensity on the major and minor axis (10 features)	Average value, Variance value, Maximum value, Minimum value, Median value
Feature value using around region of R_C (52 features)	Overlapping volume between R_{Si} and the air region, Average value, variance value, maximum value, minimum value and median value of R_{Di} and R_{Si}

R_C : Candidate region.
V_C : Volume of R_C.
SA_C : Surface area of R_C.
S_V : Sphere that has same volume and same gravity point of R_C.
r_V : Radius of S_V.
R_{Di} : Dilated region of R_C by using spherical structure element of i mm in radius. (i= 2,3,4,5)
R_{Si} : Region that subtracts R_C from R_{Di}. (i = 2,3,4,5)
BB : Bounding box of R_C.

positive and false-positive regions are computed using training datasets. A set of true-positive and false-positive regions is obtained by performing the lymph node candidate region detection process described in Sects. 3.4.2 and 3.4.3. A classifier is trained by features generated from training datasets. Figure 3.13 shows a flowchart of the false-positive reduction process using a support vector machine.

After the training process of the classifier, feature values of lymph node candidate regions are computed. These features are entered into the trained classifier to decide whether a candidate region is or is not a lymph node.

3.4.3 Examples of Lymph Node Detection on Abdominal CT Images

This section will show examples of the results of lymph node detection from abdominal CT images. Acquisition parameters of these CT images are image size

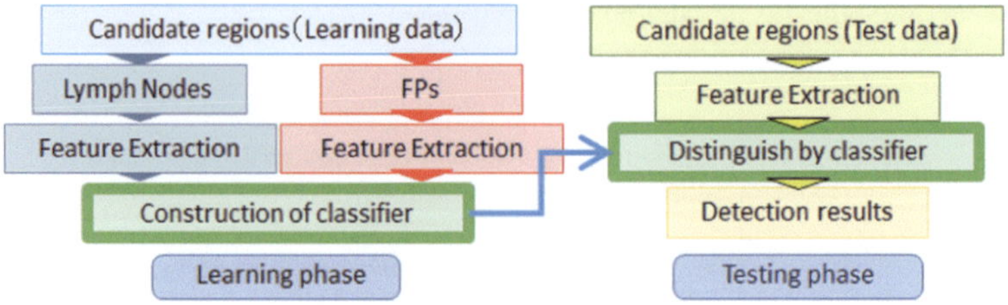

Fig. 3.13 Flowchart of false-positive reduction based on support vector machine

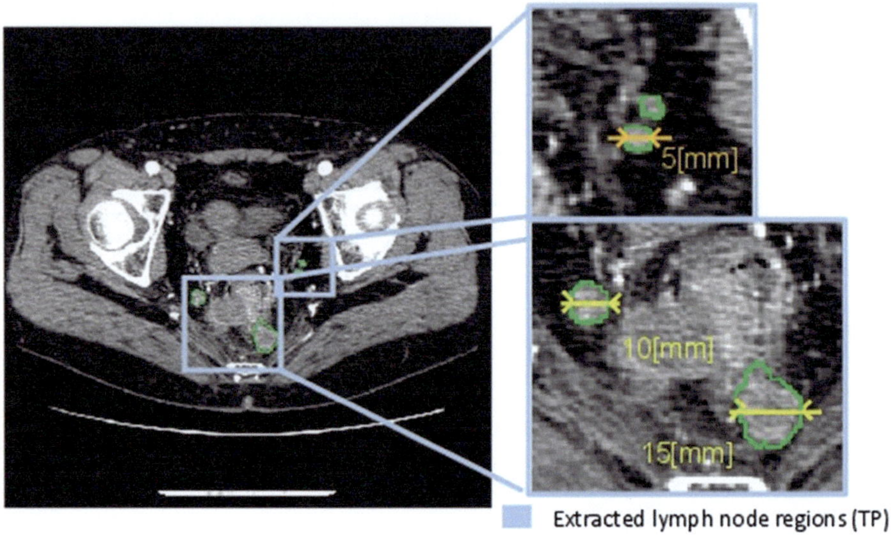

Fig. 3.14 Example of lymph node regions detected by Hessian-based analysis. This figure shows examples of detected true-positive regions

512 × 512 × 401–451 voxels, pixel size 0.586–0.702 mm^2, section thickness 1.25 mm, and reconstruction pitch 0.5–1.0 mm. The data from 28 cases of contrast-enhanced three-dimensional (3D) abdominal CT examinations were processed. These image datasets included five colorectal cancers and 23 gastric cancers. There were 95 lymph nodes included in these images. The parameters for detection were set as $T_{BSE} = 8$ and $T_{\lambda_3} = 1.1$.

Figure 3.14 shows examples of true-positive regions extracted by the method shown in Sect. 3.4.2. As shown in this figure, Hessian-based analysis can detect lymph nodes in the pelvis appropriately. Although the method can detect true-positive regions from 3D abdominal CT, it also detected several false-positive regions and false-negative regions (Figs. 3.15 and 3.16). As we can see on these

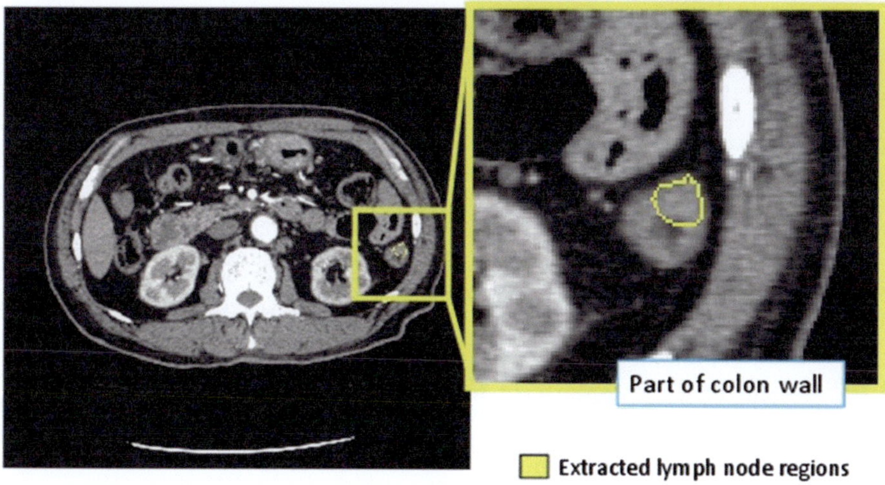

Fig. 3.15 Example of regions detected by Hessian-based lymph node detection method. This figure shows examples of false-positive regions

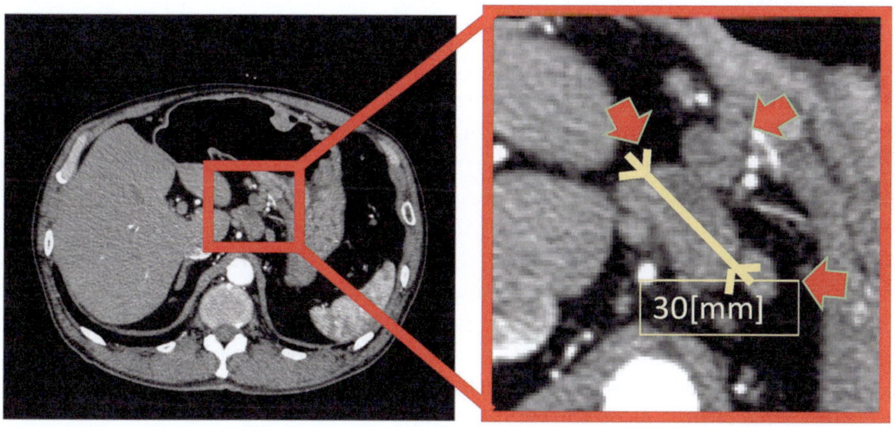

Fig. 3.16 Example of lymph nodes missed by automated method

figures, other anatomical regions such as a part of the colonic wall or residues inside the intestine tended to be extracted as false-positive regions. The overall performance of detection is shown in Fig. 3.17 as a receiver operating characteristics (ROC) curve.

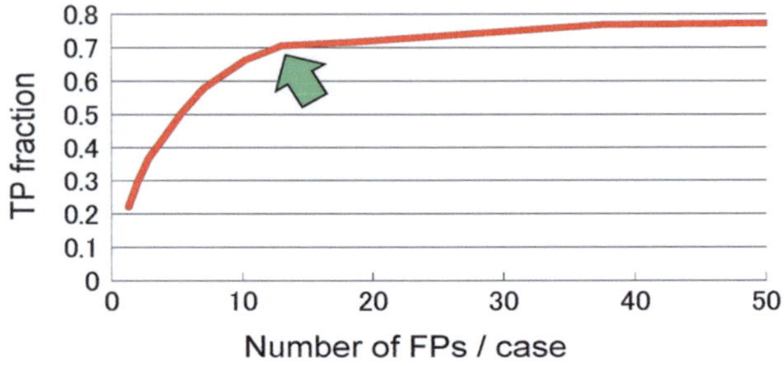

Fig. 3.17 Receiver operating characteristic (ROC) curve of lymph node detection method

3.5 Brain, Head, Neck, and Eye

Hiroshi Fukuda, Yasuyuki Taki, Kazunori Sato, Kai Wu, Yoshitaka Masutani, Takeshi Hara, Chisako Muramatsu and Akinobu Shimizu

3.5.1 *Computational Neuroanatomy*

Computational neuroanatomy is a discipline focused on analyzing and modeling the anatomy of individual brains and the structural variability across a population. The goal is not only to model normal brain structures and their variations within a population but also to identify the morphological differences between normal and pathological populations. The ultimate goal is to create a human brain structure model and classify the abnormalities of the brain from structural differences. Applications are quite important in neuroscience to minimize the influence of normal anatomic variability of the brain in functional group analysis, such as functional mapping of the brain using functional magnetic resonance imaging (fMRI). They are also important in diagnostic medical imaging to develop automatic algorithms for the diagnosis of brain diseases. Modeling the shape of the brain is difficult because of the complexity of brain morphology and the large degree of variability in normal human brain structure. These difficulties raise the need for statistics and computational methods to analyze curves, surfaces, and deformations. In this section, voxel-based morphometry (VBM) and deformation-based morphometry (DBM) will be introduced.

3.5.1.1 Voxel-Based Morphometry (VBM)

VBM was developed by Wright et al. [315] and is an automated technique that allows evaluation of structural variations in the brain across many individuals without visual assessment or operator-dependent manual procedures. VBM detects differences in the local densities (tissue probabilities) of various tissue compartments throughout the brain without any a priori knowledge of regional differences in brain structure and function [13]. Whitwell documented well on the methods of VBM [309], and the description in this section is based on that work. In VBM, T1-weighted MR images (T1WI) are analyzed by statistical parametric mapping (SPM). SPM is a statistical technique developed by Friston [89, 90] for examining differences in brain activity recorded during functional neuroimaging experiments using technologies such as fMRI or PET. It contains design matrix software for statistical analyses, statistical analyses, and visualization of the results, which was created by the Wellcome Department of Imaging Neuroscience (part of the University College London) [265]. VBM is applied not only to assess normal human brain aging [64, 108, 314] but also to detect pathological changes in brain diseases.

1. *MRI processing* (*spatial normalization and segmentation*)

To make statistical analyses across many MRIs of different individuals with different brain structure, preprocessing of the brain is necessary. The first process is known as spatial normalization or anatomical standardization. It is generally achieved by registering all images in a population into the same template image so that they are all in the same standard stereotactic space. After this process, an anatomical location in one subject's MRI corresponds to the same location in another subject's MRI. Different algorithms can be used to perform this registration [13, 64, 314]. The most commonly applied algorithm available in the SPM software (described later) includes a 12-parameter affine transformation followed by a nonlinear registration using a mean squared difference matching function [13, 265]. The template image used for the spatial normalization could be one specific MRI scan, which is selected from a population based on defined criteria, or could be created by averaging across a number of different MRI scans such as the MNI152 template that was created by averaging 152 healthy brains at the Montreal Neurological Institute (MNI). Customized templates may be created using the given study group or a group that is matched to the study group in terms of age and disease status. Such templates may improve the normalization of each subject in a study group [108].

In SPM, in which a low-parameter shape transformation is performed for spatial normalization, a step called modulation is then often necessary to correct for volume changes during the spatial normalization step [108]. Image intensities are scaled by the amount of contraction that has occurred during spatial normalization, so that the total amount of gray matter remains the same as in the original image. Then, statistical comparison of volumetric differences between scans is performed. If the spatial normalization was precise and all the segmented images appear identical, no significant differences would be detected in unmodulated data, and the analysis would reflect registration error rather than volume differences. Other techniques employ different normalization procedures that use high-dimensional elastic transformations [64], or ELAST [250]. These procedures preserve the volume of different tissues and do not require a separate modulation step.

Images are segmented into different tissue compartments (gray matter, white matter, and cerebrospinal fluid (CSF)), and statistical analysis is performed separately on either gray or white matter, depending on the target tissue to be analyzed

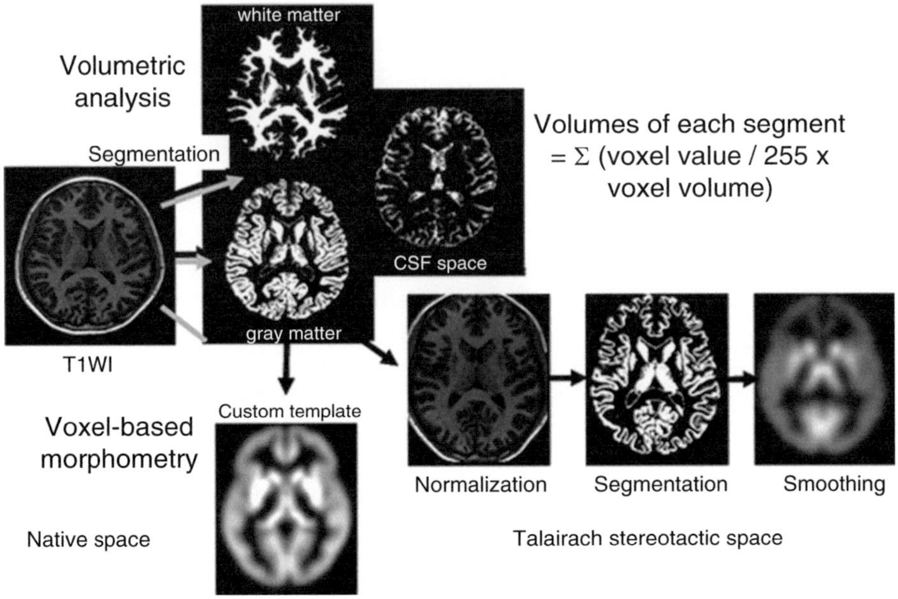

Fig. 3.18 Image processing and analyses (Figure 1 of Ref. [94])

(Fig. 3.18). There are a number of ways to perform the segmentation, including classification using voxel signal intensity combined with prior probability maps, as in SPM. Such prior probability maps may be more unbiased when generated from the specific population under study. However, weighting balance between signal intensities and prior probability should be considered, when differences in the transformation vector itself in two different populations are a matter of concern. The accuracy of the segmentation will also depend on the quality of the normalization. Iterative versions of normalization and segmentation methods have been developed which enable optimization of both processes concurrently, to improve the final segmentation [15]. In this method, the original MRI in native space is segmented, and then the segmented images are spatially normalized to gray matter and white matter templates to obtain optimized normalization parameters. The method is termed "optimized VBM" [239]. Segmentation errors can occur because of displacement of tissue and partial volume effects between gray matter and CSF. Both are more likely to occur in atrophic brains in subjects of older age or with degenerative brain diseases. The use of customized templates can help to minimize some of these potential errors [108].

2. *Statistical analysis*

Finally, the normalized and segmented images are smoothed by convoluting isotropic Gaussian kernel [13, 108], so that the intensity of each voxel is replaced by the weighted average of the surrounding voxels, in essence blurring the segmented image. The size of the smoothing kernel depends on the size of expected regional differences, which can vary across studies [150, 239, 311]. Smoothing before statistical testing has three advantages. First, the smoothing has an effect of making the data more fitted to the Gaussian field model and more normally distributed, thus increasing the validity of parametric statistical test. Second, the smoothing compensates for spatial normalization error and decreasing intersubject variability [13, 242]. Third, the smoothing reduces the effective number of statistical comparisons and thus increases the sensitivity to detect changes by reducing the variance across subjects, although excessive smoothing will diminish the ability to accurately localize changes in the brain. Although these processing steps (normalization, segmentation, and smoothing) are necessary for the analysis of data across subjects [13, 242], they can also introduce errors and variability into the analysis, which can reduce sensitivity. For example, VBM cannot distinguish between real changes in tissue volume from local misregistration of images [14, 40]. It should be noted that normalization accuracy and, thus, sensitivity will vary across regions.

Statistical analysis using the general linear model and the theory of Gaussian random fields is performed to identify regions of gray matter or white matter that are significantly related to the effects under study [13, 90, 91]. The analysis is a standard *t* test and F test extended into all voxels in 3D space. In SPM, the design matrix for statistical analysis is composed of two contrasts comparing the smoothed gray matter or white matter. These analyses generate statistical maps showing all voxels of the brain that disprove the null hypothesis and show significance to a certain p value. These maps are often shown as color maps with the *t* statistic scale (Figs. 3.19, 3.20 and 3.21). Because the statistical tests are performed across a quite large number of voxels, it is important for the results of the analyses to correct for multiple comparisons to prevent the false-positive results by chance. The classical approach to the multiple comparison problem is to control the family-wise error (FWE) rate [90], and the most common way to control the FWE rate is with Bonferroni's correction. The more conservative method is false discovery rate (FDR) correction [104]. The FWE correction controls the chance of any false positives across the entire volume, whereas FDR correction controls the expected proportion of false positives among suprathreshold voxels. A small volume correction is often used to reduce the number of comparisons to be performed and increase the chance of identifying significant results in particular regions of interest. This method typically involves defining regions of interest over particular brain structures and analyzing only these regions. The placement of these regions should be hypothesis driven and ideally based on previous work.

3.5.1.2 Deformation-Based and Tensor-Based Morphometry

DBM [13, 17] is one of the methods to detect and analyze brain shape differences among the brains of different populations. The deformation vector in each voxel is computed and a deformation vector matrix or field throughout the brain is generated by nonlinear spatial normalization, in which a brain image is transformed to a template image, and is used to describe global or local differences in brain shape.

A deformation field can be considered as a continuous 3D vector field denoted by three elements (*x, y, z*) at each voxel. When the same template (reference) image is used for the transformation of a series of image datasets, we obtain a series of deformation fields—one for each image—by which we can compare the differences between the images. For example, let us assume that there are two groups of subjects, A and B, with different characteristics (i.e., male and female, young/old, healthy/ill). Using the same template (reference) image, we can compute a series of deformation vectors for the two groups. To test the significance level in different deformations, the p value for the significance test is calculated using multivariate analysis of covariance (ManCova) and canonical variate analysis (CVA). DBM is fully automated (operator independent) and reproducible.

To localize structure differences between subject groups, tensor-based morphometry (TBM) is used to produce a statistical parametric map of regional shape differences. A deformation field that maps one image to another can be considered as a discrete vector field. The Jacobian matrix of the deformation field is generated, in which each element is a tensor describing the relative positions of the neighboring elements. Morphometric measurements derived from this tensor field can be used to localize regions with different brain structure. This is most often used to measure regional volume increase/reduction [83, 103]. Statistical parametric maps of these determinant fields can be used to compare the anatomy of subjects' groups. Other measurements are derived from the tensor fields.

3.5.1.3 Brain Image Database

To assess normal variations of the human brain across populations of different ages and genders or to detect pathological differences between normal and diseased brains, large amounts of imaging data are necessary. Recently, projects to construct a large-scale database for human brain images have been vigorously pursued by major neuroscience centers in Europe and the USA.

There are many groups involved in neuroimaging database development. Among them, the International Consortium for Brain Mapping (ICBM) [127] is well known and is one of several powerful, multicenter groups. This consortium is governed by Professor John Mazziotta (UCLA) and is composed of four core research sites: UCLA, MNI, the University of Texas at San Antonio, and the Institute of Medicine, Juelich/Heinrich Heine University in Germany. In addition, data acquisition sites in Asia (Sendai, Japan, this author's group) [128] and Europe (France, Finland, and the Netherlands) contribute to this international consortium.

We founded the Aoba Brain Imaging Center (ABIC) (with this author as the project leader) in1998. At the end of this project, ABIC had collected brain MRIs of 1600 healthy Japanese. We registered the brain images together with information on the subjects' age, sex, height, weight, blood pressure, medical history, social history, and other factors to construct a database [244]. The number of cases at the time of writing is 2743 (Table 3.3).

Table 3.3 Number of subjects in healthy Japanese brain MRI database

	Number of subjects		
Sub-database	Male	Female	Total
Aoba-1	805	786	1591
Aoba-2	184	258	442
Tsurugaya-1	92	104	196
Tsurugaya-2	118	105	223
Children	145	146	291
Total	1344	1399	2743

All subjects were living in or around Sendai City, Japan. Age of the subjects ranged from 18 to 80 in Aoba-1 and -2, 70 or older in Tsrugaya-1 and -2, and from 6 to 18 in the Children sub-database

Application of VBM to Medical Imaging: Gray Matter Volume Loss in Patients with Sub-threshold Depression VBM has been used in studies on normal brain aging and the characterization of brain pathology. Good [108] first reported age-related volume changes in the human brain using VBM. Taki also analyzed age-related brain volume changes in a large number of healthy Japanese subjects in both cross-sectional [280, 282] and longitudinal [283] designs. Baron et al. [28], using VBM, demonstrated that patients with Alzheimer disease showed gray matter loss in medial temporal structures, the posterior cingulate gyrus, and the temporoparietal association cortex. Whitwell reported gray matter loss in patients with frontotemporal lobar dementia [311] and Parkinson's disease [310]. In the following section, some results from our own studies will be introduced.

Depression is one of the most common psychiatric disorders in the elderly. There are subjects who have significant depressive symptoms, but who do not meet the criteria for major depression. However, the symptoms in these patients are associated with deterioration of physical function, worsening of physical diseases, a higher risk of mortality, and a higher risk of suicide, and thus, this state is considered to be depression syndrome and is clinically important in the elderly. It is termed "sub-threshold depression" (sD).

We assessed for differences in regional gray matter volume between community-dwelling elderly subjects with sD and age-matched nondepressed normal subject using VBM. We defined subjects with sD as those who scored ≥ 15 on the Geriatric Depression Scale and ≥ 22 on the Mini-Mental State Examination and did not fulfill the criteria for major depressive disorder (MDD) in the Diagnostic and Statistical Manual of Mental Disorders IV (DSM-IV). We collected brain MRI data of 34 subjects with sD and 109 age-matched normal subjects and assessed differences in regional gray matter volume between these two groups by applying VBM.

Only male subjects with sD had significantly smaller volumes of the medial parts of both frontal lobes and the right precentral gyrus when compared with normal male subjects (Fig. 3.19) [281]. There were no significant structural differences between female subjects with sD and normal female subjects. Our study revealed

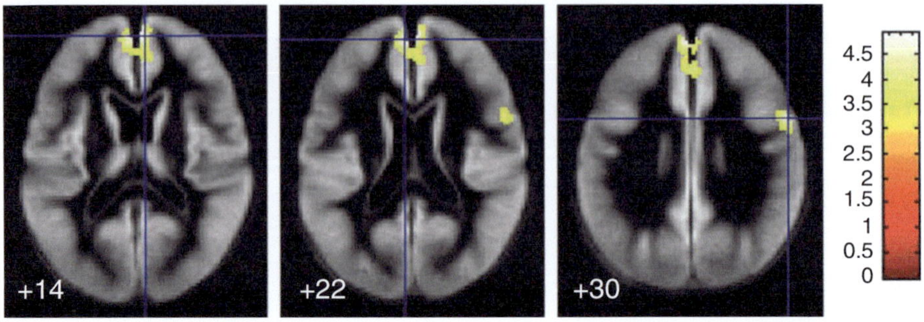

Fig. 3.19 The areas in which gray matter volume decreases in sub-threshold depression Medial frontal lobe and right prefrontal gyrus (Taki Y et al., J Affective Disorders 88:313–320, 2005. Courtesy of Elsevier. Figure 1 of Ref. [281])

that community-dwelling elderly male subjects with sD showed bilateral prefrontal gray matter volume reduction, which has been reported in elderly patients with MDD. Unlike findings in MDD, there was no significant volume reduction in the hippocampus. Our study implies the mechanism underlying the pathogenesis of brain volume loss and its relationship with sD in males.

Application of DBM in Computation of Standard Brain Models for Each Age and Sex Group DBM has been applied to assess brain shape changes in alcoholism [53], brain atrophy [102], and Alzheimer disease [88]. We sought to generate a Japanese standard brain model for each gender and age group (20–70 years old) using DBM in a large number of brain MRIs of healthy Japanese [244]. Subjects were obtained from the Aoba-1 database (Table 3.3) and were divided into age–sex groups for each decade (1920s, 1930s, 1940s, 1950s, 1960s, and 1970s). All were healthy, nondemented, and had no history of brain diseases. Abnormal brain MRIs, such as those showing cerebrovascular diseases, brain tumors, or massive white matter ischemic changes, were excluded.

Based on our previous study [244], we selected a brain MRI scan that showed the least sum energy in the process of linear transformation into other brains in each age–sex group. The brains in each age–sex group were used as reference brain (**R**) in this study. Each brain image **T** in each age–sex group was coregistered with image **R** of its group by a six-parameter affine transformation, and a coregistered image **T**′ was generated. **T**′ had the same shape and size as **T**. The deformation field, which transformed image **R** into image **T**′**,** was calculated as follows: (1) image **R** was roughly deformed to image **T**′ by a 12-parameter affine transformation, and transformed image **R**′ was generated. This affine transformation was converted to a deformation field **A**. (2) Image **R**′ was deformed to image **T**′ by an elastic transformation algorithm [250], and deformation field **B** was generated. (3) The composition of deformation fields **A** and **B** was calculated (**C** = **BA**) [245]. **C** contained both the 12-parameter affine transformation and a nonlinear elastic transformation. **C** was obtained for each brain image dataset of the age–sex group.

All the deformation field **C**s in the age–sex group were averaged by calculating the mean deformation vector in each voxel and generated mean deformation field **M**. By applying **M** to the reference image **R**, a brain image representing the mean shape of the age–sex group was generated.

Deformation fields that transformed a reference brain from the 40s age group into the brains of other age groups were calculated. In this study, a brain of a 41-year-old male was used as **R** for aging simulation in male brains. The steps used in obtaining the transformation fields were the same as those used to calculate an averaged brain for each age group. All deformation fields were averaged; this was termed a "deformation field of aging." Then, we computed a different age group brain by applying the deformation field of aging to a young reference brain ("age-simulated brain").

Figure 3.20 shows computed averaged brains for each age group (left) and an age-simulated female brain (right) [94]. By visual inspection, ventricle and sulcus enlargement, specifically of the Sylvian fissure, were observed with increasing age. Male results were similar to those for females. Figure 3.21 (upper row) shows deformation fields that transformed a reference brain into other age group brains. Although the deformation field has three components for x, y, and z, only the

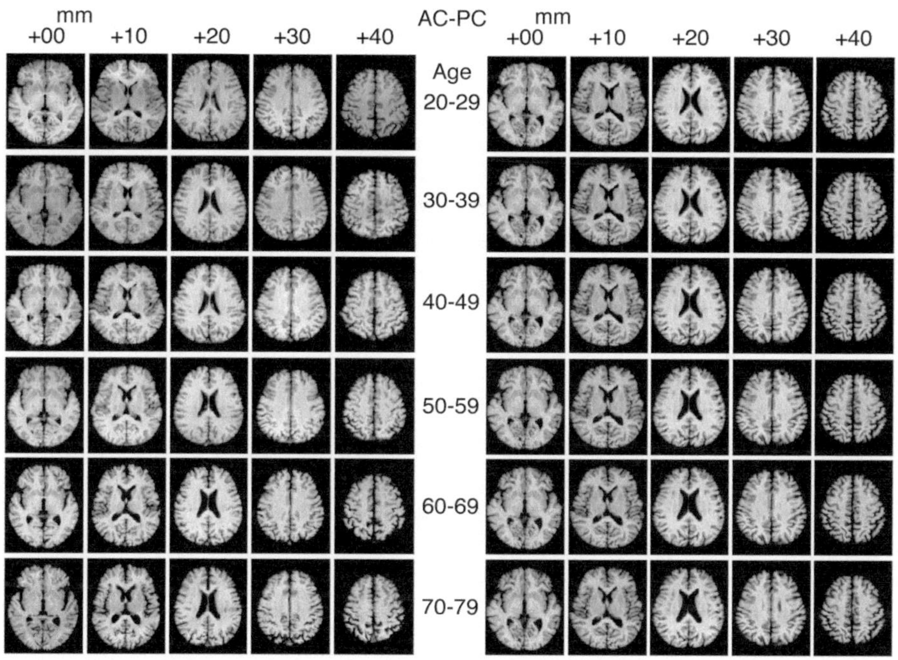

Fig. 3.20 Averaged brains for each age group and age-simulated brains for each age group for males *Left*: averaged brains for each age. *Right*: age-simulated brains for each age. The images were shown on the plane parallel to that including the AC-PC line (Fukuda H et al., In Yamaguchi T ed.: Nano-biomedical Engineering 2012:170–190, Imperial College Press) (Figure 3 of Ref. [94])

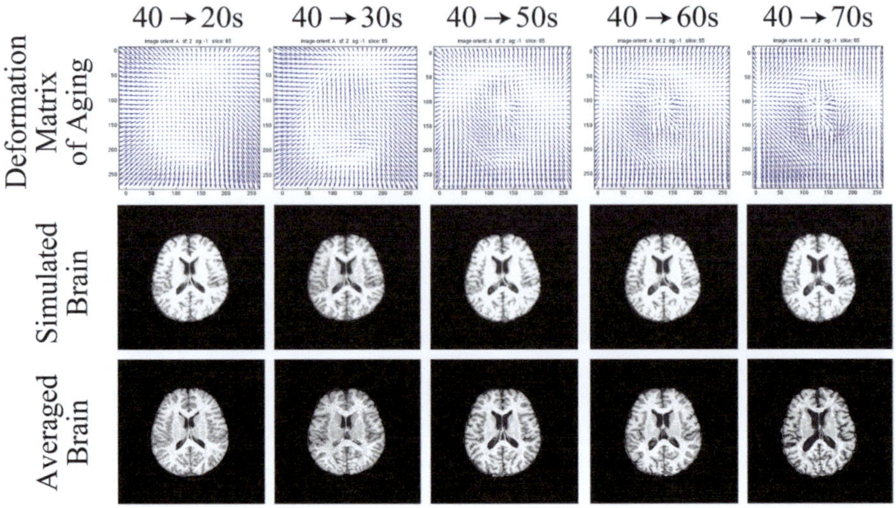

Fig. 3.21 Deformation matrices for aging simulation. Deformation fields for aging simulation "40s→20s" indicate a transformation of a brain of a 41-year-old into a 20s averaged brain. Brain images at the level of +10 mm parallel to the AC-PC line are shown (Fukuda H et al., In Yamaguchi T ed.: Nano-biomedical Engineering 2012: 179–190, Imperial College Press) (Figure 4 of Ref. [94])

horizontal plane values are shown by arrows. Figure 3.21 (middle row) demonstrates age-simulated brains at the + 10 mm intercommissural (AC-PC) line level for each age in males. A deformation field of aging transformed a young brain into an older brain with large ventricles and sulci. However, simulated aged brains were similar to averaged brains, although the transformation did not fully match the averaged brains (lower row).

Acknowledgments
Our studies were partly supported by a grant-in-aid from the Ministry of Education, Culture, Sports, Science and Technology (No 23,240,056, No 24,103,701).

3.5.2 White Matter

3.5.2.1 Brain and White Matter Anatomy

As the center of the nervous system, the brain plays an important role in human activities of both a physical and mental nature. The brain consists of two major components: gray matter and white matter. Gray matter is composed of neuronal cell bodies, neuropil, glial cells, and capillaries and is located mainly on the brain's surface with a few millimeters of thickness as the cerebral cortex. White matter is composed of mainly glial cells and myelinated axons, which serve as pathways for transmitting electrical signals within local regions of the cerebrum and from the brain to the spinal cord to exert control over the whole body.

The connective structure of the axon fiber tracts in white matter is thought to provide insight on complex brain function. However, noninvasive observation of white matter fiber structures was impossible until the end of the twentieth century using conventional imaging techniques. The advent of dMRI, which quantifies the diffusion of water molecules in biological structures, allows for noninvasive visualization of fiber tract structures in live subjects. In addition to the characterization of normal structures of white matter, dMRI can also be used to quantify pathological structures, such as tumors and infarctions via measurement of diffusion properties. Thus, the use of dMRI revolutionized techniques for brain anatomy and pathology analysis.

3.5.2.2 Diffusion MRI: A Tool for White Matter Anatomy Inference

The general term "diffusion MRI" is used for various sets of MR imaging techniques based on the diffusion-weighted imaging (DWI). DWI is the most fundamental technique of dMRI and has been used clinically since the mid-1980s [170]. A pair of DWI with identical direction of motion probing gradients (MPG), often called also as diffusion-sensitizing gradients, enables acquisition of map of diffusion coefficient in the direction. In the late 1990s, faster imaging techniques such as echo planar imaging (EPI) enabled multidirectional acquisition of DWI [49], which revealed the anisotropic characteristics of water diffusion around the fiber tracts. Such diffusion anisotropy is caused by restriction of water diffusion by the fiber structures [31]. To determine the fiber orientation, the diffusion tensor imaging (DTI) technique was developed, which is a simple approximation of the anisotropic diffusion coefficient by a second-order tensor. The eigenvector corresponding to the maximal eigenvalue of the diffusion tensor is a good estimate of fiber orientation. DTI combined with streamline visualization techniques [68] including fiber tracking is called diffusion tensor tractography (DTT) and provides a three-dimensional display of fiber tract structures [32, 59, 196].

DTT is a powerful tool for navigation during minimally invasive brain neurosurgery as it provides information for localizing fiber tracts that must be avoided, such as the corticospinal tract. For diagnostic purposes, tract-specific analysis (TSA) is used to differentiate between various diseases by comparing diffusion tensor parameters within the tract volume. One critical drawback of DTT is the well-known problem of fiber crossing [312]. The single tensor model of the anisotropic diffusion coefficient is not applicable to regions of fiber crossing where complex diffusion profiles are observed.

The most straightforward solution to overcome the fiber crossing problem involves DWI acquisition from many MPG directions to increase angular resolution of diffusion measurement. Following acquisition, non-tensor-based analysis of the orientation profile of the diffusion coefficient is conducted. Alternatively, diffusion spectrum imaging (DSI) [303] and q-ball imaging (QBI) [293] can estimate the orientation distribution function (ODF) of fibers at each location, which represents the likelihoods of fiber existence in each orientation. The DSI technique obtains the probability density function (PDF) of diffused water molecules before estimating ODF. The QBI directly estimates ODF from the profile of diffusion coefficient based on a simple approximation using the Funk-Radon transform [293]. In ODF-based tractography, fiber tracking is performed by following the local maxima of the ODF profile instead of using the orientation of the maximum eigenvalue in DTT.

Furthermore, the non-Gaussianity of diffusion prompts the characterization of tissues including pathological structures [136]. Basically, such non-Gaussianity is available by PDF analysis and can be described simply by a parameter kurtosis, which is obtained by diffusion kurtosis imaging (DKI). The dMRI techniques can be regarded as imaging techniques and models for understanding the signal values for the multiple directions and magnitudes of the MPG field. These signal models provide useful information for various purposes, which makes dMRI an indispensable tool for visualization and analysis of brain white matter fiber structures.

3.5.2.3 Tractography Techniques

Similar to "angiography," which involves images featuring blood vessels, the term "tractography" refers to images in which the fiber tracts such as white matter fiber bundles and muscular fibers are the focus of observation. Unlike angiography, which is simply obtained using a contrast agent, tractography is conducted using 3D visualization techniques. The techniques of tractography can be classified into several types, and the major categories are defined as local and global approaches. Local approaches define the structure of fiber tracts through fiber tracking based on local information at each voxel in diffusion tensor or ODF. Global approaches, however, globally optimize the fiber tract shape to guarantee smoothness and/or match it with prior information from the tract atlas. Tractography technique development was initiated by DTT [32, 59, 196], which is classified as a local approach. In DTT techniques, fiber tracking is performed iteratively in the diffusion tensor field, similar to the manner of streamline visualization [268]. An example of DTT algorithm is shown below in Fig. 3.22 and detailed as follows:

(1). *Seed point configuration*

Seed points, i.e., starting points for fiber tracking, are defined manually or based on an atlas. In the manual configuration of seed points, ROI is first specified, and then seed points are automatically placed with in the ROI.

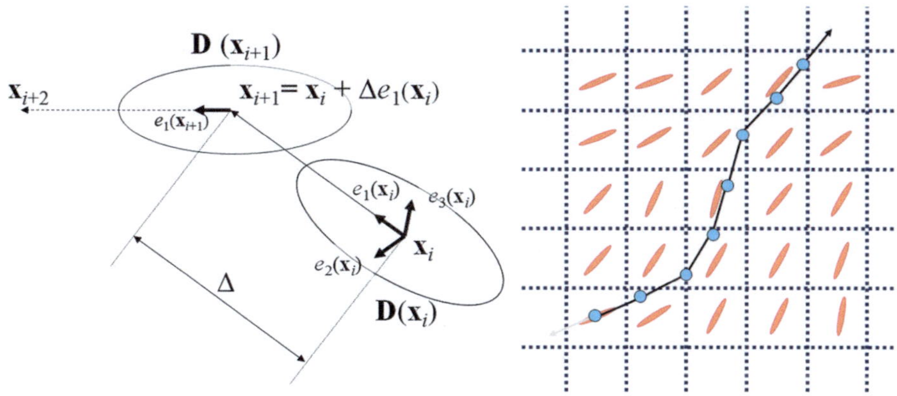

e_1, e_2, e_3: eigenvectors of diffusion tensor D

$|e_1| = |e_2| = |e_3| = 1$

Fig. 3.22 Basic scheme for DTT

(2). *Diffusion tensor calculation*

For each seed point, a diffusion tensor is calculated with DWI signal values including diagonalization for obtaining eigenvalues and eigenvectors. The DTT approaches assume that fiber orientation corresponds to the orientation of maximum diffusion coefficient obtained by the eigenvector of the maximum eigenvalues.

(3). *Short step movement*

Move in a short step distance along the direction of the maximum eigenvalue of the diffusion tensor. Note that there are two directions for initial movement at the seed point, that is, two trajectories per seed point are yielded. After the initial movement in two ways, only one direction is selected so that the trajectory is smoother.

(4). *Iteration of tracking*

Procedures (2) and (3) are repeated iteratively until tracking termination conditions are satisfied. These conditions are low diffusion tensor anisotropy, large angle of direction change, etc.

As described earlier in this chapter, the DTT based on the single tensor model is inadequate in situations involving fiber crossing. Therefore, the ODF-based tractography for DSI or QBI data is approaches are preferable as they have more options to determine tracking directions based on the local maxima of ODF. This is a natural extension of the DTT in the local approach. The local approaches of DTT and ODF-based tractography are subcategorized as "deterministic approaches," to distinguish them from another branch of local approaches known as "probabilistic approaches" [35]. In deterministic approaches, tracking is performed only once per seed point because the trajectory is identical if the tracking scheme and termination

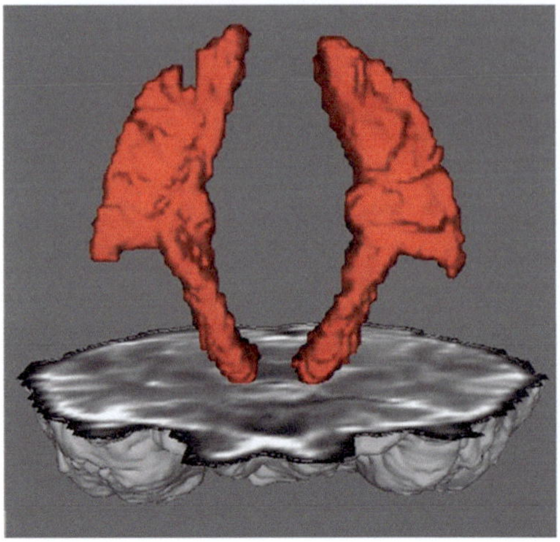

Fig. 3.23 Probabilistic tractography example

conditions are identical. In contrast, probabilistic approaches perform tracking several times from a seed point. This is because the tracking direction changes slightly each time a random sampling in the diffusion tensor or ODF is used as the probability function. Consequently, higher sensitivity for minor fiber structures occurs, resulting in false-positive structures. An example is shown in Fig. 3.23 and is based on the notion of uncertainty of fiber direction determination [71].

The global approaches, including model-based approaches using a fiber tract atlas, are quite different from local approaches. That is, in those approaches, smoothness of trajectory or similarity to the prior trajectory is considered. Figure 3.24 shows an example of global approach proposed by Reisert [237].

3.5.2.4 Clinical Applications

Clinically, DTT is used for a wide range of applications for diagnostic and therapeutic purposes. DTI remains the main dMRI technique for clinical practice to circumvent the issue of long imaging times associated with DTT. Many clinical applications were broadly performed for the analysis of brain development [125, 215] or change due to aging [1] and for diagnosing various types of brain disorders [192, 202]. Examples of three applications are shown with images [188].

Neurosurgical Planning and Intraoperative Navigation

A case involving a well-demarcated tumor without perifocal edema in the left medial temporal area is displayed in Fig. 3.25. In addition to the tumor volume, the corticospinal tract was visualized. The posterior limb of the internal capsule was used for tracking seed areas at the high-intensity area on the T2-weighted ($b = 0$)

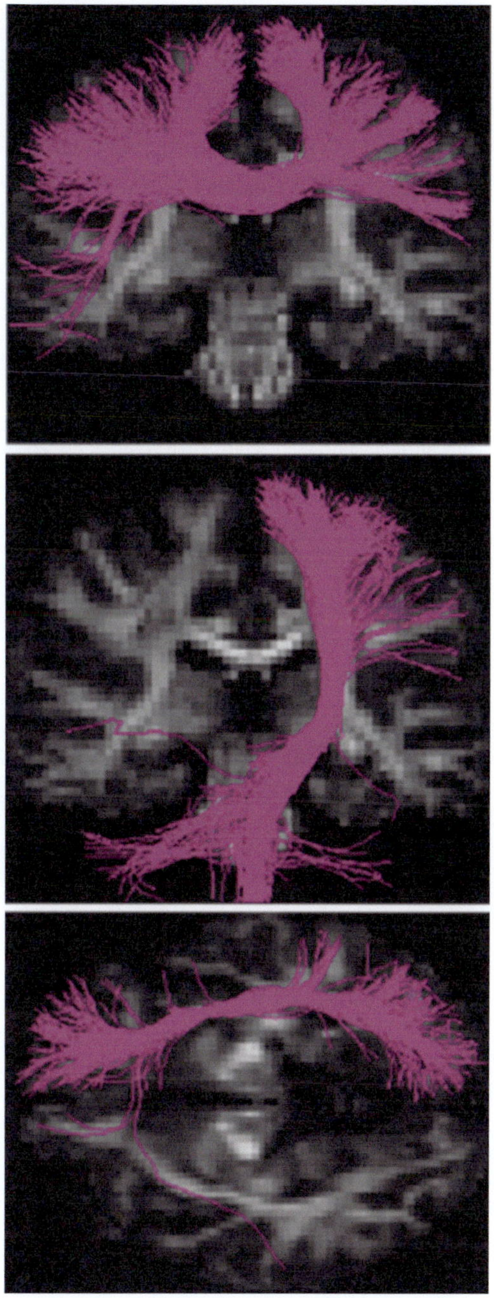

Fig. 3.24 Global tractography examples [237]

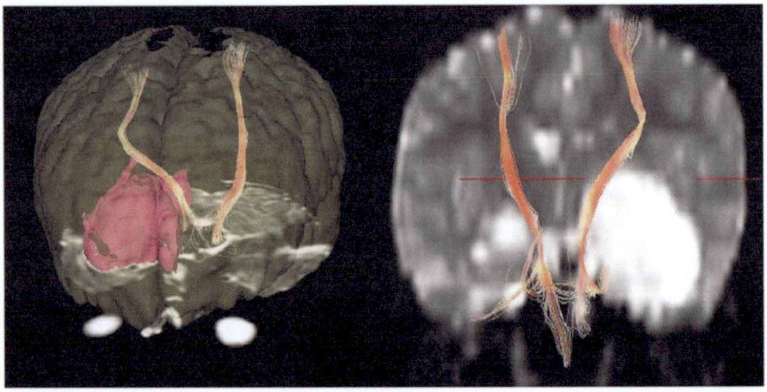

Fig. 3.25 Brain tumor. *Left*: 3D rendering of the pyramidal tracts, well-demarcated tumor, and brain surface extracted from T2-weighted image data. *Right*: Coronal section of T2-weighted image with tractography [188]

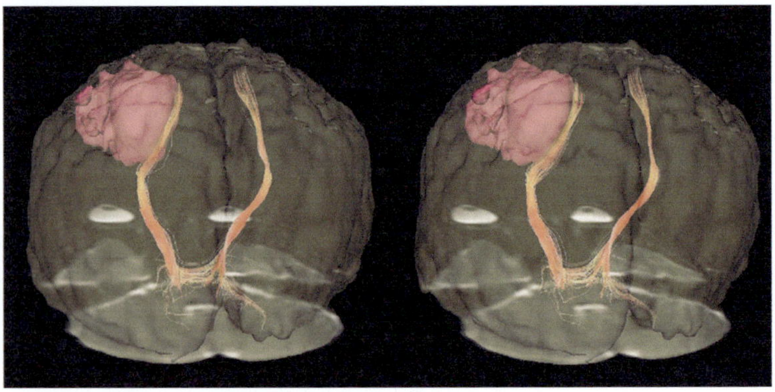

Fig. 3.26 Brain tumor of mild invasion: Tractography stereo pair [188]

image. The precentral sulcus can be recognized by the precentral knob sign, and other signs are employed as the tracking target area. Significant deformation with slight change of diffusion anisotropy in the color-encoding scheme was observed at the tract while the patient shows only mild weakness of the arm. Figure 3.26 shows a case of left superior frontal tumor without significant perifocal edema. The visualized pyramidal tract displays rounded deviation and minimal change of anisotropy.

Acute Infarction

A 15-h acute infarction in the deep white matter very close to the pyramidal tract is shown in Fig. 3.27. No significant difference in diffusion anisotropy is observed between the sides, and the patient recovered completely.

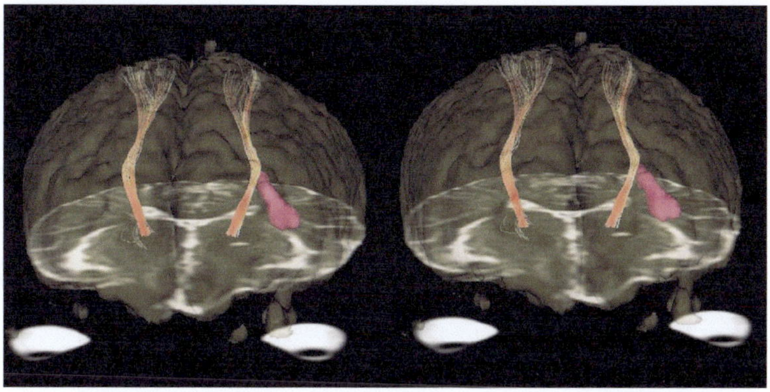

Fig. 3.27 Acute infarction: Stereo pair of tractography by 3D rendering including infarction volume [188]

Abnormality of White Matter Anatomy (Agenesis of Corpus Callosum)

A case of complete agenesis of the corpus callosum without commissure fibers is visualized via tractography in Fig. 3.28. When fiber tracking is started from the portion where the commissure fibers are located in normal subjects, thick fibers in the anteroposterior direction can be tracked while the Probst bundle running along the anteroposterior axis was visualized in this case.

3.5.2.5 Perspectives

During the past decade, dMRI has drastically changed clinical procedures, primarily neurosurgical procedures. In addition to clinical applications, the field of neuroscience strongly benefits from dMRI. Recent research involving the "connectome" [266] has relied heavily on dMRI as the source data in addition to functional MRI. The increased temporal and spatial resolution provided by dMRI has the potential to reveal more details of brain anatomy and function.

Future directions in this field involve exploring the inference of microstructures in dMRI signals. One example is the CHARMED approach to modeling (Fig. 3.29) by Assaf et al. [19], which enables us to distinguish several components of fiber structures and is extendable to inference of fiber dimensions [20]. Thus, dMRI, which originally served to capture functional information pertaining to water diffusion, is currently involved in developments which may serve to further reveal the morphology and fine structures of the brain.

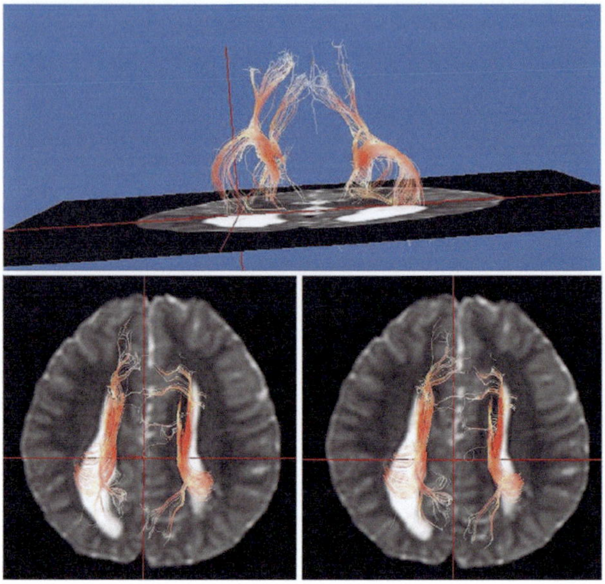

Fig. 3.28 Complete agenesis of the corpus callosum. *Top*: Posterior view of 3D tract visualization. *Bottom*: Stereo pair with an axial section of T2-weighted image in top view [188]

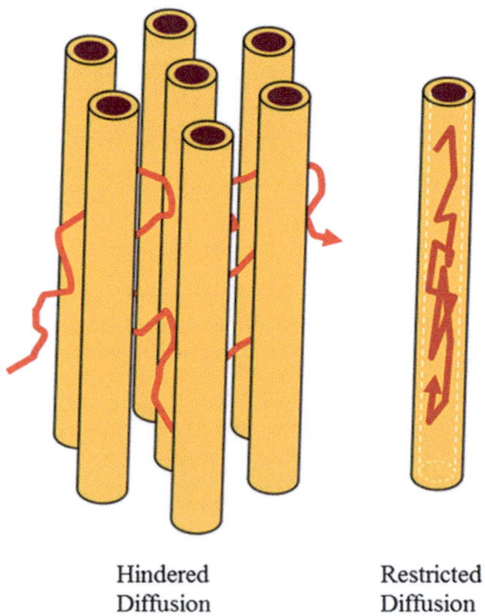

Fig. 3.29 Modeling fiber structures by CHARMED

3.5.3 Brain CT

Background Cerebrovascular accidents (CVAs) or strokes are the main cause of sudden death in emergency medicine [298]. MR images are often used for the diagnosis of nonacute brain disorders, but MR scanners are used less often in emergency situations, both because of the need to assess a patient quickly and because of the necessity of excluding all internal and external metal before allowing the patient into the scanner. CT is also quite effective in revealing acute intracranial/extraaxial bleeding and fractures promptly, and so is the modality used first in these emergency scenarios [207, 326]. Traumatic brain injury and related strokes are caused by either ischemia or intraaxial hemorrhage. Typical extraaxial bleeds are subarachnoid hemorrhage (SAH), epidural hematoma (EH), and acute subdual hematoma (SDH).

Emergency medicine staff must recognize the image changes caused by these entities on CT images and determine the treatment plans. The recognition of typical findings of early CT sign and to exclude cerebral hemorrhage are very important for prompt institution of therapy.

The analysis of brain image and function is widely recognized in the fields of brain science. The technique of SPM is a sophisticated method to compare the given brain image with a normal database voxel by voxel. Functional images using MR and PET are analyzed by the SPM method to determine the activated area in brain from visual or molecular stimuli [12, 16, 92, 191]. The regions are determined by statistical comparison with the normal database.

CAD algorithms for stroke are under development. Takahashi et al. have used a modified SPM method to obtain Z-scores, which show deviation values indicating pixel's abnormality, of brain CT values from patients' scans and developed an extraction filter to enhance cerebral infarction regions [114, 276, 277]. Nagashima et al. developed an approach to enhance bilateral changes by comparing left and right brain regions [209]. Novinski et al. reported a brain MR atlas that can be applied to brain stroke diagnosis [219].

We have constructed a standardized brain CT model using normal cases collected at an emergency hospital and developed an automated comparison technique of patients' CT scan images with the model. The comparison technique includes an automated detection method for hematoma and infarcts based on the statistical value obtained by the automated comparison method.

Normal Brain CT Model Assembling a normal brain CT model requires a voxel-by-voxel volume deformation technique to fit all normal scans to one typical brain shape after the regions are determined on each section. The deformation technique with image segmentation and characterization before assembling the normal model of brain CT is illustrated in Fig. 3.30.

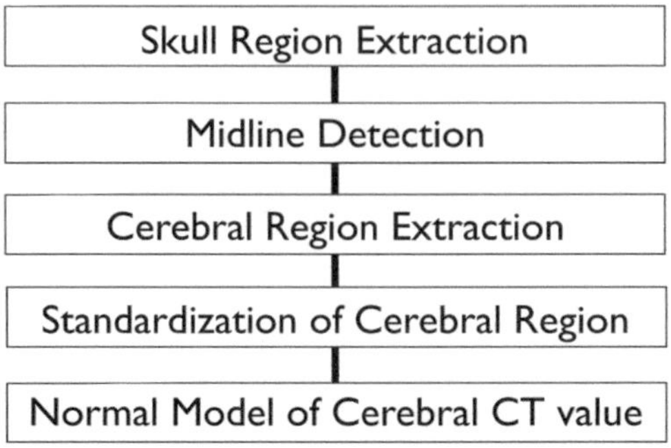

Fig. 3.30 Overview of constructing normal model of cerebral CT numbers [114]

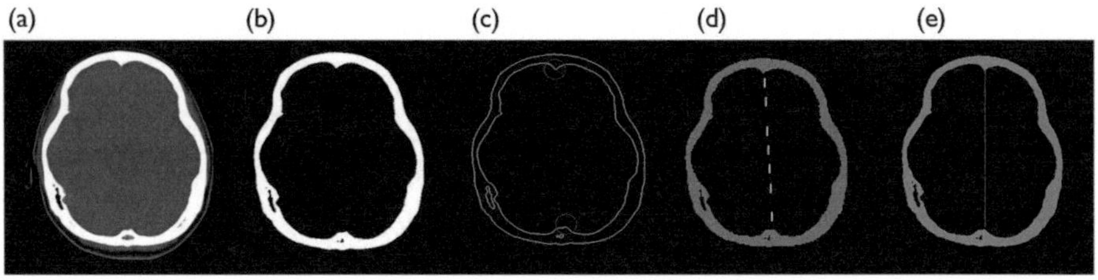

Fig. 3.31 Correction of image inclination before standardization of cerebral region. (**a**) Original brain CT image, (**b**) skull region extraction by using image thresholding, (**c**) detection of two landmarks, (**d**) midline determination based on the apices, (**e**) tilt correction [114]

The parenchymal regions in the CT images have to be segmented before the standardization of brain shape. Figure 3.31a shows the original brain CT image. The skull has a clear shape and bone density. A thresholding technique is applied to extract the skull regions. Figure 3.31b shows an example of the extracted skull at the threshold of 84 HU. The technique is applied to every section to extract the whole skull. Contours of the skull as shown in Fig. 3.31c are used to determine the midline. The landmarks of frontal crest and internal occipital protuberance are used to determine the midline. Figure 3.31d shows the midline that is estimated from the two projections. The inclination of the midline is used to align the parenchyma to upright direction section by section. The parenchymal regions are determined as intracranial regions defined within skull region after the image inclinations are aligned. A rectangular region including the cerebral parenchyma is standardized at the same size of 350×400 pixels section by section using a rigid image deformation method of affine transformation. Figure 3.31e shows an example of the section after the standardization.

The normal brain CT model is assembled as volume data to indicate the indices of mean (M) and standard deviation (SD) to show the confidence interval of CT values by voxel by voxel. The normal model includes two parameters of M and SD of CT values at every location (x, y, z). The M and the SD in each voxel are obtained by summing the normal cases voxel by voxel.

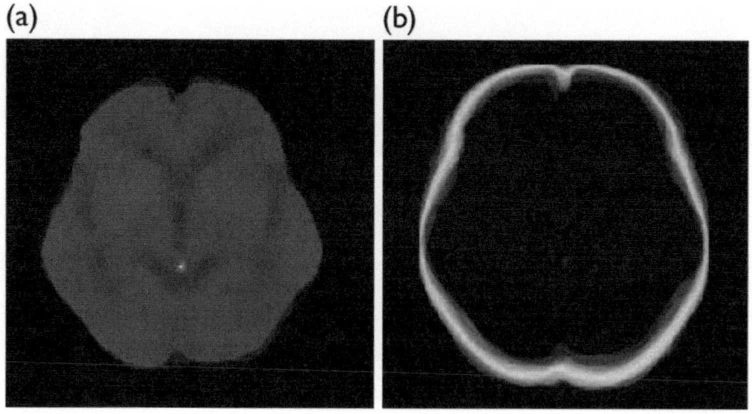

Fig. 3.32 Normal model of CT values from 50 normal brain CT cases. (**a**) Mean model, (**b**) SD model [114]

Figure 3.32 shows the example of M (x, y, z) and SD (x, y, z) images. The normal model was assembled using 50 normal cases, all acquired with the same equipment at the same emergency facility. Figure 3.32a shows a single scan of the M (x, y, z). The corpus callosum and lateral ventricles are illustrated in this scan. The white area in Fig. 3.32b indicates a misregistration region because of the boundary changes of each brain surface.

Scoring Patients' Scans Using the Brain Model The normal model expressed by M and SD can estimate a confidence interval for the brain CT value in each voxel. This means brain CT images of patients can be converted to Z-score maps after the same volume deformation procedure is applied. The Z-score of a CT value of patients at a location P(x, y, z) in the deformed images in 3D space are obtained by using Eq. (3.7), below, with M(x, y, z) and SD(x, y, z) in the corresponding location (x, y, z):

$$\text{Z-score}\,(x, y, z) = \{\mathrm{P}\,(x, y, z) - \mathrm{M}\,(x, y, z)\}\,/\mathrm{SD}\,(x, y, z) \qquad (3.7)$$

The Z-score map can depict the deviation from normal voxel by voxel. The deviation from the normal cases can be a statistical index to show the range of the abnormalities of the voxel.

Z-score Maps of Intracranial Hemorrhage Intracranial hemorrhage is an important cause of impaired consciousness. Surgery is often required to evacuate the hematoma and reduce the intracranial pressure. Fifteen abnormal CT cases were collected from a critical care medical center. Abnormal cases included SAH, EH, SDH, and cerebral contusion. Fourteen out of 15 abnormal cases had some irregular regions with a high Z-score (x, y, z). One out of 15 abnormal cases did not contain high Z-score areas.

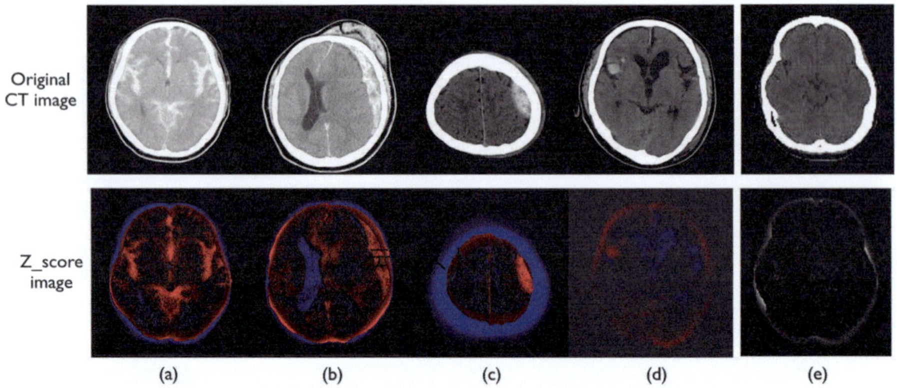

Fig. 3.33 Examples of original CT and Z-score images for brain hemorrhage and injury. (**a**) Subarachnoid hemorrhage (SAH), (**b**) traumatic acute subrural hematoma (SDH), (**c**) acute epidural hematoma (EH), (**d**) cerebral contusion, (**e**) normal case with absolute Z-score

Figure 3.33 shows the examples of Z-score (x, y, z) images in a scan from a 3D volume. Hematoma regions corresponding to each entity (Fig. 3.33a–d) were highlighted according to the Z-scores (x, y, z). The Z-scores of extraaxial region in Fig. 3.33b were not obtained because the analysis was only performed within the intraaxial region with the image normalization. In contrast, the normal case in Fig. 3.33e shows almost no signals in the Z-score (x, y, z) image in black and white image to indicate the absolute value of the Z-score. The boundary has high scores because of the misregistration in the brain CT model.

Z-score Images of Cerebral Infarction Cerebral infarction is caused by prolonged ischemia. Currently, the only effective treatment is administration of tissue plasminogen activator (tPA) to dissolve the thrombus, but if administered more than 3–4.5 h after onset, it may cause hemorrhage from the damaged parenchyma. If CT screening shows hemorrhage, non-stroke pathology, or a well-demarcated CVA in the area of interest, tPA is not administered [65, 285].

We collected 25 cases with findings verified with consensus by a group of physician with board-certified in emergency medicine, an emergency radiology technologist with board-certified in radiological technology, and a neuroradiologist.

Figure 3.34 shows the example of an ECS and the Z-score (x, y, z) images. Very faint edema is observed in original CT image on Fig. 3.34 within the circle. The Z-score map correctly indicates the region. The detection performance for abnormalities was 90.4% true positive, with 6.4 false-positive marks per case.

Observer performance studies based on receiver operating characteristics (ROC) curves were conducted to measure the effectiveness of our automated detection method for abnormalities. Twenty-five abnormal cases and 23 normal cases were used in the observer performance study. Six general physicians at an emergency medical care center took part in the study. All of the physicians were asked to register their certainty as to whether abnormalities existed on every case with and

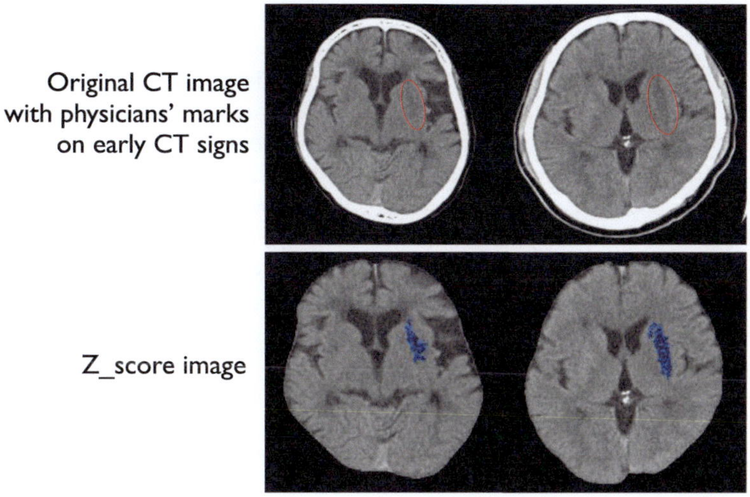

Fig. 3.34 Two examples of original CT and Z-score images for brain infarction

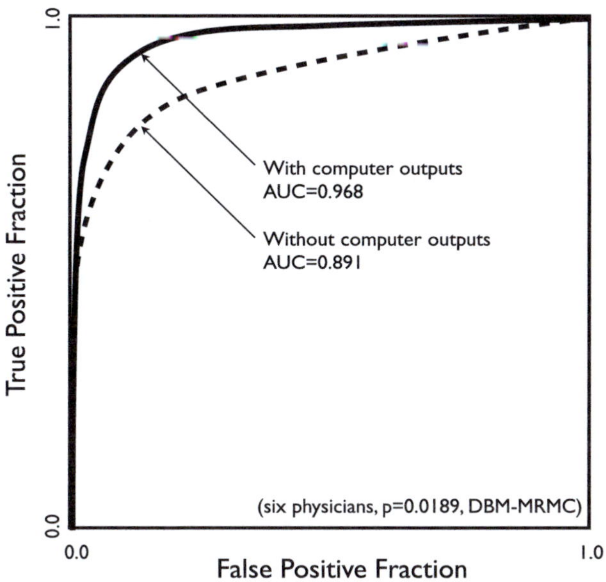

Fig. 3.35 Results of ROC analysis of six physicians without/with computer outputs

without the aid of the Z-score map. The average area under the ROC curves (AUCs) was 0.891 when the Z-score maps were not shown, but improved to 0.968 when the readers interpreted the CT images with the Z-score maps.

Figure 3.35 shows the averaged ROC curves of the six readers. We confirmed statistical significance ($p = 0.0189$) between the two results based on the multi-reader multi-case analysis method (DMB-MRMC).

Conclusions

Z-score maps of brain CT images obtained by statistical comparison with normal cases may help physicians in emergency medical situations to interpret important findings caused by stroke, such as intracranial hemorrhage and cerebral infarction. Gallagher et al. showed the statistical difference between specialists and general readers in ER [96], but the use of the automated method may improve interpretation performances not only of physicians but also of radiological technologists who will be the first medical practitioners to read the patients' images, because the AUC was significantly improved with very high discrimination performance such as 0.968 when the physicians referred the computer results during the interpretation. We believe that this will improve the chances of survival for patients with life-threatening CVAs and hemorrhage.

3.5.4 Oral/Maxillofacial Anatomy

Knowledge of oral anatomy is very important in oral and maxillofacial diagnosis and treatment as well as in orthodontic care. In addition, dental images can be used for diagnosis of systemic diseases. It has been suggested that mandibular cortical width (MCW) measured on dental panoramic radiographs (DPRs) is significantly associated with bone mineral density [69, 159], which, when significantly reduced, may indicate osteoporosis. Measurement of MCWs on images obtained for dental examination purposes may provide the beneficial information of disease risk with no additional cost. In this section, a few oral segmentation methods using CA and statistical models are briefly introduced.

Model-based segmentation of the mandible using cone beam computed tomography (CBCT) was proposed by Antila et al. [9] for the purposes of aiding dental and maxillofacial surgery planning and reconstructing panoramic radiographs from CBCT images. First they created a mean statistical mandibular model surface $\boldsymbol{S}$ using manual mandible outlines from nine dental CT reconstructions of the mandible and 31 MR scans of the head (Fig. 3.36). A parabolic approximation for the 3D centerline $\boldsymbol{C}$ of the mandibular arch surface was used to fix the coordinate system, $\boldsymbol{x}_1$, $\boldsymbol{x}_2$, $\boldsymbol{x}_3$, and $\boldsymbol{o}$, corresponding to the orthonormal basis and the origin, respectively, of the model (Fig. 3.36). This coordinate system was used for the affine transformation when applying the model. In the segmentation stage, global and local affine transformations were applied to the model to capture the rough appearance of the mandible, followed by an elastic deformation to refine the segmentation result. The global transformation of $\boldsymbol{C}$ consisted of 3D rotation and translation and $\boldsymbol{x}_1$–$\boldsymbol{x}_2$ scaling, which adjusted the size and shape of the mandibular arch. The local transformation of $\boldsymbol{S}$ consisted of rotation about the tangential vectors of $\boldsymbol{C}$ and 2D scaling in the normal planes spanned by $\boldsymbol{x}_3$ and $\boldsymbol{b}_\mathrm{j}$, which adjusted the orientation and shape of the surface cross section. For the elastic deformation, control points were placed on the surface $\boldsymbol{S}$ and were iteratively adjusted by energy minimization controlled by the surface outward normal gradient $\boldsymbol{v}\bullet\nabla I$ and value

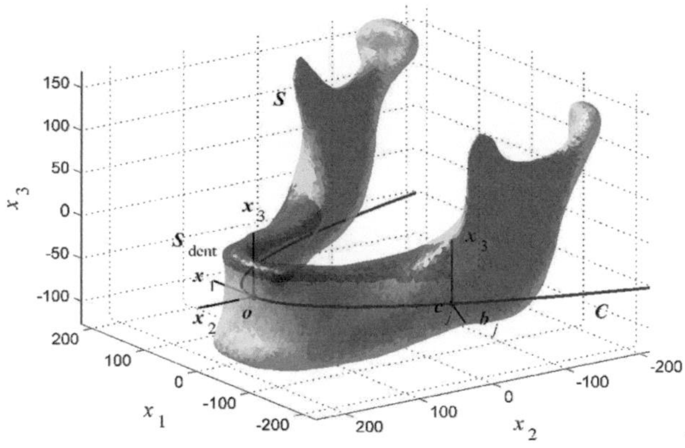

Fig. 3.36 A mean statistical mandible model surface S computed from manually delineated mandible outlines of nine dental CTs and 31 magnetic resonance (MR) scans of the head [9]

I of the grayscale intensity. With a small number of test cases, segmentation was successful in comparison with the manual references.

Other methods introduced here are based on or are modifications of the active shape model (ASM) [60]. The model can be built using feature (landmark) points of the training cases and is represented by

$$\mathbf{x} = \bar{\boldsymbol{x}} + \boldsymbol{P}\boldsymbol{b} \tag{3.8}$$

where $\bar{\boldsymbol{x}}$ is the mean shape, $\boldsymbol{P}$ is a set of orthogonal modes of variation, and $\boldsymbol{b}$ is a set of shape parameters. The mean shape can be computed by n landmark points concatenated into a $2n$ vector $\boldsymbol{x} = (x_1, x_2, \ldots, x_n, y_1, y_2, \ldots, y_n)$ of each training case. $\boldsymbol{P}$ is the matrix of t most significant eigenvectors computed using principal component analysis (PCA), whereas $\boldsymbol{b}$ is the corresponding weight vector.

Allen et al. [3] applied the ASM in determining the inferior and superior borders of mandibular cortical bone for the purpose of MCW measurement on DPRs. The model was built using 200 manually marked points with 50 equally spaced points on each of the upper and lower margins of cortical bone on the right and left sides of mandibles. Figure 3.37 illustrates the mean model and the shape variation realized by varying the weight b_1 by $\pm$ 3σ. During the application of the model to a test case, an iterative search was conducted by updating the landmarks on the basis of the gradient along the normal to the boundary at each point. Detection of the mandibular borders was relatively successful using ASM; however, the method had some limitations in having lateral shifts along the borders.

Motivated by creating a model not only describing the shape but also the texture, an active appearance model (AAM) was introduced by Cootes et al. [61].

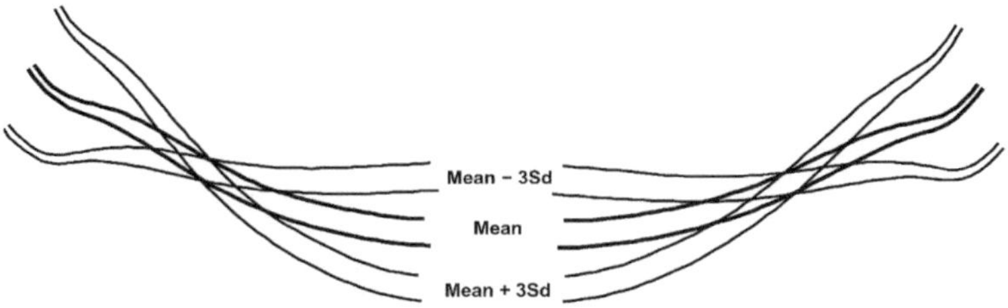

Fig. 3.37 A mean model of inferior and superior borders of mandibular cortical bone with its shape variation realized by varying the weight b1 by $\pm$ 3σ [3]

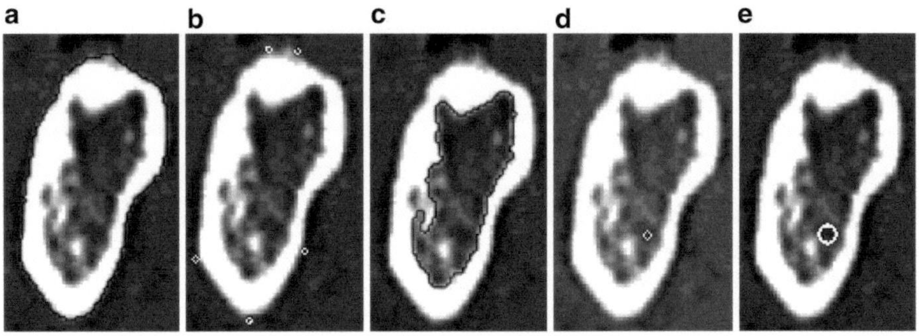

Fig. 3.38 Semiautomatic landmarking steps to extract jaw tissues based on the AAM [240]

The appearance model is represented by

$$\mathbf{x} = \overline{x} + P_s\mathbf{c} \quad (3.9)$$

$$\mathbf{g} = \overline{g} + P_g\mathbf{c} \quad (3.10)$$

where the upper equation describes the shape as in ASM and the lower one describes the texture. $\boldsymbol{c}$ is the set of parameters controlling the shape and texture, and $\overline{g}$ and $\boldsymbol{P_g}$ are the mean texture vector and the matrix describing the mode of variation in texture, respectively.

The AAM was used for automatic segmentation of jaw tissues in CT by Rueda, et al. [240] for possible utility in oral implant surgery planning. Their aim was to automatically segment cortical bone, trabecular bone, and the mandibular canal on a cross-sectional view (Fig.3.38). The model was constructed using a training set with 87 landmarks. A semiautomatic annotation is processed in five steps: (1) thresholding to find the external contour of the cortical bone (Fig. 3.38a), (2) defining five points of high curvature on the contour (Fig. 3.38b), (3) finding the contour of the trabecular core (Fig. 3.38c), (4) locating the dental nerve in the center of the mandibular canal (Fig. 3.38d), and (5) selecting the radius of the canal (Fig. 3.38e). Using these results, landmarks are placed equally, including

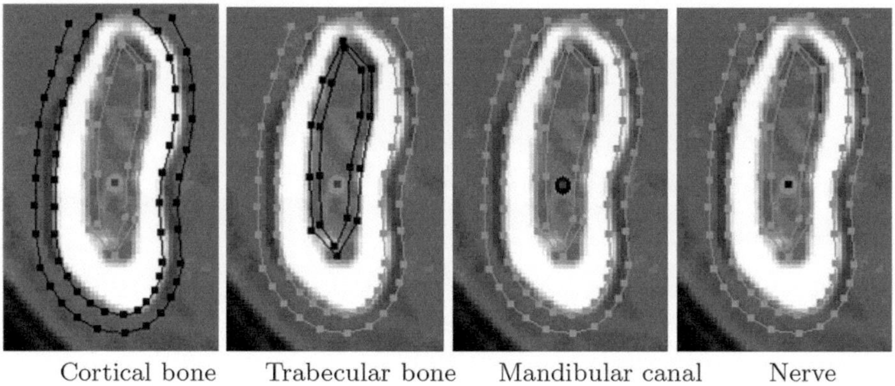

Fig. 3.39 Model landmarking for cortical bone, trabecular bone, mandibular canal, and mandibular nerve [240]

30 and 28 landmarks for the double contour of the cortical bone, 10 and 10 for the double contour of trabecular core, 8 on the mandibular canal, and 1 on the mandibular nerve inside the canal (Fig. 3.39). After the mean shape is extracted, a piecewise affine warp is applied, and the intensity is sampled from the shape-normalized images. These samples are normalized so that the effect of global intensity variation is reduced, and the texture (gray-level) vector is obtained. In the segmentation process, the model is used as an initial template, and a principal component multivariate linear regression model is used to generate new images to fit the test image. With AAM, segmentation of the cortical bones was generally successful, while segmentation of the trabecular bone was more difficult.

To improve on the AAM, Cristinacce and Cootes [63] proposed a constrained local model (CLM). The joint shape and texture model has the same form as the AAM (Eqs. 3.9 and 3.10). During the iterative search, a set of templates (patches) is generated from the model at the feature points. Based on the correlation between the current templates and a test image, a new set of feature points is predicted where a new set of templates will be generated. The CLM method was tested on different types of image datasets, including MR slices of brain, photographs of human faces, and DPRs [63]. For detecting mandibular contour on DPRs, while 78 manual points along the mandible were used for the AAM, only 22 points were used for the CLM, and it resulted in more stable performance.

Muramatsu et al. [206] have also proposed a model-based method in delineation of mandibular contour on DPRs for the purpose of automatic MCW measurement. In this method, manual contour points from the training cases (Fig. 3.40a) are used to create a mask for the lower mandibular border (Fig. 3.40b). This mask is used for detecting candidate edges with specific directions anticipated for the mandible. The individual manual contours of the training cases also serve as models from which the most similar model for a test case is selected on the basis of the similarity score with the detected edges. Using the selected model as an initial

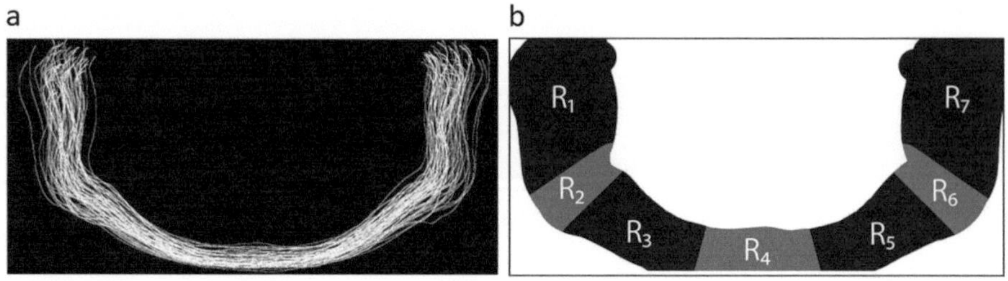

Fig. 3.40 A model-based method in delineation of mandibular contour on DPRs for the purpose of automatic MCW measurement [206]. (**a**) Manual contour points from the training cases, (**b**) a probabilistic template for the lower mandibular border

control point, the final contour is determined by fitting the points with the active contour models [151]. Using the proposed method, the mandibular contours of the test cases were successfully determined with a small number of minor partial failures in the mandibular angle.

3.5.5 Fundus Oculi

3.5.5.1 Introduction

The retinal fundus is the only part of the body where blood vessels can be observed directly and noninvasively, allowing assessment of the effects on the vasculature of both ophthalmic and systemic diseases, such as hypertension and diabetes. Ophthalmologists generally examine patients' eyes using an ophthalmoscope. It is flexible and the examination can be done in real time; however, the findings can only be stored in the form of drawing. Fundus photographs are widely and frequently obtained for diagnostic records and longitudinal comparisons. Fundal photography is well suited for mass screening of eye diseases, such as glaucoma [271], because of its simplicity and low cost.

Because the number of qualified professionals and their time are limited, computerized analysis and quantitative assessment of fundus images can be valuable. There have been numerous studies regarding computerized image analysis of these images. In this section, some of the methods for segmentation of retinal structures using CA with statistical or mathematical models are briefly introduced.

Models for Segmentation of Optic Disc and Vessels on Fundus Images The optic nerve head, also called the optic disc, is one of the main structures in the retina. It is the site where retinal ganglion cell axons converge to form the optic nerve. It is also the entry site for vessels supplying the retina. It is generally the brightest region in a fundus photograph, as shown in Fig. 3.41, and serves as a landmark. Localization and segmentation of the optic disc is essential in computerized

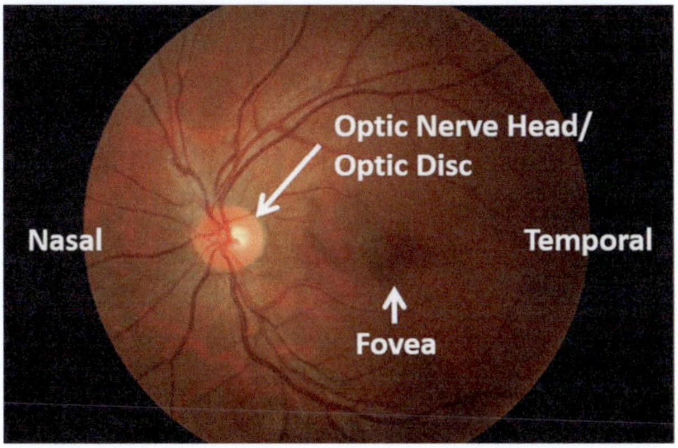

Fig. 3.41 Major anatomical structures observed in a fundus photograph

analysis for the diagnosis of glaucoma, vessel tracking, and other purposes. A number of research groups have proposed computerized methods for localization and segmentation, including those using simple thresholding, deformable models based on brightness and edge information, and pixel classification based on image features, such as gray level, edge characteristics, and texture.

One method that uses an SSM for segmentation of the optic disc was proposed by Li et al. [174]. It consists of building a point distribution model based on the idea of an ASM [60]. For disc boundary detection, 48 landmark points, 34 of which are evenly spread around the optic disc boundary, and the rest of which are on the main blood vessels inside the disc, are selected. Figure 3.42 illustrates the landmark points. The shape model was constructed by using the landmarks of eight training cases and applying PCA. In their study, only the four largest eigenvectors were used, which represent about 93% of the total variance of the training shapes. In the application of the model to segmentation, the method consists of initialization, matching point detection, and shape parameter update. The disc center location, which is detected on the basis of the similarity to multi-scale eigendiscs, and the mean shape (of the training cases) are used to initialize the model. For each landmark, its matching point is searched by the first derivative of the intensity distribution along the normal profile of the model. At the disc margin, a single pulse can be observed, while a negative pulse followed by a positive pulse can be found at blood vessels. In this study, two aspects were proposed for improving the original ASM: inclusion of self-adjusting weight and exclusion of misplaced landmark points in the update of shape parameters. Weights were adjusted by whether or not a matching point was detected and how far it was from the model landmark point. If it was too far, the point was not included in the parameter update. Landmarks were iteratively updated until the error was converged.

Li et al. also used ASM for the purpose of detecting the fovea, which is located near the center of the retina and is responsible for high-resolution central vision. In

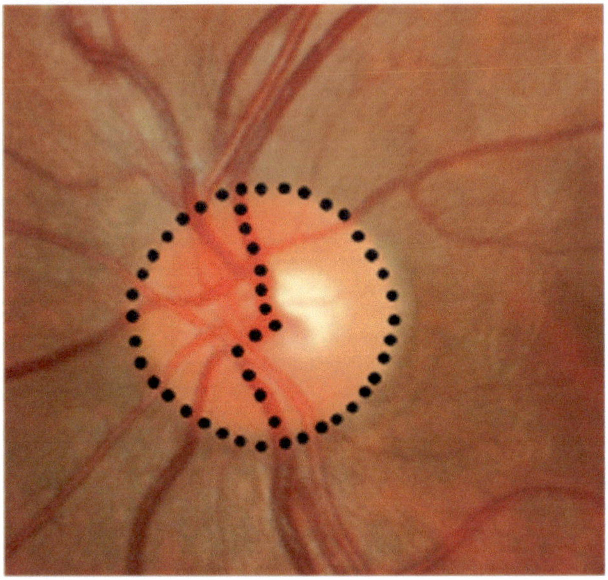

Fig. 3.42 Landmark points in ASM [174]

fundus images, it is observed as a dark brown circular/oval region located about two disc diameters temporal to the disc (Fig. 3.41). In their method [174], 30 landmarks are placed on a main vessel arch that traverses the optic disc, which are eventually fit by a parabola to estimate the location of the fovea. The model construction and fitting are processed in the same manner as for the disc itself. An example of the landmarks and the fitted parabola is illustrated in Fig. 3.43. Despite the small number of training cases, the results of a small dataset were generally satisfactory using the modified ASM.

Fujita et al. [93] used probabilistic models for localization of the optic disc. The idea is based on the fact that the disc is visible as a bright oval region, slightly brighter in the temporal half than in the nasal half (Fig.3.41), and that the central retinal artery and vein emerge from the disc and branch out to supply and drain the retina, respectively. Using training cases, an intensity model and a vessel likelihood model were created by registering disc centers and averaging the intensity images and vessel detection images, as shown in Fig. 3.44. In the localization step, the vessel score at each pixel is determined by shifting the vessel likelihood model and multiplying it with a test image. At the pixel with the highest score, local matching with the intensity model is performed for refinement. The method, in general, works well on the test databases that are publicly available; however, it tends to fail on images in which the optic disc and the major vessels are partially visible and on images with large abnormalities.

The blood vessels are another important anatomical feature in the retina. Since the blood vessel network in individuals is highly distinctive, it is used for personal

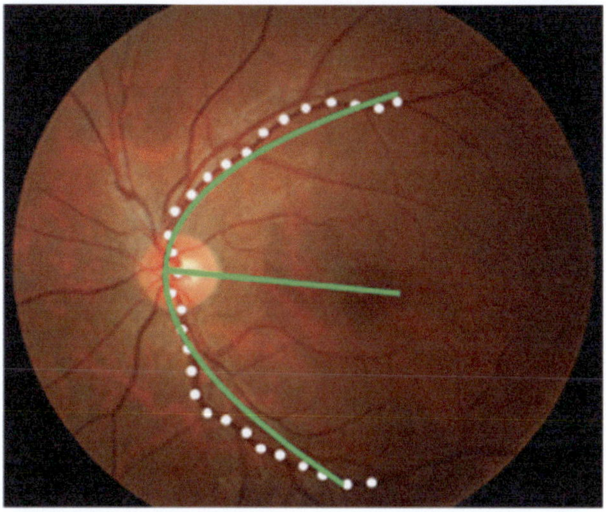

Fig. 3.43 Landmark points placed on a main vessel arch and a parabola fitted to the landmark points [174]

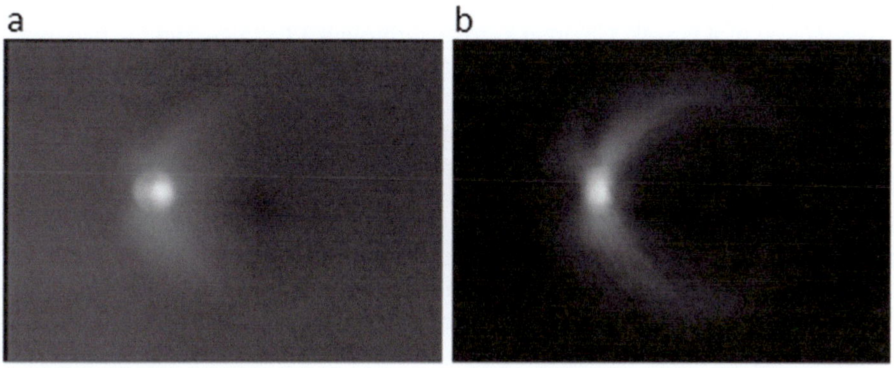

Fig. 3.44 An intensity model (**a**) and a vessel likelihood model (**b**) learned from a training dataset [93]

identification, and sequential examination images are often stored and registered to look for changes over time. In terms of disease diagnosis, vessel caliber is important quantitative information because of its association with cardiovascular disease risk. Therefore, not only is the detection of vessels essential but accurate determination of vessel diameter is also important. A large number of computerized methods for retinal vessel segmentation have been proposed. Some of these methods use mathematical models for vessel segmentation, tracking, and measurement.

In principle, the models are based on a Gaussian function. Chaudhuri et al. [56] proposed a matched filter that is shaped as a Gaussian curve corresponding to the

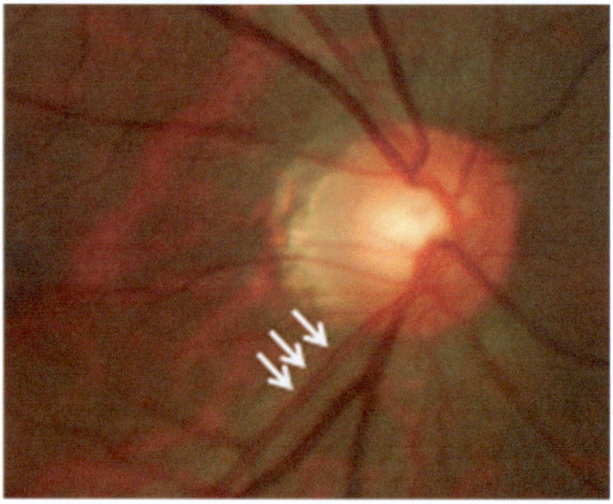

Fig. 3.45 An example of *bright lines* observed along the central axis of a blood vessel (*arrows*) caused by the reflection of a flash of light

cross section of a vessel. The filter is rotated through 12 angles, and the maximum response is used for segmentation. Similarly, an amplitude-modified second-order Gaussian filter was proposed by Gang et al. [97] for detection and measurement of vessels. A dual Gaussian model was used by several groups [82, 99, 175, 183, 214] to take into account the central light reflex, (Fig. 3.45), which appears as a bright line along the axis of blood vessels caused by the reflection of a flash from a fundus camera. It is often more apparent in arteries than veins and may affect vessel segmentation. The model can be constructed by the subtraction of two Gaussian functions, in which one with a smaller σ, proportional to the width of the central light reflex, is subtracted from one with a larger σ, proportional to the width of a vessel. An example of such a model is shown in Fig. 3.46. These models are used for segmentation, measurement of vessel width and length, and classification of arteries and veins. To control the model with a smaller number of parameters, a multiresolution Hermite polynomial model was proposed [302], which is represented as

$$H = \left(1 + a\left(x^2 - 1\right)\right) G \tag{3.11}$$

where G corresponds to the Gaussian function. The shape is quite similar to a dual Gaussian model. When $a = 0$, $H \equiv G$ and the two peaks get further apart as a becomes larger. By using the models that take into account the central light reflex, segmentation results are generally better than those using a simple Gaussian model.

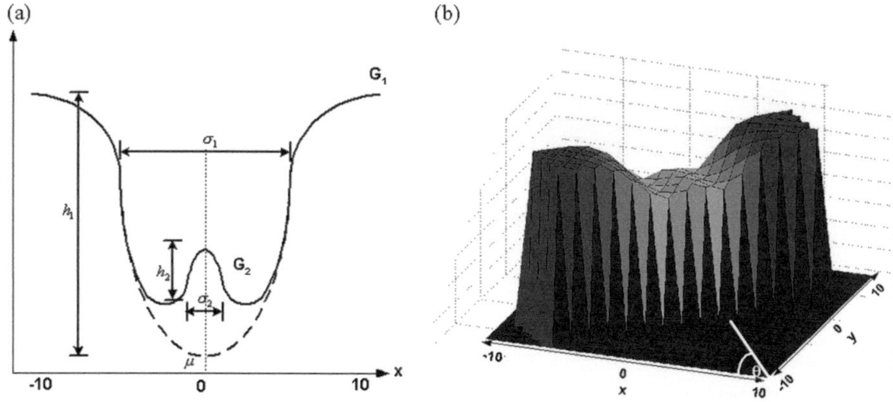

Fig. 3.46 Dual Gaussian model constructed by the subtraction of two Gaussian functions [183]

3.5.6 Retinal OCT

Since the invention of optical coherence tomography (OCT), a number of scanning mechanisms have been proposed to improve the resolution and scan speed [70]. In recent years, OCT has been increasingly used in ophthalmologic examinations for diagnosis of diseases, including, but not limited to, macular degeneration and glaucoma. The spectrum of clinical applications of OCT scanning has rapidly widened. An endoscope, a laparoscope, and a catheter have been combined with an OCT scanner, and their clinical usefulness was proved not only in ophthalmology but also in cardiovascular and digestive surgery.

This section focuses on OCT for fundus examination. Quantitative measurement of intraretinal layers in an OCT volume can be useful in the diagnosis of diseases, such as age-related macular degeneration (AMD), glaucoma, and symptomatic exudate-associated derangement (SEAD). Consequently, most of the image processing algorithms developed so far for an OCT volume extract intraretinal layers in their initial process. This section explains anatomy in a retinal OCT image and image processing algorithms followed by CAD with OCT (see review paper [67] for other topics).

Retinal Anatomy on OCT Figure 3.47 is a retinal tomography image centered on the macula, scanned by an OCT scanner during a fundus examination [162]. The retinal layers include a nerve fiber layer (NFL), ganglion cell layer (GCL), inner plexiform layer (IPL), inner nuclear layer (INL), outer plexiform layer (OPL), outer nuclear layer (ONL), external limiting membrane (ELM), photoreceptor inner and outer segments (PR IS, PR OS), and retinal pigment epithelium (RPE). The concave part at the center of the image is the fovea. The OCT image depicts the anatomy of about ten retinal layers to which the horizontal axis of the image is roughly parallel. Note that the relationship between a layer in an OCT image and an anatomical layer might not be one-to-one correspondence because of the imaging limitations of

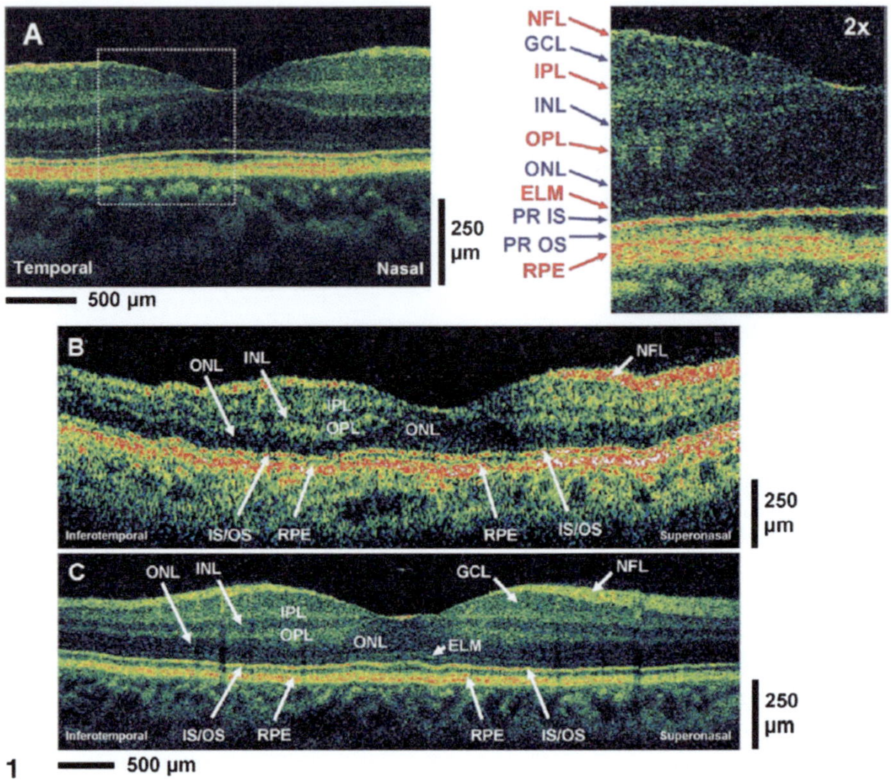

Fig. 3.47 (**a**) Ultrahigh resolution (UHR) OCT, (**b**, **c**) macular OCT images (Figure 1 of Ref. [162])

resolution and signal-to-noise ratio (SNR). The gray values of layers are similar to each other. In addition, vessels and hard exudates absorb and/or reflect near-infrared light and decrease the gray values of the regions deep to them, resulting in shadows and artifacts in the image. Consequently, it is a difficult task to recognize individual retinal layers on OCT images. Note that Fig. 3.47 is a pseudo-color display of an OCT image frequently used in clinical situations.

Intraretinal Layer Segmentation in OCT Retinal layer segmentation is the most popular topic in the field of image analysis of OCT. It is, however, difficult to carry out retinal layer segmentation owing to the low SNR. To overcome the problems and recognize thin layers, noise reduction and prior knowledge of CA are essential.

The pioneering study of retinal layer segmentation was done by Hee [117], who applied a one-dimensional edge detector to the A-scan direction of an OCT image and proposed an algorithm to measure the thickness of the retinal nerve fiber layer and the whole retina. This process is sensitive to noise because no anatomical features were introduced. Another study [165] presented an algorithm with an MRF to increase its robustness against noise. Subsequently, multilayer segmentation algorithms based on an adaptive thresholding technique, an edge detection process, or a texture analysis have been proposed [25, 50, 74, 109, 133, 319]. To reduce the noise, denoising algorithms, such as a complex diffusion process [105] or a coherence-enhancing diffusion filtering [305], have been employed.

In contrast, a spline-based active contour model [204] and a level set-based segmentation algorithm [322] achieved higher performance in segmentation than conventional algorithms, where minimization of a shape energy defined by curvature easily achieved a smooth surface, which is a favorable feature from the point of view of anatomy. The mean Dice coefficient between a true and an extracted layer was 0.85.

Kajic et al. [142] proposed the use of an AAM for multilayer segmentation of the eight layers from NFL to RPE. First, layer boundaries with relatively strong edges, i.e., upper boundaries of NFL and CL and the lower boundary of RBE, were found using an adaptive thresholding method followed by polynomial fitting. Using manually segmented training cases, the shape model was constructed by sparsely sampled distances (26 points) of eight boundaries from the top (NFL) boundary. The texture model consisted of four features from each of eight layers, including the mean pixel value in the original image, mean and standard deviation of the median filtered image, and the mean of multiple scale edges sampled along the boundaries. Rather than PCA, a neural network was used for dimensionality reduction of the original shape feature space from 208 to 12 and the texture feature space from 32 to 2. The original AAM would generate a new image with a texture learned from the texture variation and compute the distance between the synthesized image and the test image. Their method, instead of a pixel-wise comparison, used the layer boundaries produced by the model to compute texture features during optimization and compared it to the expected texture properties of each layer. The objective function includes a term that penalizes deviations from the boundaries determined by the adaptive thresholding and a term from the AAM to constrain the optimization process. In addition, instead of starting from the mean model, the most similar model was selected based on the distance between the top and bottom boundaries as well as the ratio of foveal pit distance to the greatest thickness, and it was used as the initial model.

In their subsequent study, a variant of AAM was applied for segmentation of the choroidal layer [143]. The model was constructed in a similar way but used distances from the lower boundary of RPE rather than distances from NFL. In addition, a blob detector was employed in the objective function. The multiple thresholding technique was applied to detect blobs that correspond to vessel cross sections. For optimizing the shape, the algorithm tried to maximize the ratio of the choroidal area covered by blobs to the total area of the choroid and the post-choroidal region. The authors concluded the method was successful for this relatively difficult task.

Rathke et al. [236] proposed the combination of the local appearance model and shape model for segmentation of nine boundaries, b^1, ..., b^9, or NFL, GCL + IPL, INL, OPL, outer nuclear layer and inner segment (ONL + IS), connecting cilia (CL), OS, RPE, and choroid. In constructing the appearance model, sample patches $s(i, j)$ of 3 × 15 pixels were drawn from labeled images for each of 19 classes corresponding to the ten layers and nine boundaries. Using the training cases, a class-specific density $N(s; \mu_k, \Theta_k^{-1})$ for each k can be estimated with mean parameter μ_k and sparse precision matrix Θ_k by applying a lasso penalty [87]. Given an image I, class-conditional likelihood of I_{ij} is defined as

$$p\left(I_{ij}|m_{ij}=k\right) := N\left(s\left(i,j\right)|\mu_k, \Theta_k^{-1}\right), k \in \left\{l^1, \ldots, l^{10}, b^1, \ldots, b^9\right\}. \tag{3.12}$$

For each pixel, the local class variable m_{ij} can then be determined by

$$p\left(m_{ij}|I_{ij}\right) = \frac{p\left(I_{ij}, m_{ij}\right)}{p\left(I_{ij}\right)} \tag{3.13}$$

using uniform $p(m_{ij})$. The shape model is constructed by the continuous height values of all boundaries for all image columns j. When applying the model, initial estimations of the boundary locations are made by using the appearance model and the shape prior of column j, i.e., marginalizing out all other columns. However, they showed that when the conditional probability was iteratively updated by using the predictions of b_j^n for other columns, the result could be improved.

Recent important studies have been based on graph cuts, which can find a global minimum instead of the local minimum from which the active contour model and level set models suffer. Garvin et al. [101] constructed a 3D graph in which the spatial relationship between neighboring layers was embedded. Multiple layers were segmented simultaneously by minimizing a defined cost function based on edge/regional image information, a priori surface smoothness, and interaction constraints. Six layers, or seven boundaries, were extracted from a healthy eye, and the average surface error between extracted and true boundaries was 5.69 μm. The same research group presented a multi-scale graph cut approach which extracted ten layers, or 11 boundaries, with a 5.75 μm surface error (Fig. 3.48) [233].

Computer-Aided Diagnosis of Retinal OCT and Other Topics The aforementioned multilayer segmentation algorithms are useful for CAD of retinal OCT, for example, computer-aided staging and evaluation of AMD treatment, and assessment of SEAD and drusen [74, 75, 233]. Some studies have suggested an association of layer's thickness with the diseases. Dufour et al. [72] proposed the use of SSM for detecting pathology. Drusen found in patients with AMD can be detected as a displacement of the OS layer. The model was constructed by using a simple grid around the fovea without requiring anatomical landmarks. The model was deformed to fit a new segmentation result, allowing 99% of the variation encountered in the normal training cases. For every landmark, the residual fitting errors between the deformed model and the segmentation result were calculated, and the map of

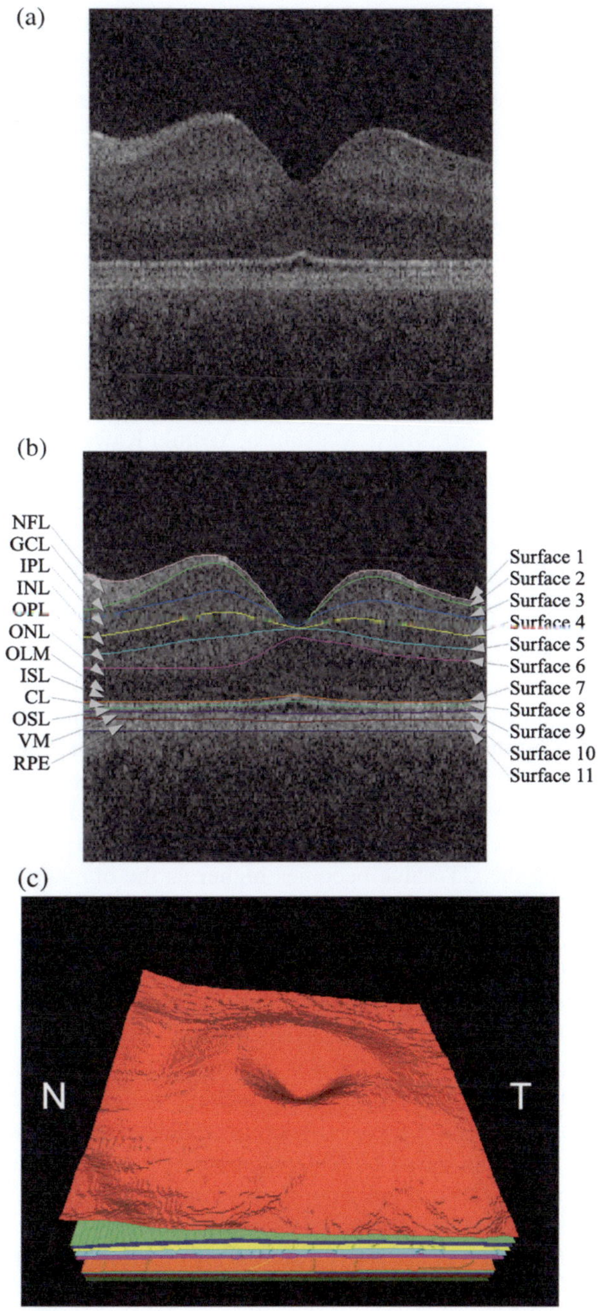

Fig. 3.48 Segmentation results of 11 retinal surfaces (ten layers). (**a**) X–Z image of the OCT volume. (**b**) Segmentation results, nerve fiber layer (*NFL*), ganglion cell layer (*GCL*), inner plexiform layer (*IPL*), inner nuclear layer (*INL*), outer plexiform layer (*OPL*), outer nuclear layer (*ONL*), outer limiting membrane (*OLM*), inner segment layer (*ISL*), connecting cilia (*CL*), outer segment layer (*OSL*), Verhoeff's membrane (*VM*), and retinal pigment epithelium (*RPE*). The stated anatomical labeling is based on observed relationships with histology although no general agreement exists among experts about precise correspondence of some layers, especially the outermost layers. (**c**) 3D rendering of the segmented surfaces (*N*, nasal; *T*, temporal). (Figure 2 in Ref. [233])

abnormality measures was determined by the fitting error normalized by the natural residual error in the normal samples.

Besides retinal layers, we can find optic nerve head and vessels as important anatomical structures in a retinal OCT. Herzog et al. [119] proposed an extraction algorithm for the optic nerve head in OCT images. The cup–disc ratio was evaluated based on the extracted optic nerve head [44]. Lee et al. and Wehbe et al. [173, 304] proposed vessel segmentation algorithms from an OCT image, in which they measured blood flow velocity based on the segmentation result [304].

3.6 Thoracic Organs

Kensaku Mori, Mikio Matsuhiro, Yoshiki Kawata and Noboru Niki

3.6.1 Bronchus and Vessels

3.6.1.1 Overview

The tracheobronchial tree is the macroscopic framework of the respiratory system. The trachea bifurcates into two branches, the right main stem bronchus and left main stem bronchus. Many branchings continue down to the bronchioli, which allow air to diffuse into the alveoli, the site of actual gas exchange. Because the bronchial tree is filled with air, it has negative density on CT. Typical CT values of the bronchial lumen are in the range −1000 H.U. to −900 H.U. Because of the partial volume effect, the CT value becomes higher as the branches become thinner. To segment the tracheobronchial tree from chest CT examinations, the basic method involves tracing the negative-density regions from the trachea, which can be easily identified, in the direction from the center to the peripheral. There are several methods for extracting bronchial regions. These methods can be classified into two categories: (a) the region-growing-based method and (b) the machine learning-based method. In the category (a), basic region growing (e.g., one threshold value in region growing) [195] or adaptive region growing (e.g., changing threshold branch by branch) [156] is utilized. The category (b) determines whether each voxel belongs to the bronchial lumen or not based on features computed at each voxel. This classification is done by morphological operation or machine learning [23]. Then the selected voxels are connected to portray the tracheobronchial tree.

Lo et al. (2012) [181] discussed the comparison of world-representing methods for bronchus extraction from 3D chest CT images. Their findings were based on results of a bronchus region extraction competition held in MICCAI in 2009.

3.6.1.2 Region-Growing-Based Method

Simple Thresholding This method extracts the tracheobronchial tree by tracing dark regions on chest CT images. The seed point is set inside the trachea and the growing process commences. When growing is performed, it is necessary to define some conditions to determine whether the growing area is inside a lumen or not. A simple threshold value is all that is necessary in this region. If a voxel has CT values lower than a given threshold value t, the growing process is performed. If not, the voxel will not be included in the queue. One issue with this method is a selection of a threshold value. If t is too low, insufficient regions will be extracted, and not all branches will be segmented. In contrast, higher t will cause extraction of aerated lung outside the tracheobronchial tree. In such case, *extraction explosion* occurs [195]. When drawing a graph of the number of extracted voxels, the number rapidly increases at a certain threshold. It is possible to find such a threshold value t by monitoring the number of extracted voxels. Figure 3.49 shows examples of segmentation results obtained by an optimal threshold value and a threshold value causing *explosion*. Because CT values of the bronchus wall become low because of the partial volume effect, explosion easily happens. A bronchus wall enhancement filter is sometimes applied. In this case, small leakage of regions growing, which is much less than *explosion*, will happen.

Adaptive Thresholding The method shown in the previous section uses one threshold value in the extraction of the entire tracheobronchial tree. However, as stated before, the threshold values should be adjusted branch by branch because of the partial volume effect of the surrounding aerated lung. The method shown in [156] defines a volume of interest (VOI) for a bronchial branch. The growing process is performed inside this VOI. Because the method is designed to find the running direction of a branch, the VOI is alliteratively extended in the running direction of a branch. Luminal region extraction is simultaneously performed inside the VOI. If the method detects bifurcation, two VOIs are set for secondary branches of a target branch. The threshold value of region growing is computed for each VOI. This expansion process is performed using the breadth-first search method. Branch-

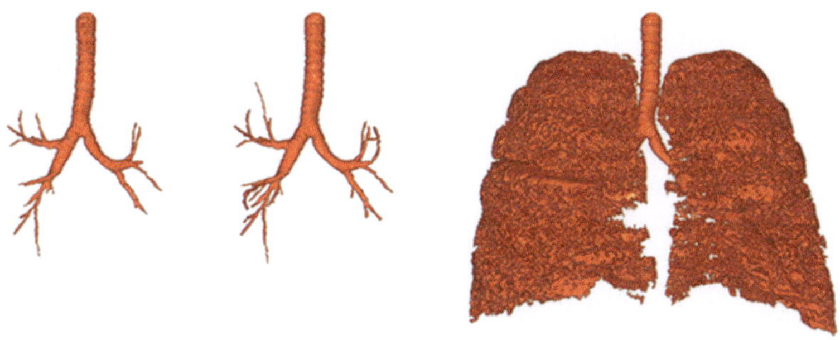

Fig. 3.49 Extraction of bronchus regions with appropriate threshold and inappropriate thresholds

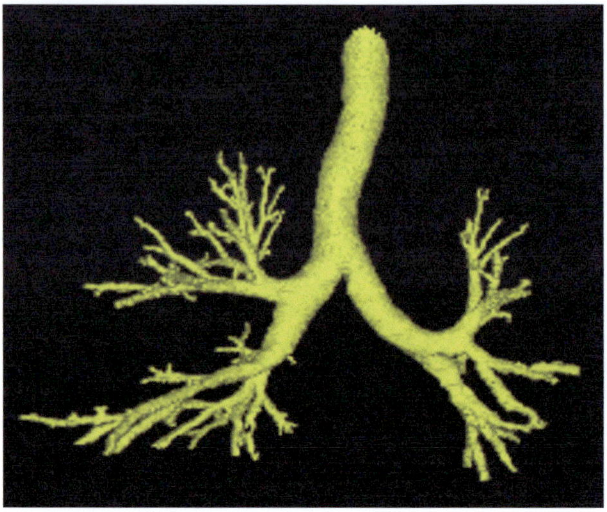

Fig. 3.50 Segmentation result of region growing with adaptive thresholding [156]

by-branch growing has another advantage in that it can detect not only bronchus regions but also branching structures. An example of extraction obtained by this method is shown in Fig. 3.50.

3.6.1.3 Anatomical Labeling

When constructing an application system for bronchoscope navigation or lung nodule diagnosis, it is important to segment the tracheobronchial tree from CT while knowing the anatomical names of each bronchial branch Fig. 3.51, which are the terms used by clinicians.

Because the bronchi bifurcate in an almost fixed manner, it is possible to assign anatomical names using graph matching between the input tree structure and the graph structure of a bronchial bifurcation atlas (learning tree structure). Mori et al. (2000/2002) [197] reported on an anatomical labeling process based on graph matching. The matching is performed based on running direction information and constraint of parent branches. Labeling of bronchial branches has also been performed based on graph matching [290]. An example of bronchi branching models is shown in Fig. 3.52.

There are many variations in branching patterns of bronchi. It is necessary to have branching pattern atlases and to develop an algorithm to use such atlases showing variations. Mori et al. (2005) [198] showed a method for selecting suitable atlases in anatomical labeling. This method divides the tracheobronchial tree into five parts:

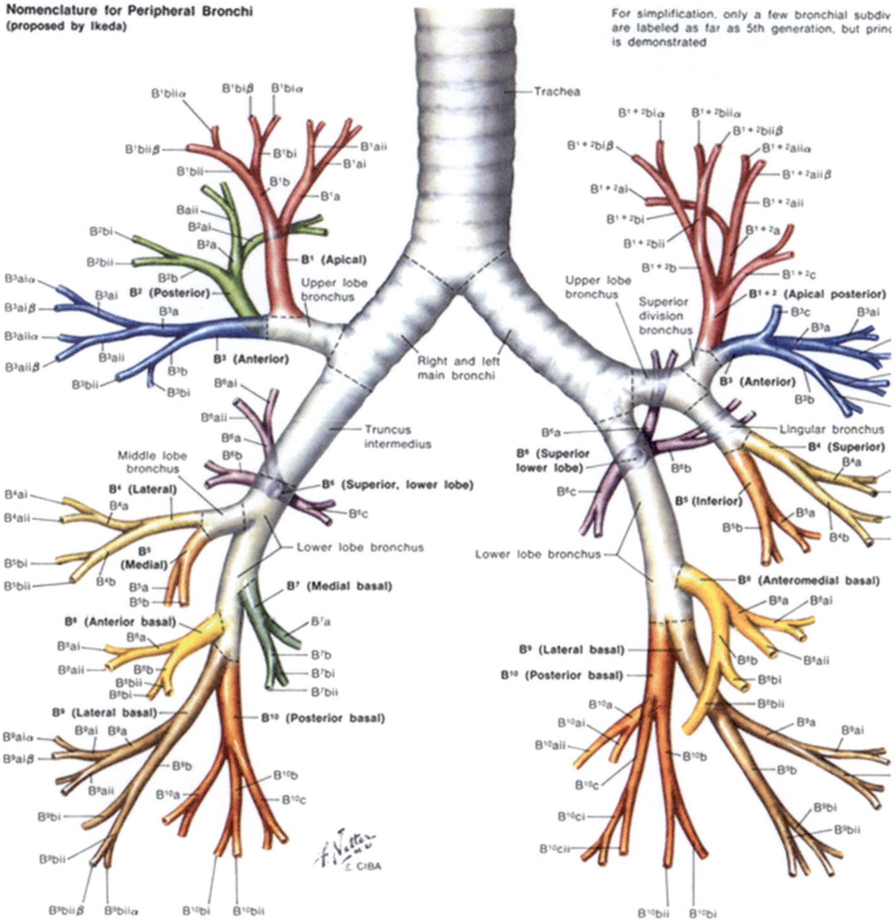

Fig. 3.51 Atlas of bronchus [217]

(a) trachea, (b) right upper lobe area, (c) right middle and lower lobe area, (d) left upper lobe area, and (d) left lower lobe area. Branching pattern databases are created for each area. Graph matching is performed in each divided area and an atlas removal process is introduced for choosing the best atlas in each area. Figure 3.53 shows an example of such an atlas.

Another method is based on a machine learning approach [199]. This method computes many features for each branch name and constructs the classifier that outputs anatomical names from input features.

Figure 3.54 shows an example of anatomical labeling.

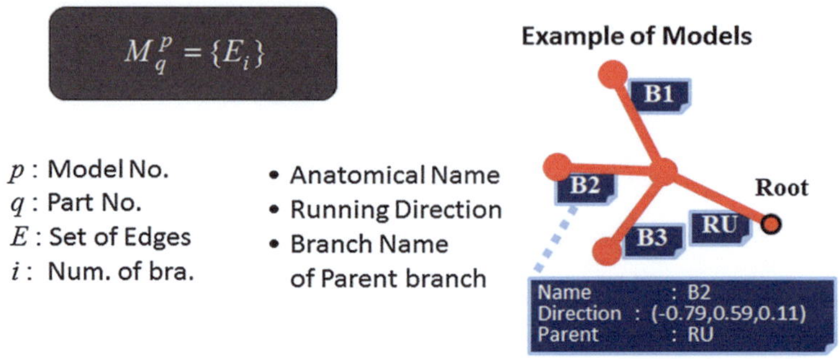

Fig. 3.52 Examples of bronchus branching model

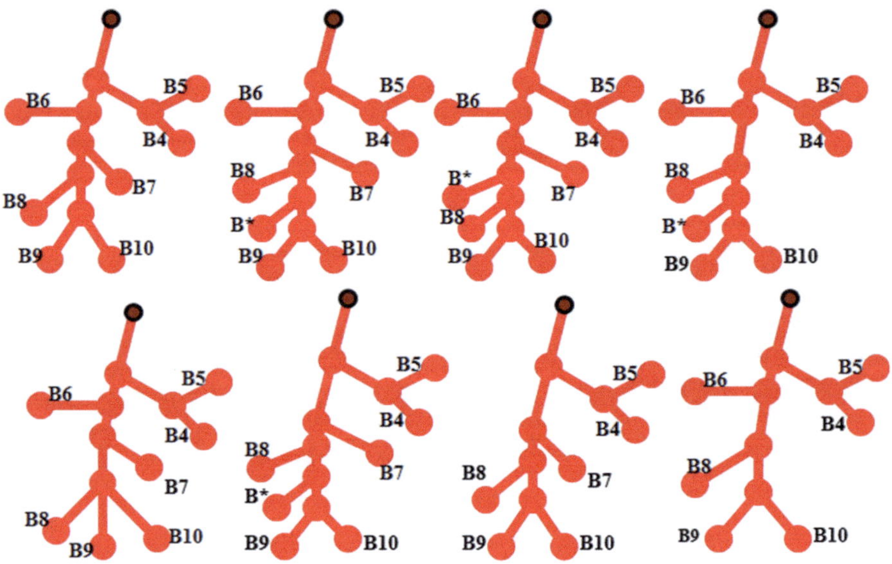

Fig. 3.53 Examples of bronchus branching model atlas. This figure shows atlases of the right middle and lower lobe area

3.6.2 *Pulmonary Blood Vessels*

3.6.2.1 Overview

Detailed segmentation and analysis of the pulmonary circulation are needed for computer-assisted diagnosis or computer-assisted surgery as described in Chap. 4. For example, adjacent pulmonary blood vessel anatomy is an important information to include in the analysis of lung nodules detected on CT. In bronchoscopic navigation, it is important to prevent inserting a biopsy needle into a pulmonary artery branch during transbronchial needle biopsy (TBLB).

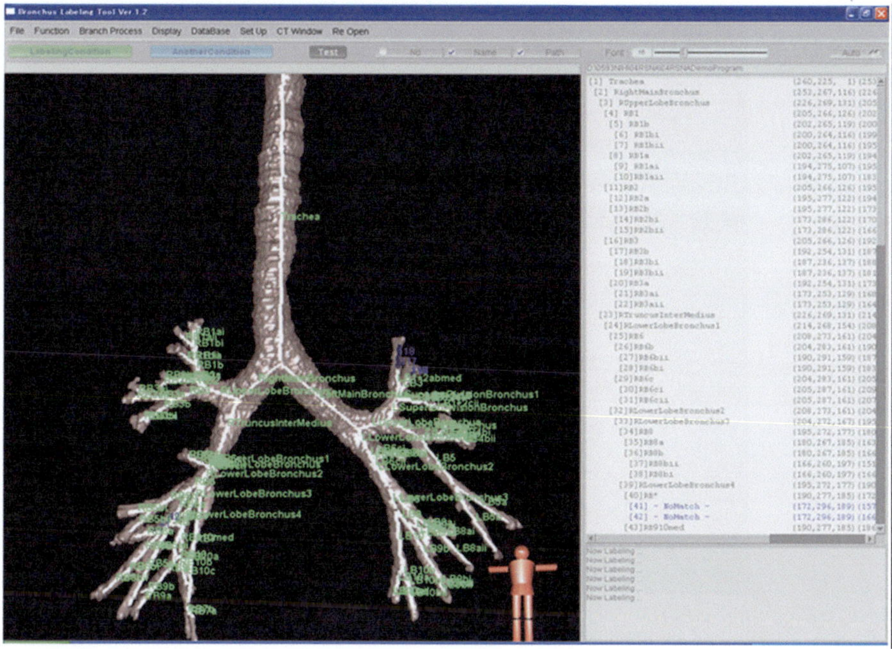

Fig. 3.54 Examples of anatomical labeling of bronchi

Because pulmonary blood vessels are depicted as white regions on chest CT images, it is possible to segment lower-order vessels by simple thresholding. However, because of gravitational pull, the side of the lungs that is close to the ground has relatively higher density than that on the upper side. Adaptive thresholding or another more sophisticated approach is necessary.

3.6.2.2 Simple Thresholding

This method extracts pulmonary blood vessels based on simple thresholding. Because pulmonary blood vessels have CT values of around −500 H.U. or more, it would be possible to extract pulmonary blood vessels by performing a thresholding operation inside the lung area. The lung area is also easily segmented from CT images by simple thresholding and connected component analysis.

3.6.2.3 Line Enhancement Filter

As stated previously, Sect. 3.6.2.1 CT values of pulmonary vessels differ because of the partial volume effect or gravity. One solution to these issues is to use a differential filtering technique. Frangi's "vesselness filter" [80] or Sato's filter [243]

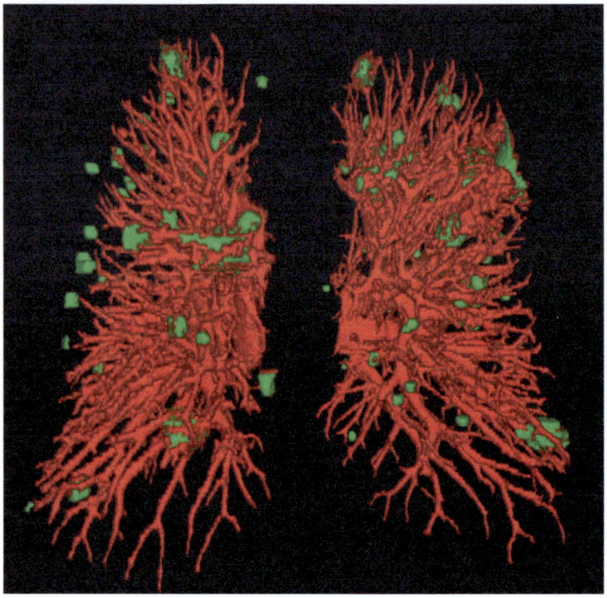

Fig. 3.55 Example of pulmonary blood vessels extracted by Level set segmentation. Pulmonary blood vessels are colored in *red* in this figure

are often used to enhance pulmonary blood vessels. This detection process enables easy extraction of pulmonary vessels.

3.6.2.4 Level Set-Based Approach

This approach extracts pulmonary blood vessels by using level set segmentation. The core part of level set segmentation is the definition of the speed function. One approach is to use the output of the vesselness filter shown in Sect. 3.9.8. The speed function is designed to extract pulmonary blood vessels in 3D chest CT images. Figure 3.55 shows an example of pulmonary blood vessels extracted by level set segmentation.

3.6.3 Lung and Pleura

With the development of CT technology, complete volumetric chest images can be acquired over a single breath hold. With increasing resolution, the data load has substantially increased. A lung CAD system may help radiologists to deal with these data loads more effectively. Accurate lung segmentation is fundamental for quantitative analysis of lungs by CAD systems.

The lung is divided into lobes by interlobar fissures that are potential spaces lined by visceral pleura. Extraction of the interlobar fissures is essential, because it make it possible for a lung CAD system to determine the lung lobe of the pathology and calculate volume rate of pathological region in the lung lobe. Clinically, extraction of the interlobar fissure is often important to determine whether a disease affects one or more lobes, when lobar resection is considered.

Anatomy of Lungs and Interlobar Fissures The right and left lungs are covered by a tightly attached layer of visceral pleura. The veins, arteries, airways, and lymphatics comprising the lung roots attach to the mediastinum in the center of the chest (Fig. 3.56a). The entire thoracic cavity is lined by an outer layer of pleura called the parietal pleura. A potential space is located between these two pleural layers and can be enlarged by fluid (pleural effusion), air (pneumothorax), or disease. The lungs consist of sections called lobes. The left lung usually contains two lobes, the upper and lower, while the right lung usually consists of three, the upper, middle, and lower lobes. An accessory right upper (azygous) lobe is not uncommon, and there are other variations that are less common. These lobes are separated by interlobar fissures lined by visceral pleura. The fissures can be of varying depth, extending down to the lung root, or incompletely dividing the parenchyma. Both lungs have a major fissure dividing the upper and lower lobes, and the right lung has a minor fissure dividing the upper and middle lobes. Examples of lung lobes are shown in Fig.3.57. The lobes are divided into pulmonary segments, and each segment contains multiple secondary pulmonary lobules divided by thin connective tissue septa. In pathological cases, such as pleural effusion, lung cancer, interstitial pneumonia, and severe emphysema, features of the lungs and pleura are usually altered. Pleural effusion in a CT image is shown in Fig. 3.56b.

Lung Segmentation Many methods for automatically extracting the lung regions based on 3D CT volumes have been proposed [263]. As visualized on CT images,

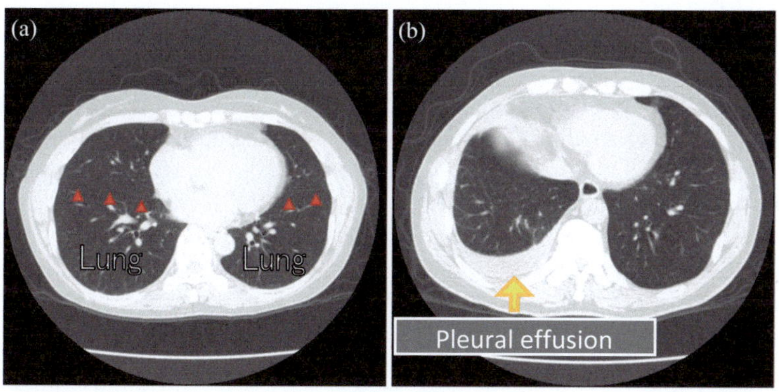

Fig. 3.56 CT images of (**a**) normal chest and (**b**) right pleural effusion. *Red* triangles point to interlobar fissure

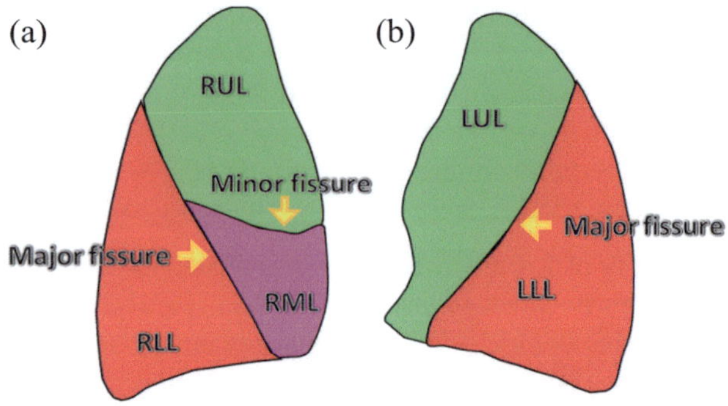

Fig. 3.57 Image of lungs viewed from outside. *RUL* right upper lobe, *RML* right middle lobe, *RLL* right lower lobe, *LUL* left upper lobe, and *LLL* left lower lobe. (**a**) Right lung, (**b**) left lung

the lung parenchyma, which is mostly air, is dark. This contrast between lung and surrounding tissues is the basis of most segmentation methods [11, 45, 122, 147, 329]. In lung segmentation methods, air is extracted by gray-level thresholding. Therefore, the lungs are identified by imposing restrictions on size and location. Alternatively, the lung volumes can be determined by region-growing segmentation from the trachea. The trachea is recognized as 2D circular air regions in the first slices of the scan or as a 3D tubular air region located centrally in the upper part of the scan. The main lung volume is separated into left and right lungs. The trachea and main stem bronchi are removed. Morphological processes work out filling holes and smooth borders of segmented lungs.

Such methods are known to be simple and effective for normal cases. However, they often fail to extract lungs affected by pathologies, especially when the pathologies involve the pleura. Pleural effusion, like that illustrated in Fig. 3.56b, has higher density than normal lung tissue, and therefore segmentation methods using contrast fail.

For pathological cases, segmentation methods often use the shape of the lung. Sluimer et al. proposed a registration-based approach in which a shape template is registered to an input CT volume [262]. They achieved significant improvements in the segmentation of lungs with pathologies. Kido and Tsunomori proposed another registration-based method using a template obtained from normal cases [153]. Two-step matching improved the performance in a case with severe pleural effusion. Hua et al. proposed a method that combines the classification process with a graph-search algorithm [123]. The method has been shown to be effective in cases with pathology. Nakagomi et al. proposed a graph cut-based method that can take into account the multiple shapes generated from an SSM of the lungs. The method used in energy term introduces neighboring structures of the lungs, e.g., the aorta and the body cavity [210].

Interlobar Fissure Extraction The interlobar fissures are very thin and low contrast on CT images. Simple methods such as gray-level threshold cannot extract them. Many extraction methods have been proposed. Kubo et al. proposed a method that extracts sheet shapes by morphology operation from the emphasis of 2D linear shadow images [166]. Saita et al. suggested a method that uses blood vessel information [241]. This method classifies lobar blood vessels, whereas the region containing the interlobar fissure is identified on the basis of the 3D distance from the lobar blood vessel. The interlobar fissures are extracted by the emphasized sheet shadow from the identified region. Zhang's [328] method extracts the major fissures using an anatomic pulmonary atlas. A ridgeness measure is applied to the original CT image to enhance the fissure. A fuzzy reasoning system is used in the fissure search to analyze information from three sources: the image intensity, an anatomic smoothness constraint, and the atlas-based search initialization. Van Rikxoort's method uses supervised enhancement filters [297]. These filters, which enhance interlobar fissures and suppress others, are constructed using training data. Ukil and Reinhardt proposed a method based on the information provided by the segmentation and analysis of the airway and vascular trees [295]. An ROI is generated using this information, and the interlobar fissures are extracted by enhancement using a ridgeness measure in the ROI. Pu et al. pointed out existing methods that were problematic in the presence of pathologies and susceptible to interindividual differences; they proposed a method using the membranous properties of the fissures. The method creates polygons by the marching cube algorithm. Surface shapes are extracted using multiple Laplacian smoothing from polygons [231]. The interlobar fissures are extracted by normal vectors of surfaces that are made by extended Gaussian image. This interlobar fissure is filled blank spaces by average of normal vector of the surface [232]. Agarwala et al. proposed an atlas-based segmentation approach [2]. An atlas is a set of two images: the intensity image and its segmentation. Because segmentation obtained using an atlas-based approach may have local errors because of local failures of the image registration algorithm, they applied a local version of selective and iterative methods for performance level estimation that uses local weights for fusion of the input segmentations. Lassen et al. employ a method that uses information on blood vessels and sheet shapes extraction using a Hessian matrix [169]. Matsuhiro proposed a method that uses features of the interlobar fissures' film shape that can extract interlobar fissures from pathological cases, e.g., lung cancer, interstitial pneumonia, and severe emphysema [189]. This method contains three extraction phases: coarse extraction, fine extraction, and correction. Coarse extraction enhances images by 4D curvature. Film shapes are extracted from contrast-enhanced images. The interlobar fissures' shape extraction parameters, e.g., angles and sizes, are trained by training case data. In these data, interlobar fissures are predetermined manually. Coarse interlobar fissures are extracted from film shapes by these parameters. Fine extraction is implemented in iterative enhancement around fissures. Correction is implemented in interpolation of interlobar fissures by normal vectors. The extraction results are illustrated in Fig. 3.58.

This section describes lung segmentation and interlobar fissure extraction methods. Developing the performance of those methods is expected. Methods that segment pulmonary segment and secondary pulmonary lobule are expected to be developed, too.

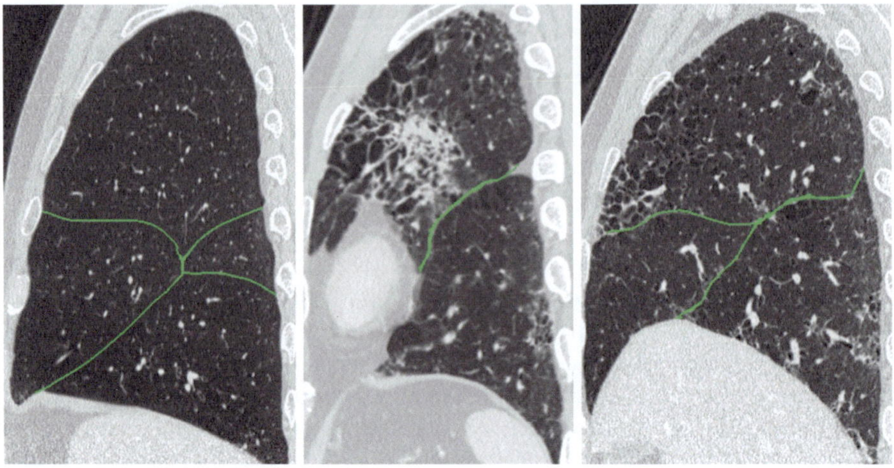

(a) (b) (c)

Fig. 3.58 Extraction results of interlobar fissures. (**a**) Interstitial pneumonia, (**b**) lung cancer, (**c**) hypersegmentation

3.7 Breast

Daisuke Fukuoka, Tomoko Matsubara

3.7.1 Ultrasound Imaging: Classification Methods for Masses

Breast cancer is a common cancer among women all over the world. Combined with mammography, ultrasonography is used to suggest the degree of suspicion of malignancy to help determine whether follow-up or biopsy are necessary. Sonographic features that are assessed include shape, margin, echotexture, posterior acoustic shadowing, and orientation.

Recently, various classification methods using shape and other ultrasonographic features have been reported. Horsch et al. [121] reported a CAD method that is based on the automatic segmentation of lesions and the automatic extraction of four features related to the lesion: shape, margin, texture, and posterior acoustic behavior. Chen et al. [58] reported on methodology based on fractal analysis and k-means clustering. Other classification methods have been proposed that use the statistical properties of echo signals [100, 253, 254, 278, 291, 292]. K-distribution and Nakagami distribution are used for modeling the echo signals. Takemura et al. [278] compiled a total of 208 features for discrimination, including those based on a parameter of a log-compressed K-distribution. Their proposed system classifies types of diseases as cancer, cyst, or fibroadenoma (a common benign entity) using an ensemble classifier based on the AdaBoost algorithm with feature selection.

Tsui et al. [291] examined five contour feature parameters (tumor circularity, standard deviation of the normalized radial length, area ratio, roughness index, and standard deviation of the shortest distance) and calculated the Nakagami parameters estimated from the ultrasonic backscattered signals. The Nakagami parameters are only dependent on the statistical distribution of the echo waveform and are not affected by the echo amplitude. In another paper, Tsui et al. [292] suggested that the Nakagami image can visualize the scattering properties of breast lesions. Figure 3.59 shows the B-mode image and the corresponding Nakagami image of a benign breast tumor.

SSMs of Breast Ultrasound Images

Few studies have focused on SSMs of breast ultrasound images, because of the paucity of landmarks, the deformability of the tissue, and operator dependence. To overcome such problems, several automated whole breast ultrasound scanners have been developed [275, 307]. Figure 3.60 shows an example of a volumetric whole breast image, where a $16 \times 16\ \text{cm}^2$ area is scanned automatically. It is assumed that whole breast scanning techniques will enable the construction of an SSM for the

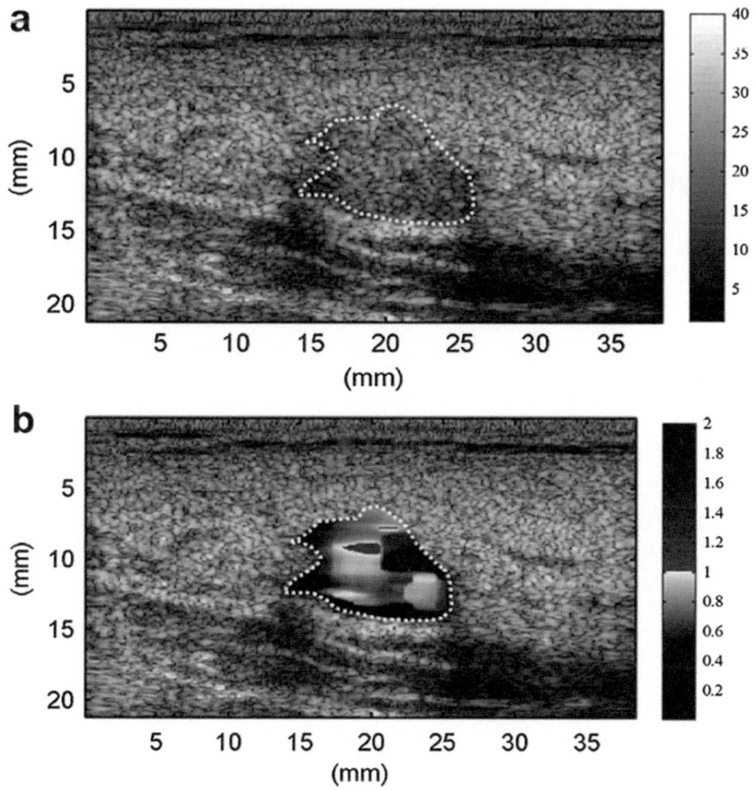

Fig. 3.59 The B-mode (**a**) and Nakagami (**b**) images of a benign breast tumor [292]

(a) (b)

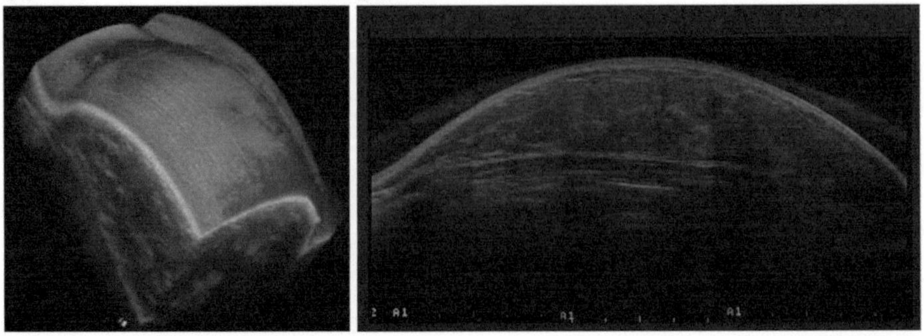

Fig. 3.60 An example of volumetric whole breast image (**a**) volumetric data in a 16 × 16 cm^2 area. (**b**) A whole breast section view [131]

(a) (b)

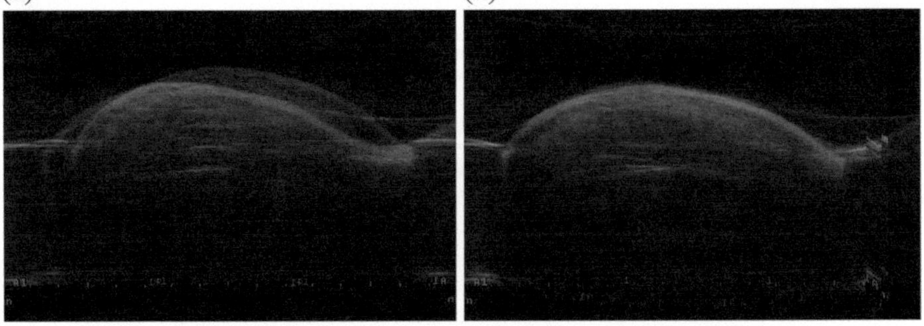

Fig. 3.61 Bilateral breast images are registered with reference to the nipple positions and skin lines. (**a**) Fusion image before registration. (**b**) Fusion image after registration [131]

purpose of assessing symmetry, looking at lesions over time, and fusing ultrasound images with mammograms and MR images.

Ikedo et al. [131] have reported bilateral breast comparison and registration methods for reduction of false positives in a detection system for masses. Normal left and right breasts in the same subject are usually architecturally symmetrical. A mass-like area in a region is classified as normal tissue if the same position in the other breast has a similar feature. Example sections after implementation of the registration are shown in Fig. 3.61.

3.7.2 Mammography

Early detection of breast cancer is of fundamental importance for improved prognosis in the treatment of the disease. Screening mammography is generically regarded as the most effective tool for interpreting early breast cancer. Several ongoing studies to improve the clinical performance of mammography require the simulation of realistic patient breast models. Screening mammography typically involves taking two views of the compressed breast, from above (craniocaudal view (CC)) and from an oblique or angled view (mediolateral oblique (MLO)). The breast shape in the CC view mammograms is sometimes characterized as a semicircle or semiellipse. Although this simplifier has proved appropriate, more sophisticated models that include clinically realistic breast shape variations would be useful. Feng et al. developed models of compressed breasts based on objective analysis. To catalogue the breast shapes of clinically acquired CC and MLO view mammograms, an automated edge detection algorithm is first used. Principal component analysis (PCA) is performed on detected edge shapes. The principal components are described by the eigenvectors $\mathbf{e_j}$, which can be obtained by calculating the following eigenvalue problem:

$$cov\left(\boldsymbol{X}\right)\boldsymbol{e}_j = v_j\boldsymbol{e}_j, \tag{3.14}$$

where $cov(X)$ is a square covariance matrix and v_j is the variance of each principal component. The n principal components with the greatest v_j are selected and their respective $\boldsymbol{e}_j$ assembles into a $100 \times n$ principal component matrix $\mathbf{E}$. The PCA parameter vector r_i for image i can be defined as follows:

$$\boldsymbol{r}_i = \left(\boldsymbol{E}^T\boldsymbol{E}\right)^{-1}\boldsymbol{E}^T\left(\boldsymbol{x}_i - \overline{x}\right), \tag{3.15}$$

The identified principal components are used for developing the breast shape models. The PCA breast shape models of the CC and MLO mammographic views have been found to be able to reproduce breast shape with strong fidelity. The average distance errors (ADEs) of the PCA models of CC and MLO views based on six principal components are 0.90 mm and 1.43 mm, respectively. Pared t-tests of the ADE values of each image between the six-, four-, and two-component models showed that these differences were statistically significant, and the ADE based on the six principal components was the smallest. Histograms of the PCA parameters for six principal components are fitted Gaussian distributions. The authors confirmed that the six-component PCA model could generate realistic breast shapes[1].

Practically all breast cancers originate in the ductal or lobular epithelium, with very few arising in the connective or adipose tissue. Therefore, the ductal network is an important element of breast anatomy. Bakic et al. proposed the realistic simulation of the breast ductal network as part of a computer three-dimensional breast phantom. The synthetic tree model is implemented based upon the description of ductal branching patterns of tree line structures given by ramification matrices (R matrices). Random binary tree (RBT) is used to simulate the ductal network,

[1]The sentences in this paragraph are reproduced from ref. [76] by permission of 2013 American Association of Physicists in Medicine.

consisting of multiple lobes. The RBT model is evaluated by comparing manually traced ductal networks in clinical galactograms and manually tracked networks of synthetic RBTs. As a result, a root-mean-square (rms) fraction error between the R-matrix elements corresponding to clinical and synthetic images is 41%. In addition, it is also analyzed that the synthetic trees generated using R matrices. A comparison of the synthetic and clinical images yields an rms fraction error of 11%. The authors reported that these results suggest the possibility that a more appropriate model of the ductal branching morphology may be developed[2].

3.7.3 Breast MRI

MRI of the breast is increasingly used to monitor high-risk patients. Tracking tumors in different states is useful for obtaining important clinical information. Computational models of the deformation of the breast in different positions are being developed. Malcolm et al. investigated the use of partial least squares regression (PLSR) to predict breast deformation from the prone to the supine positions [185]. Meshes of prone breast images are fitted to data segmented from T1WI. Because of unavailability of supine MRI data in their database, the associated supine geometrics are generated using finite element models (FEM) mechanics simulation. PLSR trains a statistical model of deformation using the population of associated prone and supine models from these FEM simulation results. It is clear that the PLSR approach has the potential to be a reliable alternative to FEM because the volume averaged geometric and relative errors between the PLSR predicted supine models and the associated FEM solutions were 1.9 ± 0.7 mm and $12 \pm 7\%$, respectively. In addition, the PLSR predictions were five orders of magnitude faster than the FEM solution.

Breast density is an important risk factor for developing breast cancer. Khalvati et al. developed multi-atlas-based breast MRI segmentation [152]. To have a diverse atlas, the training images with the manual segmentation for the whole breast were first clustered based on the similarity of the corresponding phase congruency maps (PCMs). The phase congruency of an image at location $\boldsymbol{x}$ ($PC(\boldsymbol{x})$) can be calculated as follows:

$$PC(x) = \frac{\lfloor E(x) - T \rfloor}{\sum_n A_n(x) + \varepsilon}, \tag{3.16}$$

where E($\boldsymbol{x}$) is the local energy of the image, T is a threshold to suppress the effect of noise on the local energy of the image at that location, A_n presents the amplitude of the nth Fourier component, and ε is set to a small number to avoid division by 0. Afterward, group-wise registration is performed on the original images in each

[2]The sentences in this paragraph are reproduced from ref. [26] by permission of 2003 American Association of Physicists in Medicine.

class based on the distance map of the corresponding PCMs to create the mean image and the corresponding mean label. For segmentation of a target image, it is first determined which class a target image belongs to. To generate the segmentation results, the target image and centroid of the class are aligned in order by a nonrigid registration. The segmentation results for the entire volume were compared with the ground truth results to evaluate the performance of this algorithm, and an average Dice similarity coefficient (DSC) of 0.93 was achieved.

Wu et al. proposed a fully automated segmentation of fibroglandular tissue and estimation of volumetric density using an atlas-aided fuzzy C-means (FCM-Atlas) method. Firstly, an initial voxel-wise likelihood map of fibroglandular tissue is produced by applying FCM clustering to the intensity space of each 2D MR slice. To achieve enhanced segmentation, a prior learned fibroglandular tissue likelihood atlas is incorporated to refine the initial FCM likelihood map. An updated likelihood map u_{ij}^{*r} can be calculated as follows:

$$u_{ij}^{*r} = u_{ij}^{*} \cdot (W \cdot A), \tag{3.17}$$

where u_{ij}^{*} is the FCM-generated likelihood map, W represents the warping transformation that deforms the standard atlas to the shape of the specific breast being processed, and A is an overall fibroglandular likelihood atlas. The absolute volume of the fibroglandular tissue (|FGT|) and the amount of the |FGT| relative to the whole breast volume (FGT%) of this proposed method were compared with that of manual segmentation obtained by two experienced breast radiologists. The automated segmentation achieved a correlation of $r = 0.92$ for FGT% and $r = 0.93$ for |FGT|, which were not significantly different from the manual segmentation. In addition, it was also clear that the segmentation performance was stable both with respect to selecting different cases and to varying the number of cases needed to construct the prior probability atlas by the additional robustness analysis[3].

Segmentation of pectoral muscles is important for volumetric breast density estimation and for pharmacokinetic analysis of dynamic contrast enhancement. Gubern-Mérida et al. developed two atlas-based pectoral muscle segmentation methods in breast MRI [111]. One method is based on a probabilistic model and the other method is a multi-atlas registration-based approach. The probabilities of the atlas are first mapped by registration process composed of two stages. The first stage is a translation transform and the second stage is a nonrigid transform based on B-splines registration. Subsequently, two atlas-based segmentation methods are performed. In the probabilistic atlas-based segmentation, method 1, a probabilistic atlas is used in a Bayesian framework and is created by computing the frequency with which each location is labeled as pectoral muscle. The probabilistic atlas, the tissue models, and the target are supplied to the Bayesian framework as a prior probability $P(X)$, conditional probability $P(Y|X)$, and a set of intensity values Y, respectively. The Bayesian framework estimates the segmentation X that maximizes

[3]The sentences in this paragraph are reproduced from ref. [316] by permission of 2013 American Association of Physicists in Medicine.

$P(X)P(Y|X)$ and also includes a Markov random field regularization. Multi-atlas segmentation, method 2, approaches consist of two steps after mapping all the atlases onto the target space. First, the deformed anatomic images are compared with the target to select the most similar atlases. The selection is based on the normalized cross correlation similarity measure, and a ratio is calculated as follows:

$$r_i = \frac{NCC\,(T, A_i \circ M_i)}{\max_j NCC\,(T, A_j \circ M_j)}, \tag{3.18}$$

where M is the mapping between the target and an atlas, j is the deformed atlas with maximum similarity, and T is target volume. The selected deformed atlas labels are fused to yield a single final segmentation of the patient or target image. The probabilistic and the multi-atlas segmentation frameworks were evaluated in a leave-one-out experiment. The multi-atlas approach performed slightly better, with an average DSC of 0.74, while, with the much faster probabilistic method, a DSC of 0.72 resulted. The authors stated that both atlas-based segmentation methods have high reliability because of their DSC values being higher than the computed interobserver variability.

3.8 Cardiac

Hidenobu Suzuki, Yoshiki Kawata, Noboru Niki, Ryo Haraguchi and Katsuda Toshizo

3.8.1 *Morphologic and Functional Modeling of the Heart*

The heart functions by cycling through contraction and expansion. There is a close association between morphology and function. Electrophysiology is also important in cardiac function. The presence of a conduction system allows the heart to modify its output to react to different requirements. For computational modeling of the heart, it is necessary to consider both the morphological and functional anatomies; however, building a model that integrates all scales (from molecular to organ) and all phenomena (mechanics, fluid, electrophysiology, and molecular dynamics, among others) is impossible. Clarification of the functions of interest is important.

In clinical practice, assessment of the contractile function of the left ventricle (LV) is important for diagnosing cardiovascular disease. The most widely used index of the LV contractile function is the LV ejection fraction (LVEF). The LVEF is derived from end-diastolic volume (EDV) and end-systolic volume (ESV). The LVEF can be calculated using various noninvasive cardiac imaging modalities, including MRI, CT, SPECT, and echocardiography. Therefore, numerous segmentation techniques for the LV myocardium, including statistical techniques, have been proposed.

Much work is still required to gain an integrated understanding of the normal heart. The computational modeling of the diseased heart has just begun. Because of the complexities of cardiac structure, such as rotational myocardial fibers, and the conduction system, a shape model is not always sufficient to represent various disease states.

In this section, we provide some examples of morphologic and functional modeling of the heart.

Processing Algorithms and Imaging Specific to the Heart There are several noninvasive cardiac imaging modalities. In most modalities (CT, MRI, PET, and SPECT), data acquisition is performed over a few heartbeats with electrocardiographic (ECG) gating. Moving cardiac images can then be reconstructed using the gated data.

For segmentation of cardiac images, it is necessary to consider several points specific to the heart. Epicardial delineation is more difficult than endocardial delineation because of poor contrast and fuzzy boundaries between the heart and other tissues. Endocardial delineation needs intelligent processing because of the presence of the papillary muscles that control the valve leaflets, and myocardial trabeculation, which gives the endocardium its irregular surface. The right ventricle (RV) wall is thinner than the LV wall, making segmentation more difficult. In echocardiography, the limited field of view prevents acquisition of data including the entire organ.

Many methods have been proposed for computational morphologic and functional modeling of the heart. These methods are classified into four categories: (1) statistical models, (2) deformable models/level set, (3) biophysical models, and (4) nonrigid registration using basis functions [149, 284]. The use of statistical models for segmentation has some advantages, such as robustness for regional low contrast, intelligent processing that excludes the papillary muscles, and interpolation for outside the field of view.

Depending on the clinical question, other MR sequences such as DTI (assessing orientation of the myofibers), tagging imaging (assessing myocardial contraction), and velocity encoding imaging (motion of the blood/myocardium) can be performed.

Example 1: Active Appearance Motion Models Bosch et al. [41] proposed the active appearance motion model (AAMM) technique, which allows fully automated continuous delineation of LV endocardial contours over the heart cycle from echocardiographic images. The AAMM describes both image appearance and object shape within the dynamics of the heart cycle. The authors used 129 infarct patients' echocardiographic transthoracic four-chamber sequences with manually defined LV contours. They split the datasets randomly into a training set of 65 patients and a testing set of 64 patients. The AAMM was generated from the training dataset. The generated AAMM was applied to segmentation of the 64 sequences and successfully matched 62 patients (97%). The example results are shown in Fig. 3.62. Statistical models built by machine learning (e.g., AAMM) are useful for automatic and robust segmentation of cardiac images.

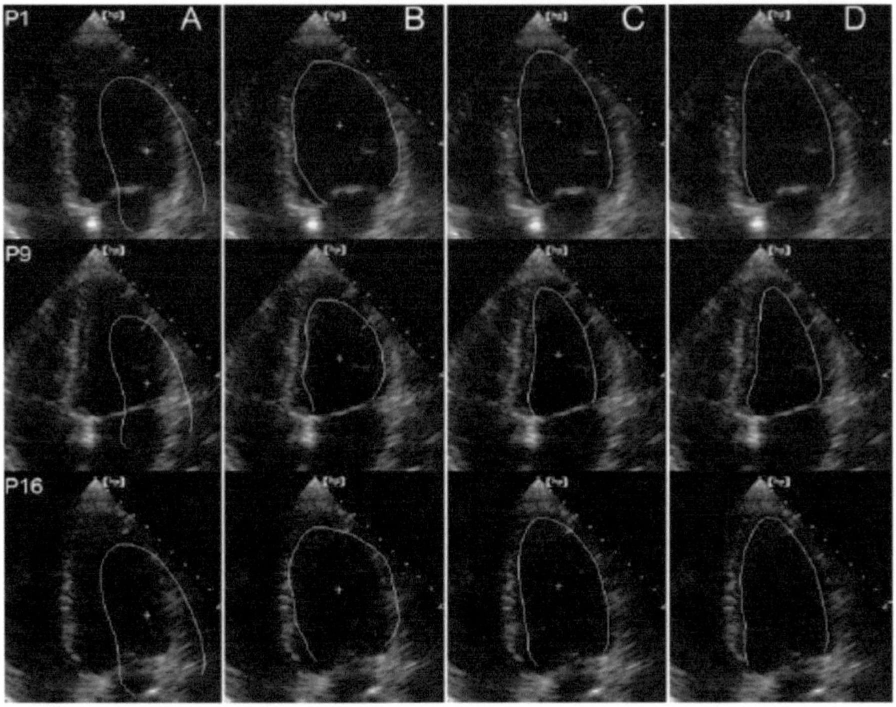

Fig. 3.62 Fully automated segmentation results obtained by applying the AAMM to an echocardiographic image sequence. (**a**) Initial AAMM model positioned on phase images 1, 9, and 16. (**b**) AAMM match after five iterations. (**c**) Final match after 20 iterations. (**d**) Manual contours for comparison [41]

Example 2: Modeling Contractility Wenk et al. [306] built a finite element model (FEM) of a patient's beating infarcted LV and measured regional myocardial deformation with three-dimensional tagging MRI (Fig. 3.63). They showed evidence of depressed contractility in the border zone of a myocardial infarction by combining MR tagging imaging and computer simulation. The combination of biomechanical computational modeling and noninvasive functional imaging techniques can be a powerful methodology to clarify the mechanisms of cardiac diseases.

Example 3: Modeling Myofibers The spatial arrangement of myofibers within the myocardium, which is termed "fiber orientation," must be taken into account for better understanding of cardiac electrophysiology patterns, mechanical function, and remodeling processes in the living/modeled heart. Lombaert et al. [182] built a statistical atlas of myofiber architecture with a human dataset of ten healthy ex vivo hearts (Fig. 3.64). The myofibers were imaged with diffusion tensor magnetic resonance imaging (DT-MRI). They used isolated hearts filled with hydrophilic gel to preserve the diastolic volume. In vivo DT-MRI of a beating heart is under

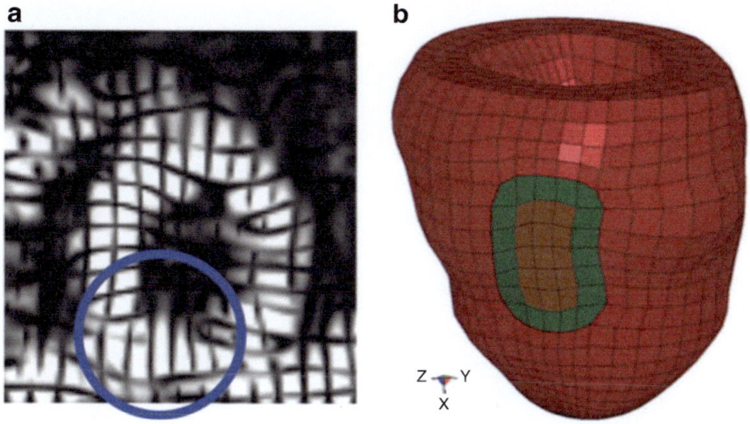

Fig. 3.63 (**a**) Short-axis view from 3D tagging MR image of patient LV. (**b**) Finite element model with infarct (*brown*) and border zone (*green*) regions [306]

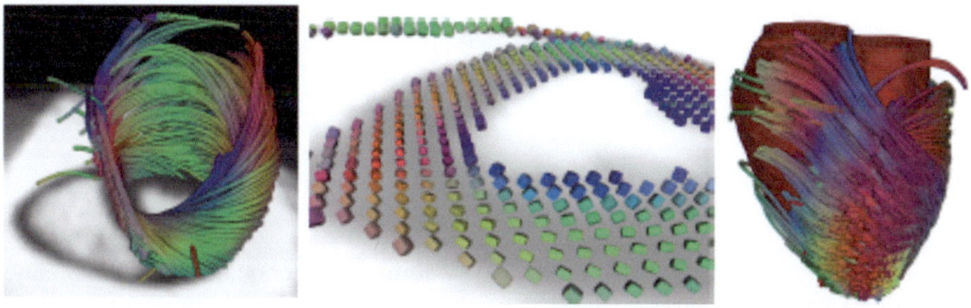

Fig. 3.64 A statistical atlas of human myofiber architecture, (*left*) fiber tractography of the left ventricle, (*center*) close-up of the fiber orientations of a short-axis slice, (*right*) fiber tractography around the left ventricular blood pool [182]

development, though the availability of isolated human hearts is extremely rare. Therefore, their statistical myofiber model is valuable.

Example 4: Interactive Modeling for Congenital Heart Disease Congenital heart diseases (CHD) involve developmental abnormalities of the heart and/or great vessels present since before birth. Early treatment is often necessary. Because of its real-time imaging capabilities, 2D echocardiography is often used to diagnose CHD. CT and MRI are unsuitable for real-time diagnosis. 3D echocardiography is sometimes insufficient for detailed imaging of CHD. Only an experienced physician could diagnose from the 2D echocardiographic images based on a spatial perception of the 3D heart; however, it is difficult to transfer the specialist's spatial perception of the 3D heart structure to other medical staff. There is no effective method of communicating the condition of an individual CHD patient, and sharing the special perception is difficult. Haraguchi and Nakao and Nakao et al. [115, 213] proposed a

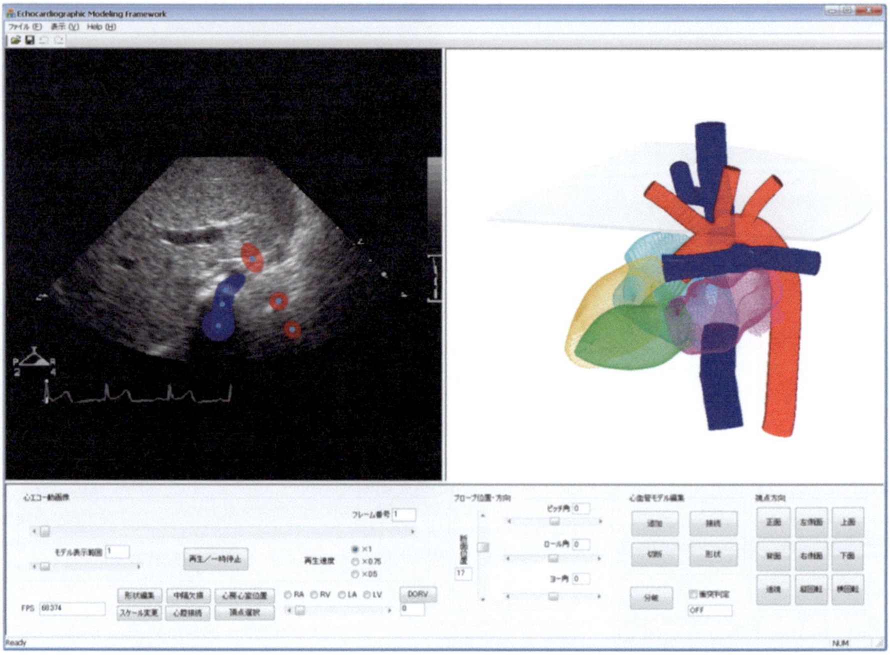

Fig. 3.65 The heart and great vessels interactive modeling system for CHD

3D heart and great vessels rapid modeling system using echocardiographic images with added simple interaction with the model by the operator (Fig. 3.65). They focused on the expression in a 3D format of the disease state as seen in a mental image by an experienced physician. Physicians can interactively construct patient-specific heart and great vessel models within a practical time frame and share the complex topology. This model cannot represent precise morphological information (e.g., vessel diameters). Nevertheless, it is an example of interactive modeling that may be useful in a particular application.

In the Future Several statistical models of the heart have been proposed for segmentation and registration. These computational models are useful for automation in determining clinically significant indices describing the disease state.

A shape model is not always sufficient to represent the various disease states (e.g., ischemic heart disease and dysrhythmia). The heart has a complex internal structure (e.g., myocardial fiber direction and trabeculation) and involves various phenomena (e.g. mechanics, fluid, and electrophysiology). Developing a biophysical model may help produce an integrated understanding of normal and diseased hearts [18, 129, 296].

To improve cardiac imaging, radiological technologist skills are also required [273, 274]. For example, in MRI, it is necessary to consider many parameters, physical conditions (e.g., quantity of the heterotopic fat) of the patient, heart rate, and breathing-related complicated movement, among other factors.

3.8.2 Coronary Arteries

Coronary artery disease is a major cause of death worldwide [238]. If a coronary artery becomes narrowed or occluded owing to the buildup of plaque (e.g., calcium, fat, and cholesterol), or the formation of a thrombus, the blood flow to the myocardium will be reduced. Restriction of oxygenated blood flow is called ischemia, and the narrowing of a vessel is called stenosis.

In current clinical practice, conventional coronary angiography (CCA) via cardiac catheterization is considered to be the gold standard imaging technique to diagnose coronary artery disease [51]. Computed tomographic angiography (CTA) is a potential alternative to CCA [203]. CTA is a noninvasive technique that allows assessment of the coronary lumen and the evaluation for the presence of coronary calcifications and other causes of stenosis [171].

Recently, non-contrast CT has been used for mass screening for lung cancer [148, 286]. CT images are also useful in the quantification of coronary calcification [30]. Coronary artery calcium is graded by Agatston score, volume, mass score, or density [37, 62] for risk stratification for future cardiac morbidity and mortality [10].

Coronary Artery Anatomy The coronary arteries supply oxygenated blood to the myocardium. The right and left main coronary arteries exit the ascending aorta from ostia just above the right and left aortic valve leaflets [5]. These two branches subdivide and traverse the epicardium. The American Heart Association (AHA) Writing Group on Myocardial Segmentation and Registration for Cardiac Imaging divides the coronary arteries into 15–16 segments as shown in Fig. 3.66 [78].

Detection of Coronary Arterial Stenoses with CTA This section describes algorithms for segmentation of coronary arteries and detection of coronary arterial stenoses with CTA. Research on coronary artery segmentation have implemented several methodological solutions: topological thinning [57], particle filtering [79], graph-based analysis [272], fuzzy connectedness [300], vessel tracking and active contours [172], minimal cost path computation [205, 226], mathematical morphology [43], hybrid strategy using multi-scale filtering and Bayesian probabilistic approach with level set model [318], multi-scale enhancement and dynamic balloon tracking [333], and two-stage shape regression [247].

After segmentation, assessment for stenoses is performed. The approaches include a 3D level set [7, 8], skeletonization and geometric analysis of a branch [57], morphological filtering and interactive masking [42], and fuzzy distance transform [317].

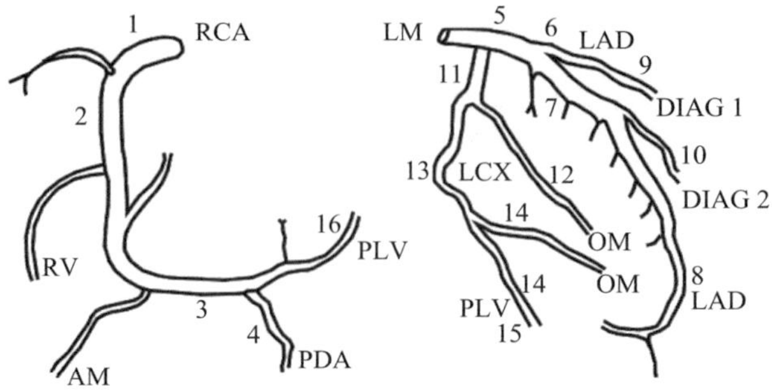

Fig. 3.66 Segments of coronary arteries. [78] *RCA* right coronary artery, *RV* right ventricular branch, *AM* acute marginal branch, *PLV* posterolateral ventricular branch, *PDA* posterior descending artery, *LCA* left coronary artery, *LM* left main coronary artery, *LAD* left anterior descending artery, *DIAG1* first diagonal branch, *DIAG2* second diagonal branch, *LCX* left circumflex artery, *OM* obtuse marginal branches

The segmentation algorithm by Schaap et al. is a coarse-to-fine robust shape regression approach [247]. First, the method is initialized with an approximate centerline [246]. The cross-sectional planes of vessels are generated based on the centerline. Next, vessels are represented by a combination of local SSMs of the vessels' appearance and shape. Then, the coarse shape of the vessel is estimated with linear multivariate regression. Finally, the vessel shape is refined with nonlinear regression by optimizing the landmark coordinates to their most likely position.

Detection of Coronary Calcification from Non-contrast-Enhanced CT Images This section describes detection algorithms for coronary calcifications from non-contrast CT images which are used in mass screening for lung cancer. The approaches of previously published methods involve a neural network [294] and two-stage classification with feature selection [132].

For the detection of coronary calcifications from thick-section CT images (10 mm slice thickness), Ukai et al. proposed an algorithm that is composed of four processes [294]. First, a CT image set is divided into three volume sections (upper, middle, and lower thirds of the heart) by a neural network (four inputs: slice position, heart shape, scapula, CT value uniformity), which is trained using the back propagation algorithm. Second, each section is segmented, using the information from the adjacent lung and vertebral body. Third, the candidate regions for the coronary calcifications are detected using a weight coefficient map consisting of a prior probability of the location for the coronary artery as shown in Fig. 3.67.

This probability is determined based on the distribution of the coronary arteries, which were manually segmented from 80 patients' heart regions normalized in terms of height and width. Finally, the artifact regions included in the candidate regions are excluded by the diagnostic rule based on a neural network.

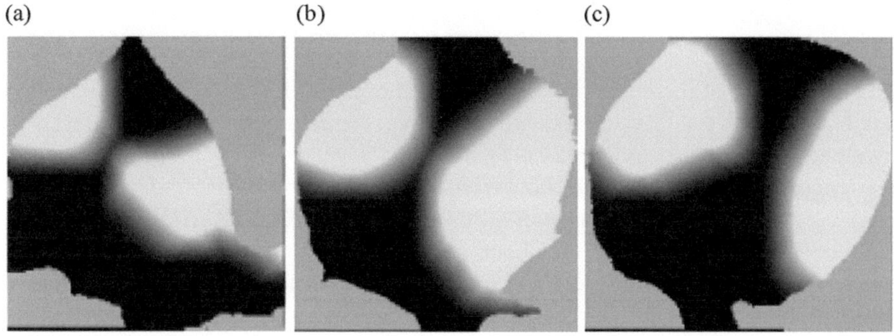

Fig. 3.67 Weight coefficient map. (**a**) Weight coefficient map in upper part of the heart, (**b**) weight coefficient map in middle part of the heart, (**c**) weight coefficient map in lower part of the heart. The value is a prior probability of the location of the coronary artery

For the detection of coronary calcifications from thin-section CT images (3 mm slice thickness), Isgum et al. proposed an algorithm that is composed of three processes [132]. First, candidate objects are extracted using threshold of the intensity and size. Second, a set of features is calculated. This set is composed of volume, shape features (three eigenvectors, λ_1, λ_2, λ_3), spatial features (the coordinate system is defined by determining the smallest box around the heart), and appearance features (maximum intensity, average intensity, derivatives (L_x, L_y, L_z, L_{xx}, L_{xy}, L_{xz}, L_{yy}, L_{yz}, L_{zz}). Finally, candidate objects are classified into either positive or negative objects using a two-stage classification system with a k-nearest neighbor classifier and a feature selection scheme (sequential floating forward feature selection[SFFS]) [135]. In this study, 14 efficient features (six spatial and eight appearance features) were employed by the feature selection.

This section describes technologies based on computational anatomy model for the diagnosis of coronary artery disease. Clinical applications including these algorithms can improve the diagnostic performance of coronary artery disease.

3.9 Abdomen

Yoshinobu Sato, Akinobu Shimizu, Kensaku Mori and Takayuki Kitasaka

This section describes algorithms for analysis of the abdominal organs using CT and MR data. Segmentation algorithms of the organs are described, and then extraction and analysis of internal structures, that is, vessels and abnormal regions, are discussed for the purpose of diagnosis and surgical planning assistance.

3.9.1 Liver

The liver is one of the most-studied abdominal organs with respect to segmentation and further analysis. Early investigations of liver segmentation were based on slice-by-slice 2D image processing [24, 98]. Since the development of multi-detector row CT (MDCT) in the late 1990s, however, volume imaging of the abdomen, that is, image acquisition with a thin-section thickness, regarded as 3D isotropic imaging, has become popular. Therefore, liver segmentation methods based on fully 3D image processing were developed [186, 264]. They demonstrated potential clinical utility in tumor detection and surgical planning. Later on, two noteworthy papers on liver segmentation were published [168, 229], in which computational anatomy (CA) models, that is, a probabilistic atlas [229] and an SSM [168], were used as prior knowledge of the liver shape and location to formulate the segmentation problem as a MAP estimation based on Bayesian theorem. Representative methods for liver segmentation typically use an SSM or a probabilistic atlas [118, 141, 331]. The remaining part of this section describes automated liver segmentation algorithm using CA models from upper abdominal CT data and then mentions applications to diagnosis and surgical planning assistance.

Anatomy of Organs and Tissues Adjacent to the Liver The liver is the largest abdominal organ, located in the right upper quadrant of the abdominal cavity, just under the diaphragm, which separates the thoracic and abdominal cavities. The shapes of the liver dome and right lung base are strongly constrained by the diaphragm's variations in contour during the respiratory cycle. The liver consists of large right and left lobes and small caudate and quadrate lobes. The shape of the right liver dome correlates closely with the right lung base via the intervening diaphragm. The contour of the inferior surface of the liver contains impressions from the right kidney, duodenum, gallbladder, and right colon. The liver is also surrounded by the lower ribs and abdominal musculature. Although CA models should deal with the above multi-organ relations systematically by mathematical models, we focus on the liver and utilize these relations in a manually specified manner in this subsection. The topic of modeling multi-organ relations will be addressed in a subsequent subsection.

CA Models of the Liver and Their Application to Segmentation from CT Images CA models, typically probabilistic atlases and SSMs, are represented in the reference frame (coordinate system), which needs to be determined from input CT volume when they are utilized. There are two main approaches to define the reference frame, that is, an organ-centered frame and an external frame. The organ-centered frame is defined by features inherent in the target organ, and shape priors are represented by probabilistic atlas and the SSM. The external frame is defined by features of external structures (such as the lungs and ribs), and both shape and location priors are represented, where the locations of the target organ are represented relative to the external structures.

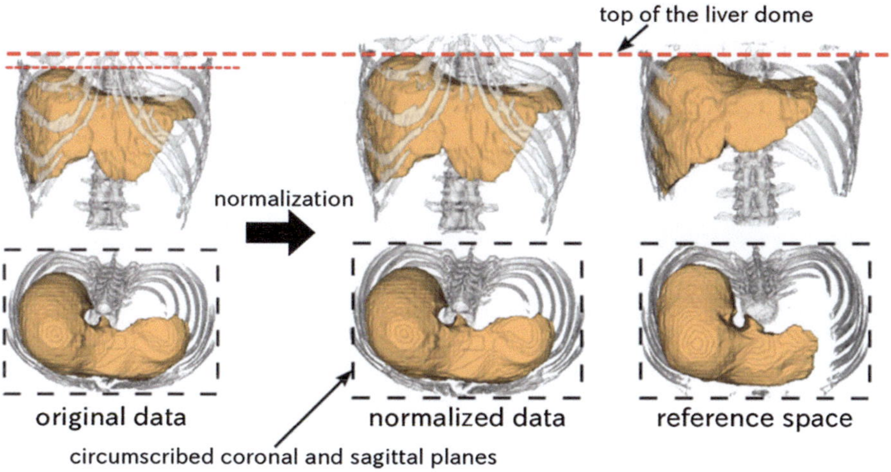

Fig. 3.68 Example of external reference frame for spatial normalization [222]

Figure 3.68 shows the determination processes of an example of the external reference frame [221, 222]. The reference space is defined based on the upper abdominal cavity, whose shape and location are constrained by the diaphragm, bones, and abdominal musculature. To determine the reference space, an approximated bounding box of the abdominal cavity is utilized, whose top plane is constrained by the diaphragm and side planes by the bones. More specifically, the upper plane corresponds to the axial plane tangential to the right dome of the diaphragm (which covers the liver surface), and the four side planes correspond to the sagittal right- and leftmost planes and the coronal front and back planes of the musculature and ribs. The lateral and anteroposterior dimensions of the reference space are normalized to the mean dimensions of all the patients. The right dome of the diaphragm separates the dome of the right lobe of the liver and the base of the right lung. Thus, it can be extracted by locating the right lung base. The lung and bone regions are not affected by contrast agents and are well delineated from other tissues in CT data because of their very low and very high densities, respectively. Therefore, these regions can be segmented in a stable manner from CT data irrespective of differences in contrast enhancement protocols. All the patient CT volumes are translated and scaled so as to be aligned to the normalized reference space, which defines the reference frame. This initialization process is called *spatial normalization* or *spatial standardization*.

Once spatial normalization has been completed, CA models such as probabilistic atlases and SSMs can be applied to the target CT data to perform segmentation. Figure 3.69 shows a probabilistic atlas and SSM of the liver, which are represented in the reference frame based on the bounding box approximating the upper abdominal cavity. Segmentation using CA models is formulated as a MAP estimation by combining data fidelity terms, which are typically derived from intensity models

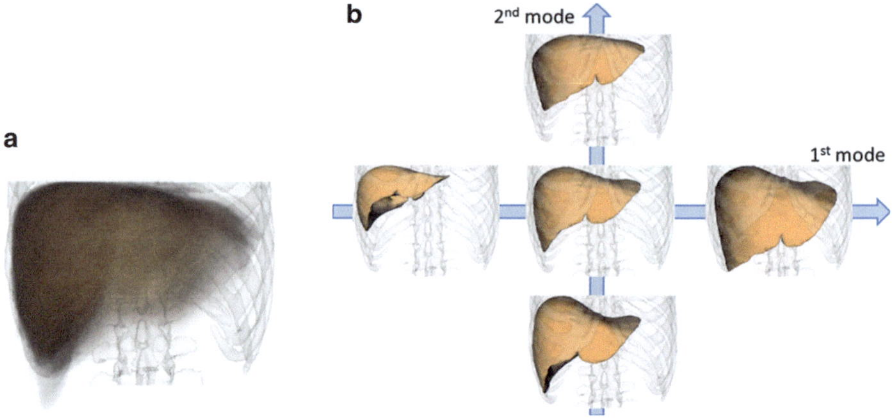

Fig. 3.69 Computational anatomy (CA) models of the liver. (**a**) Probabilistic atlas. Volume rendering of the probabilistic atlas is displayed by assigning opacities proportional to probabilities to voxels. (**b**) SSM. The shape located at the center is the mean shape, and the shape variations of the first and second modes are indicated

or edge localization error models, as discussed in Chap. 2. Segmentation using a probabilistic atlas is regarded as voxel-wise MAP estimation, which does not require any initialization and nonlinear optimization (see Fig. 2.17 for typical segmentation processes in Chap. 2). Conversely, SSM requires initial values for the shape parameters for their nonlinear optimization. Although the average shape is often used for the initial values for an SSM, it is desirable to combine some methods to avoid stacking in poor local minima. Typically, edges to be fitted to SSMs are searched for near SSM surfaces while SSMs are being fitted to input images. The initial SSM needs to be sufficiently close to the true organ surfaces for successful edge search. The mean shape, however, may sometimes largely deviate from the true shape and fail to capture the true edges. To overcome this problem, probabilistic atlas-based segmentation results can be used for initial values for SSM-based segmentation, instead of the mean shape. Then these parameters are used for the initial values for subsequent SSM-based segmentation [222].

The criteria for evaluating the performance of SSM are known as *specificity* (property of maintaining the characteristics specific to the organ shape) and *generality* (property of representing any shapes accurately). Specificity and generality are a trade-off. To overcome this, a hierarchical SSM (H-SSM) has been developed [221, 222]. A coarse-to-fine strategy is adopted for the H-SSM, in which the top level of SSM is first fitted, and then its estimated shape parameters are used as initial values for subsequent fitting of sub-shape SSMs. Figure 3.70 shows an H-SSM of the liver, in which the whole liver shape is gradually decomposed into sub-shapes and SSMs of the sub-shapes, in addition to the whole liver shape, are constructed. By using hierarchical SSMs, generality and specificity are better balanced. While the top level of SSM is more specific to the liver, the lower levels of SSMs are more accurate in terms of generality.

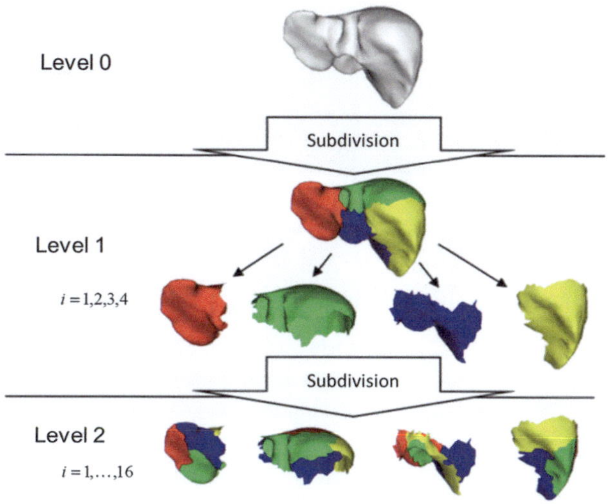

Fig. 3.70 Shape decomposition for hierarchical SSMs (H-SSMs) [222]

Figure 3.71a shows the result of probabilistic atlas-based segmentation, in which initialization was not required after the spatial normalization. Figure 3.71b shows the result of subsequent H-SSM-based segmentation, in which the initialization was provided by the probabilistic atlas-based segmentation result. Figure 3.71c shows a result of graph cut refinement (described in Chap. 2) for the H-SSM-based segmentation result. In Fig. 3.71, clear improvements of segmentation accuracy are observed as the segmentation progresses from (a) to (c).

So far, the CA models of the liver have been generic; that is, they are assumed to cover the variabilities in various patients except for spatial normalization. If additional information on a patient is provided, however, we can make the CA models to represent a patient-specific variability, in which generality and specificity will improve by assuming conditions specific to the patient of interest. One way to construct a patient-specific CA model is to use intermediate results during the segmentation process [287]. In the previously mentioned method, probabilistic atlas-based segmentation was first performed, and its result was used as an initial state for subsequent SSM-based segmentation. In this method [287], the probabilistic atlas-based segmentation result provides conditions specific to the patient, which are used to generate a conditional SSM [66] (described in the previous section) adaptive to the patient. The feature parameters are calculated on the gross shape of the liver observed from the probabilistic atlas-based segmentation result, i.e., the object volume, the area of the projected object in the coronal plane, the 50th percentile point of the x-coordinate, and so on. Figure 3.72 shows a schematic diagram of the approach. Given the observed conditions as features obtained from the target CT data, the conditional SSM is generated specifically to the given conditions. The right bottom frame of Fig. 3.72 shows a simplified example of a

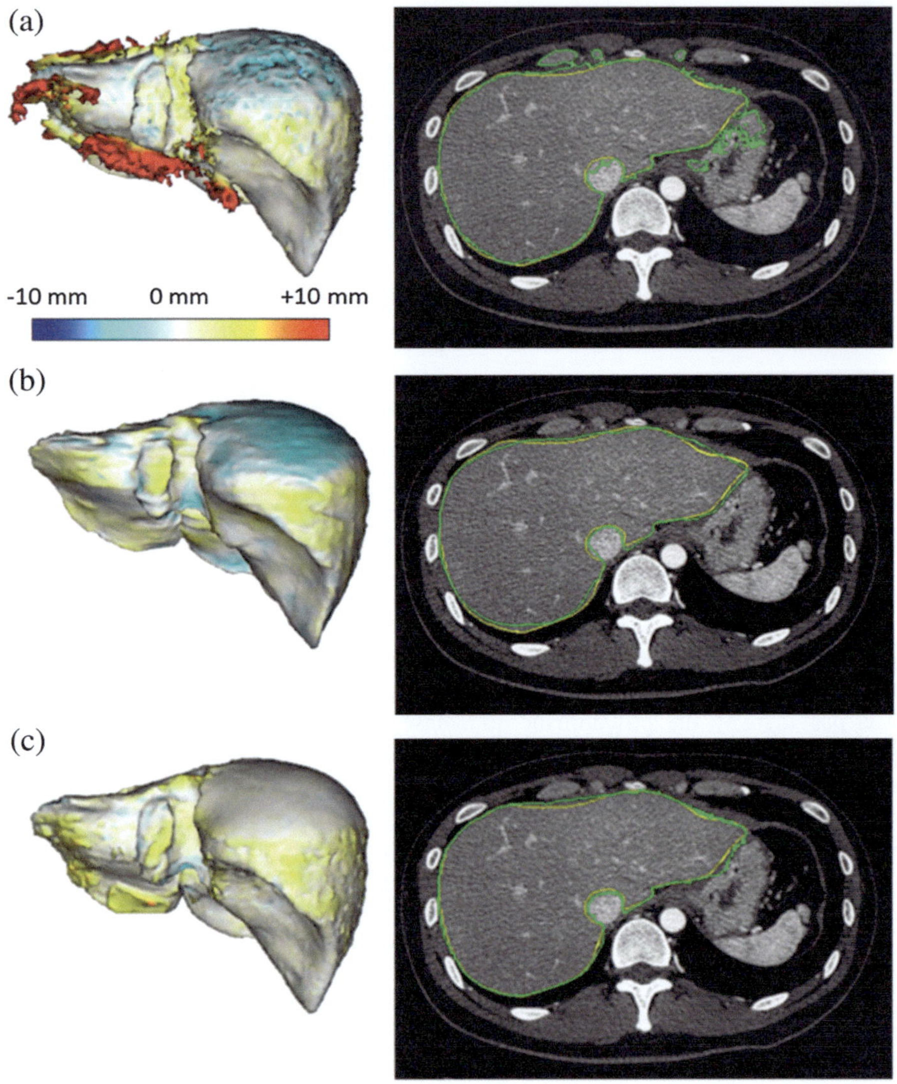

Fig. 3.71 Segmentation of the liver from CT volume. *Left*: 3D shape of segmented liver region. Distance error from the manual trace (ground truth) surface is color-coded according to the indicated color bar. *Right*: Yellow and green contours denote manual trace and automated segmentation, respectively. (**a**) Probabilistic atlas-based segmentation. (**b**) H-SSM-based segmentation. (**c**) Graph cut refinement

conditional SSM of the liver, where the feature $\boldsymbol{x}_0$ is the lateral dimension of the bounding box of the liver region obtained by probabilistic atlas-based segmentation. Because the error from the width of the true liver shape is unavoidable in $\boldsymbol{x}_0$, the error model of $\boldsymbol{x}_0$ is combined with the conventional conditional SSM so as to

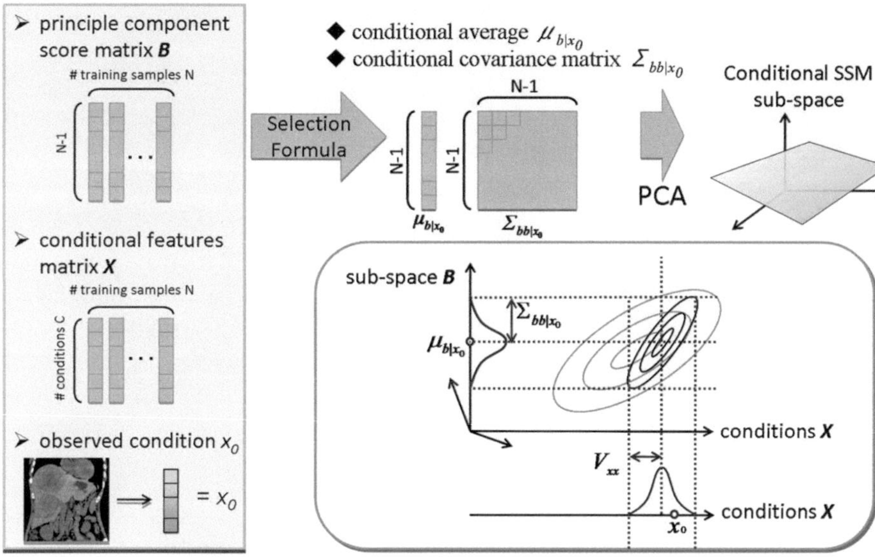

Fig. 3.72 Schematic diagram of conditional SSM generation from observed condition [287]

incorporate specificity inherent in the patient while maintaining sufficient generality. Therefore, this conditional SSM represents the remaining ambiguity and expected error after probabilistic atlas-based segmentation rather than inter-patient variability. These generated conditional SSMs are shown to be particularly useful for accurate segmentation of livers with largely deformed shapes [287].

Role in Diagnosis and Therapy of Liver Diseases Computational anatomy approaches including machine learning are useful for characterizing diffuse and focal liver abnormalities such as cirrhosis and tumors. Surgical planning for tumor resection needs precise patient-specific anatomy, including relations of tumors to vessels, and locations in anatomical segments of the liver. In the following, modeling and application for diagnostic and therapeutic assistance are described.

Computer assistance is most commonly used to target tumors. There are two main approaches for tumor detection and segmentation. One approach assumes pre-segmentation of the liver region and the other does not. The former fails if the pre-segmented liver misses the tumor region(s), which often occurs because tumors usually have different intensity properties compared with normal tissues. The nonsegmented approach may suffer from more false positives due to larger search areas. In the former approach [179], inaccurate segmentation areas in the initial conventional liver segmentation are detected by incorporating the proposed shape ambiguity measure in subsequent level set segmentation. Figure 3.73 shows a typical

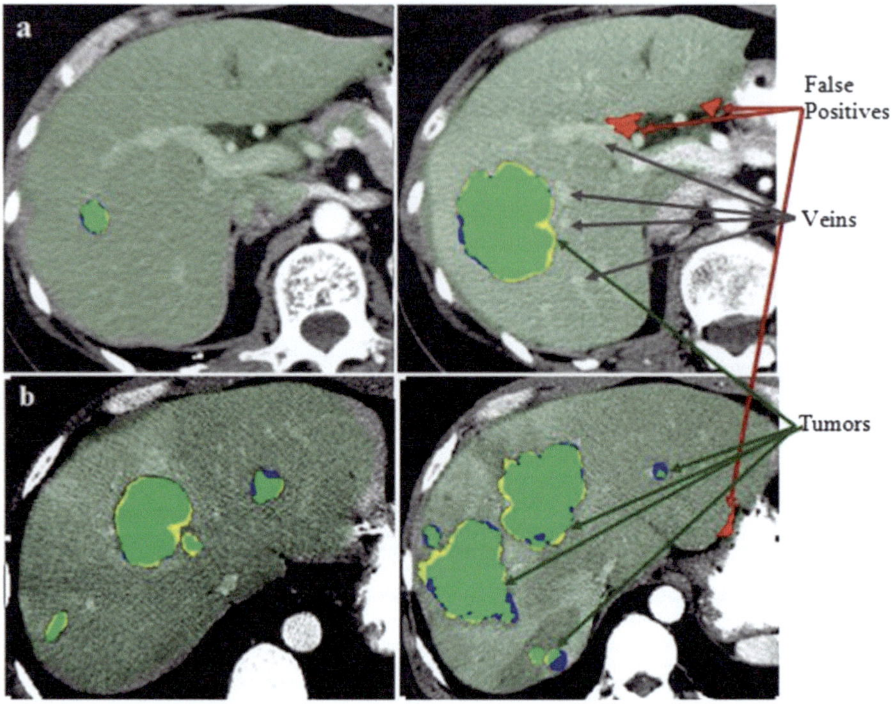

Fig. 3.73 Segmentation of the liver and tumors [179]. *Blue* and *yellow* regions are manually and automatically traced regions, respectively, and *green* regions are the overlapped regions. *Red* regions denote false-positive detections by the automatic tumor segmentation

result of segmentation of the liver that includes tumors. The tumor segmentation is performed inside the segmented liver regions. In one representative method of the latter, ensemble learning was used for segmentation of tumor regions [257]. Recently, a robust statistics mechanism was incorporated in ensemble learning to significantly improve tumor segmentation accuracy [260].

Liver cirrhosis/fibrosis is one of the important diseases of the liver. While biopsy is still regarded as the method for definitive diagnosis, some noninvasive diagnostic methods, such as ultrasound/MR elastography and blood tests, are showing increased progress in diagnostic accuracy. Liver morphology can provide useful diagnostic information because cirrhotic livers, which may initially enlarge, subsequently shrink in size and are known to show characteristic shape deformations [86] as shown in Fig. 3.74, which shows 3D visualizations of healthy (fibrosis stage 0) and cirrhotic (fibrosis stage 4) livers. The traditional quantitative imaging method based on the ratio of the left to right lobe volumes [22] was successful to some extent. SSMs are expected to well capture the characteristics of the organ shape, and some successful results are reported in the brain [269]. One method to use shape deformations for fibrosis quantification is to relate the shape parameters of SSMs

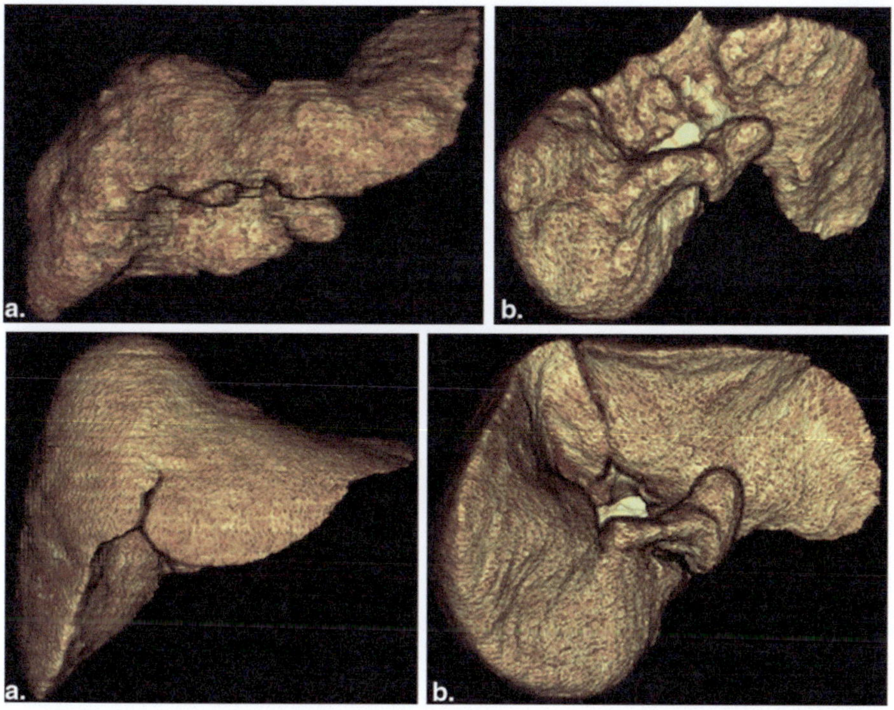

Fig. 3.74 3D visualizations of liver shape (**a**) side view, (**b**) bottom view [120]. *Top*: Typical cirrhotic liver (fibrosis stage 4). *Bottom*: Healthy liver (fibrosis stage 0)

with the fibrosis stage by using support vector regression (SVR) [120]. Accuracy in fibrosis stage estimation improved by adopting the method using SSMs and SVR (approximately 90% in sensitivity and specificity) [120] in comparison with the traditional method based on the left and right lobe volumes (approximately 60–80% in sensitivity and specificity).

Understanding patient-specific segmental anatomy of the liver is important, especially for surgical planning for tumor resection. The computational approach to approximating patient-specific segmental anatomy from CT images was addressed by two seminal papers in the early 2000s [36, 252]. Selle et al. [252] demonstrated that accurate and detailed vessel extraction and classification were critical for accurate approximation. Several efforts for virtual reality systematization [36] and automating the segmental anatomy approximation [228] according to the Couinaud nomenclature have been made since these papers were published. Figure 3.75 shows examples of automatically classified vessels and generated Couinaud liver segments based on them [228].

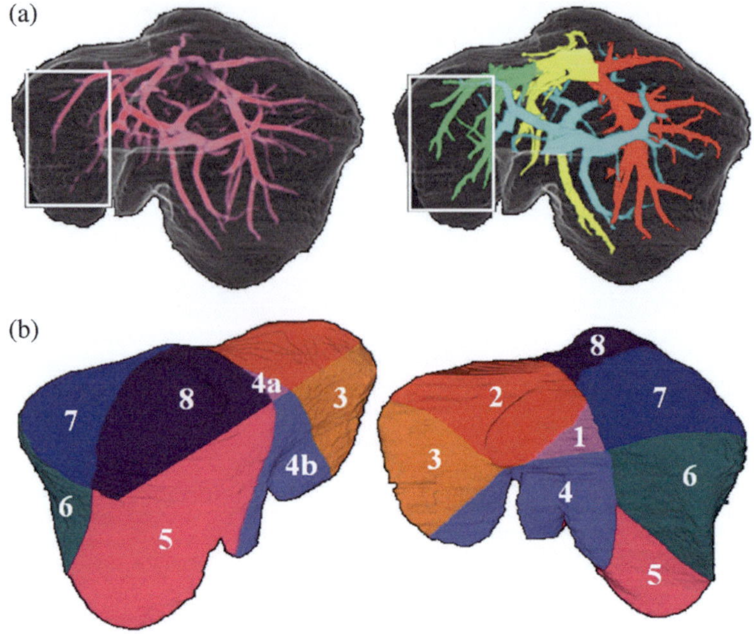

Fig. 3.75 Vessel extraction and classification of Couinaud liver segments [228]. (**a**) Vessel enhancement (*left*) and labeling. (**b**) Anterior (*left*) and posterior (*right*) views of resulted segmented anatomy

3.9.2 Pancreas

Anatomy The pancreas is located in the lesser sac deep in the upper abdomen. It runs horizontally from liver to spleen, as shown in Fig. 3.76, and its shape and location differ greatly between subjects. It has a specific relationship with the surrounding vessels. The splenic vein runs along its length, and the superior mesenteric vein runs posterior and perpendicular to the pancreas as it joins the portal vein. All these vessels enhance with contrast (see Fig. 3.77).

Overview of Pancreas Segmentation Algorithms Pancreatic segmentation from a CT volume is crucial for subsequent detailed image analysis of pancreatic pathology in a CAD system. So far, several algorithms have been proposed. One study [211] used an algorithm that analyzed contrast-enhanced images; however, part of the algorithm was executed manually, and it was based primarily on two-dimensional image processing, which can degrade segmentation accuracy. A more sophisticated automated pancreas segmentation algorithm has been presented [158], but it also suffers from low segmentation accuracy due to large variations in the location and shape of the pancreas. A third report [255] presented an algorithm that extracted the pancreas as well as 11 surrounding organs simultaneously from a non-contrast CT volume. A multi-atlas-based algorithm was proposed to extract multiple abdominal

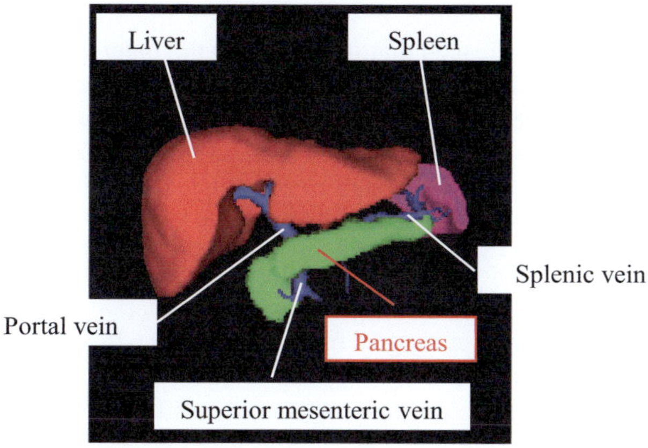

Fig. 3.76 Anatomy of neighboring organs and vessels closely associated with the pancreas

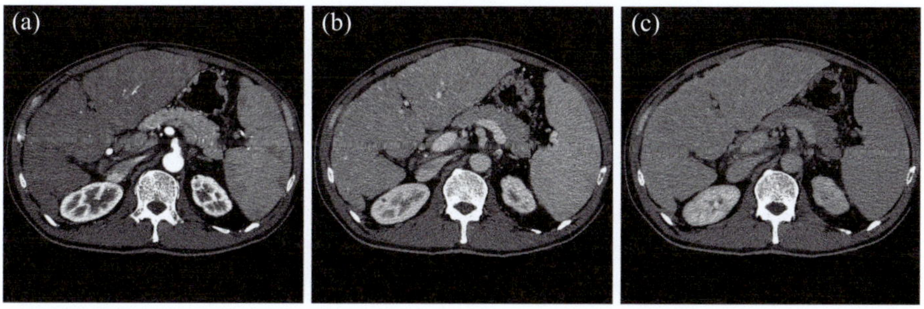

Fig. 3.77 Examples of input contrast-enhanced CT volumes of the pancreas segmentation algorithm (**a**) Early phase, (**b**) portal venous phase, (**c**) late phase

organs including pancreas [313]. Although multi-organ segmentation algorithms might be useful for pancreas segmentation, such algorithms are explained in Sect. 3.9.6. This section focuses on a single-organ segmentation algorithm, or an automated pancreas segmentation algorithm uses a CA model, or an SSM, from contrast-enhanced multiphase CT volumes [258] [259].

CA Models of the Pancreas and Their Application to Segmentation from CT Images The inputs are three-phase CT volume data: early/arterial-, portal-, and venous-phase volumes are presented in Fig. 3.77. Once the three-phase volumes are aligned by a registration algorithm based on normalized mutual information and radial basis function, segmentation of the liver and spleen is performed. The segmentation process assigns a label to each voxel, based on MAP using a probabilistic atlas and parameters estimated by an expectation maximization algorithm. The portal, splenic, and superior mesenteric veins are then extracted by a region-growing-based algorithm using location information for the extracted liver and spleen to establish landmarks for pancreas registration between different subjects. An input patient volume is subsequently warped in a nonlinear fashion with a radial basis function, such that landmarks in the input volume coincide with those in the reference volume data. The second stage roughly extracts the pancreas in the warped CT volume. This is based on the MAP method with a patient-specific probabilistic atlas generated from an SSM of the pancreas or a level set distribution model. After this rough segmentation, a morphological operation with a classifier ensemble is performed to refine the boundary further.

The algorithm was trained using three-phase CT volumes from 98 cases whose size was 512 × 512 × 161–261 voxels at a section interval of 1 mm. The pixel interval in the axial direction ranged from 0.546 to 0.625 mm. To validate the performance of the algorithm, it was applied to three-phase CT volumes from 20 test cases. Figure 3.78 shows examples of the segmentation for a test case where the JI between the extracted pancreas and the true one was 0.560 using a MAP segmentation step and 0.699 using a fine segmentation step. In summary, the average JI of 20 cases was 0.579 using the fine segmentation step.

Compared with the segmentation performance in other organs, such as the liver (see Sect. 3.9.1), the JI is relatively low. One possible reason is that since the pancreas has a slightly lobulated texture, is closely applied to surrounding structures, and is small in volume, the interobserver variability of segmentation is large. In fact, the JI between two true pancreas regions manually delineated by two independent observers (computer engineers) who have carried out studies on medical image processing was 0.760 on average, which is much lower than that for liver segmentation. Although the low JI of pancreas segmentation may be explained by interobserver variability, it does not fully explain some differences between the

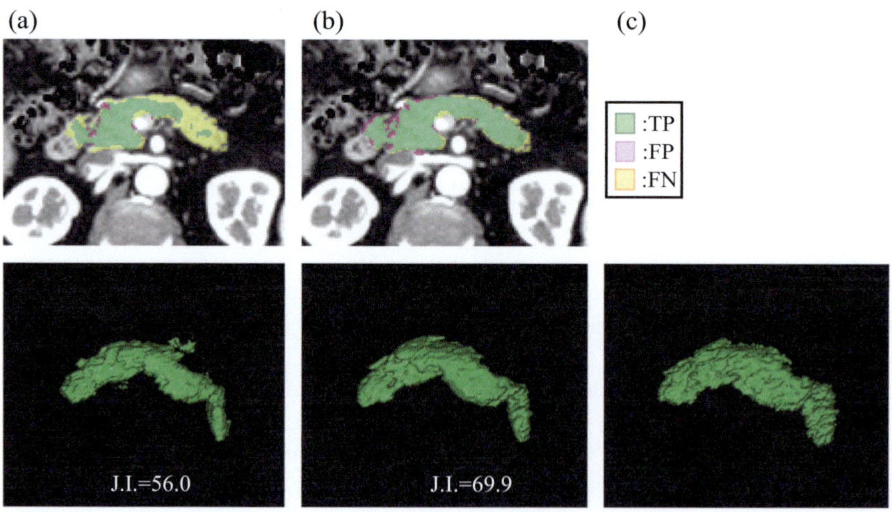

Fig. 3.78 Examples of MAP segmentation, fine segmentation, and its corresponding ground truth. (**a**) Rough segment on result, (**b**) fine segmentation result, (**c**) ground truth

segmentation performance and the JI. Variability in location, shape, and CT value of the pancreas, as well as essential difficulties in segmentation of a deeply indented and lobulated surface, are also contributing factors.

Computer-Aided Diagnosis in the Pancreas Pancreatic cancer is one of the major causes of cancer-related mortality in Japan, accounting for 29,916 deaths in 2012 [55]. CT is the most widely used imaging examination for detection and staging. CAD systems would be helpful to support the radiological interpretations and findings.

A few algorithms for diagnosis of pancreatic lesions have been presented. One group [130] presented an algorithm to discriminate between pancreatic ductal adenocarcinoma and mass-forming pancreatitis based on the radiological findings extracted by a radiologist, in which no medical image processing was performed. A fuzzy c-means clustering-based tumor extraction algorithm was presented by a second group [144], in which a tumor region was identified by combining the clustering result with a simple manual input by a user. The algorithm was applied to a few data, and the results were validated visually. A quantitative validation study using a large database will be an important future goal in this field.

3.9.3 Spleen

Introduction The spleen is located in the left upper quadrant of the abdomen (see Fig. 3.76) and is an important part of the reticuloendothelial system. Most of the spleen segmentation algorithms from a CT volume have been reported as part of the multi-organ segmentation scenario (Sect. 3.9.6). This subsection focuses on a segmentation algorithm with a CA model, or an SSM, customized to the spleen [261].

Machine learning is a popular technique in the field of medical image segmentation. In principle, an arbitrary machine learning algorithm, such as a support vector machine and the AdaBoost algorithm [38, 84], can be applied to organ segmentation where each voxel in a 3D image is classified by the algorithm as organ or background. Because of the powerful classification performance of state-of-the-art machine learning algorithms, this algorithm achieves higher segmentation accuracy than conventional algorithms. The inherent weakness of machine learning-based segmentation is, however, that voxel classification is carried out voxel by voxel, independently, producing unnatural shapes from the point of view of anatomy. To solve this problem, shape information has been incorporated in a machine learning-based segmentation algorithm.

This section focuses on ensemble learning as a promising machine learning technique, in particular a boosting algorithm, such as AdaBoost, whose performance has been proved to be favorable in terms of not only classification accuracy but also computational cost [38, 84]. Combining MRF with AdaBoost decreased false isolated connected components and holes in an extracted target region [279]. Several papers have presented algorithms that consider shape features of a target region, such as SpatialBoost [21] and Spatial AdaBoost [218]. Shape features used in these algorithms are local features which do not capture global shape features. Global shape features were introduced in the literature [4, 201]. There is, however, no boosting algorithm that can take into account statistical shape variations of an organ. This section presents ShapeBoost [261], which minimizes not only error loss but also shape loss, and evaluates the accuracy of an extracted shape based on a subspace of an SSM of the spleen.

Proposed Shape Loss Function Given a sample $\boldsymbol{x}_i$ and its class label y_i, the conventional boosting algorithm AdaBoost [38, 84] generates a strong classification function $F(\boldsymbol{x})$, which consists of a series of weak classifiers by minimizing the following loss function of $F(\boldsymbol{x})$:

$$L\left(F(x)\right) = \sum\nolimits_{i \in \Omega} \exp\left(-y_i F\left(x_i\right)\right) \tag{3.19}$$

where a set Ω represents the entire area of the image to be processed. Because the algorithm focuses on the error in each voxel, a surface extracted by a constructed classifier often includes undesirable irregularity. We propose a new loss function, given in Eq. (3.20), to make an extracted surface reasonable from the point of view of CA:

$$\begin{aligned} L\left(F(x)\right) &= \sum\nolimits_{i \in \Omega} \exp\left\{-y_i F\left(x_i\right) + \lambda \ln\left(G\left(F\left(x_i\right)\right) + 1\right)\right\} \\ G\left(F\left(x_i\right)\right) &= \left\{w_i - \mathrm{O}_i\left(\mathrm{U}\mathrm{U}^{\mathrm{T}}\mathrm{W}\right)\right\}^2 \\ w_i &= \phi\left(\mathrm{H}\left(F\left(x_i\right)\right)\right) \end{aligned} \tag{3.20}$$

where H is a Heaviside function and ϕ denotes a signed distance function. U is a matrix which consists of eigenvectors of an SSM of an organ, or a level set distribution model of the spleen in this study. **W** is a vector of signed distance function of a training spleen label which includes w_i. O_i is an operator that extracts the ith element of a vector. The coefficient λ represents the weighting between error loss and shape loss. Intuitively speaking, shape loss G in Eq. 3.20 represents distance between an extracted shape and a subspace of SSM, and its minimization causes the segmentation results to be closer to the subspace, or the natural shape, in terms of anatomy. Eventually, GradientBoost [85] was employed to minimize the loss of function that is not a convex function, and a strong classifier for spleen segmentation was obtained after minimization using a set of training labels.

Materials for validation were three-phase contrast-enhanced CT volumes, or early/arterial, portal, and venous phases (see Fig. 3.77) scanned from 80 cases. The size of the CT volume was 512 × 512 × 253–675 (voxels). A registration algorithm based on a radial basis function with normalized mutual information was used to align the different phase CT volumes. Spatial standardization based on the abdominal cavity was employed to reduce variation in location of the spleen [255]. Forty-two features were measured for weak classifiers of the ShapeBoost.

Cross validation tests using 80 cases were carried out to validate the performance, in which 40 cases were used for training a level set distribution model and a strong classifier, and the constructed classifier was tested on the remaining 40 cases. The same procedure was repeated after switching the roles of training and test.

Figure 3.79 shows examples of segmentation results where the resultant shape, in particular, the shape of the splenic hilum (denoted by arrows), by the ShapeBoost was more natural than that of AdaBoost, resulting in a higher JI between an extracted spleen and a true one. The error rate decreased from 3.09 to 2.66% on average ($p < 0.01$; Wilcoxon signed-rank test). After extracting a connected component with maximum volume, average JI by ShapeBoost reached 0.764, which was significantly higher than that by AdaBoost ($p < 0.01$; Wilcoxon signed-rank test). Note that this process is a rough segmentation process which can be refined by a graph cut-based algorithm as is done in lung and liver segmentation [210, 288].

3.9.4 Kidneys

This section deals with kidney segmentation algorithms followed by computer-aided detection or diagnosis of kidney pathologies.

Overview of Kidney Segmentation Algorithms Several kidney segmentation algorithms have been proposed for a non-contrast or a contrast-enhanced CT volume, such as a region-growing-based algorithm [176, 230], an algorithm based on knowledge of organs' location and CT values [163]. Three-dimensional deformable model-based approaches have been proposed [137, 289] to achieve more sophisticated boundaries. This section presents a 3D deformable model-based segmentation algorithm with CA, or statistical shape feature [289].

A NURBS-based surface was employed to represent the surface of the kidney because of its high flexibility in shape representation (see Fig. 3.80).

The following energy function was used for the segmentation:

$$\begin{aligned} \boldsymbol{E} = \boldsymbol{E}_{ext} + \omega_{\text{int}}\boldsymbol{E}_{\text{int}} = &- \sum_{U=0}^{M_U-1}\sum_{V=0}^{M_V-1} \left|Dir\left(\boldsymbol{S}\left(\boldsymbol{U},\boldsymbol{V}\right)\right)\boldsymbol{G}_{\sigma} * \nabla_d \boldsymbol{I}\left(\boldsymbol{S}(\boldsymbol{U},\boldsymbol{V})\right)\right| \\ &+ \omega_{\text{int}} \sum_{U=0}^{M_U-1}\sum_{V=0}^{M_V-1} (x-\overline{x})^t \Sigma_{UV}^{-1} (\boldsymbol{x}-\overline{\boldsymbol{x}}) \end{aligned} \tag{3.21}$$

where ∇_d represents a gradient operator of gray values with difference distance d, G_σ is a Gaussian function with standard deviation σ, I denotes an original image, and * is a convolution operator. Function Dir defines the similarity between direction of a gradient vector and that of a normal vector of the NURBS-based surface. The second term is an internal energy which is the sum of Mahalanobis distances from the shape feature vectors to its average. Shape feature vector $\boldsymbol{x}$ is defined at each sampling point on the surface marked by black dots in Fig. 3.81,

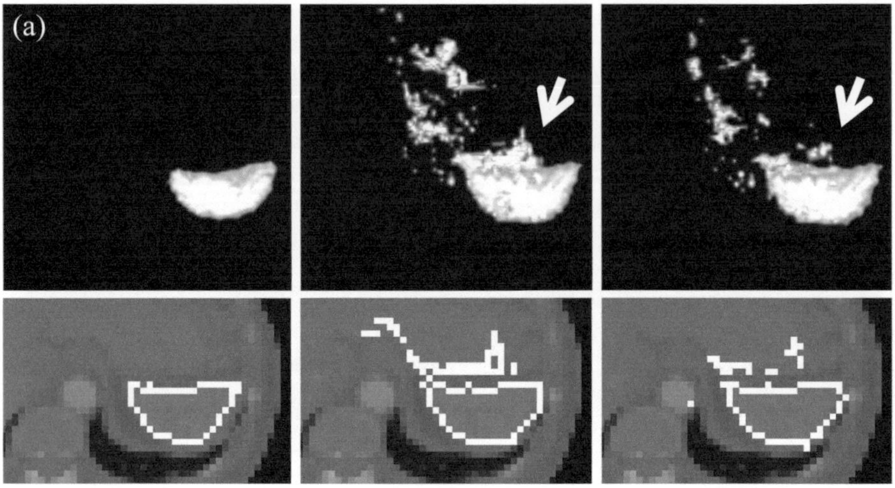

True

AdaBoost
JI = 0.516
error[%] = 2.78

ShapeBoost
JI = 0.599
error[%] = 1.97

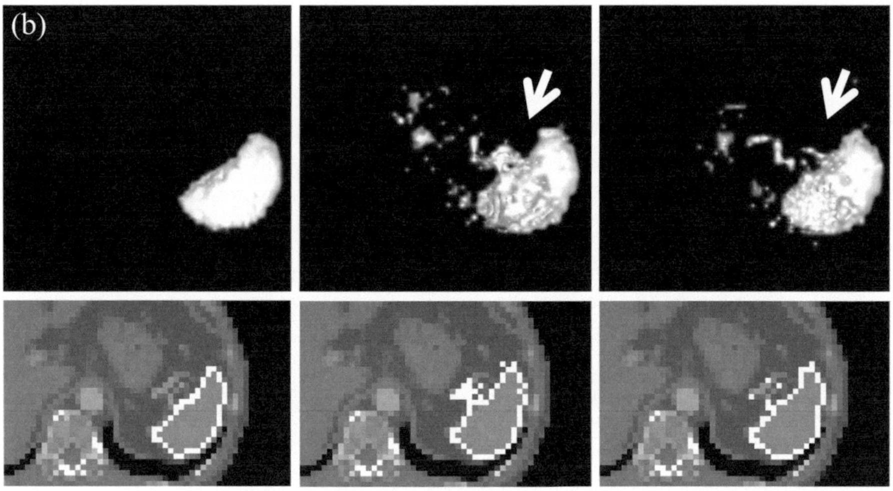

True

AdaBoost
JI = 0.715
error[%] = 1.83

ShapeBoost
JI = 0.732
error[%] = 1.67

Fig. 3.79 Examples of segmentation results by ShapeBoost and AdaBoost (**a**) case 5, (**b**) case 14

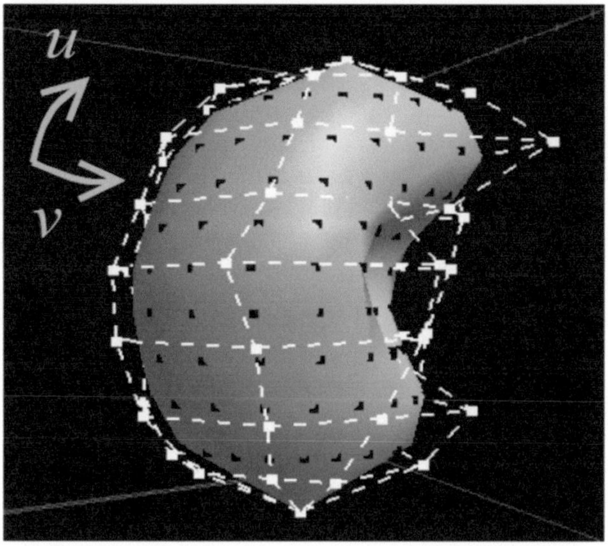

Fig. 3.80 Kidney surface represented by a NURBS function with control points (*white dots*) and sampling points used for measuring shape features (*black dots*) (Figure 3 of Ref. [289])

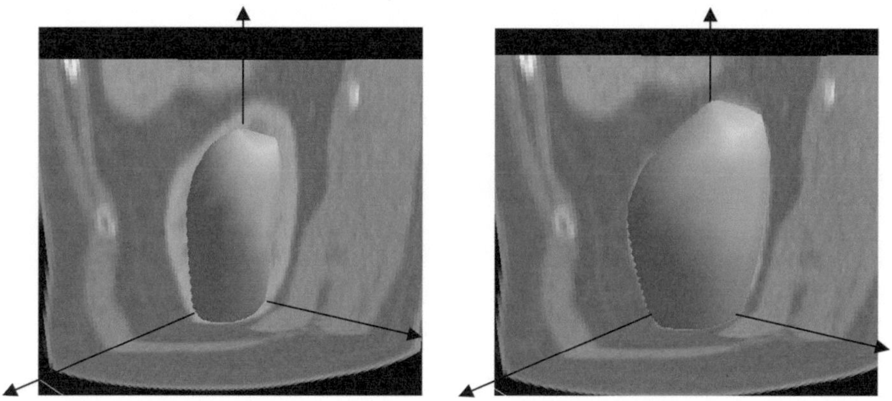

Fig. 3.81 An example of kidney segmentation; initial boundary (*left*) and final boundary extracted by the proposed algorithm (*right*)

and components of the shape feature vector are principal curvatures, or minimum and maximum curvatures, at the sampling points. Statistical information about the shape variation is incorporated into the deformable model via an internal energy whose average vector $\overline{\mathbf{x}}$ and covariance matrix $\boldsymbol{\Sigma}_{\mathrm{UV}}$ were computed from training dataset of kidneys.

Thirty-three upper abdominal CT volumes were used to evaluate the performance of the algorithm. The size of the volume was 512 × 512 × 18–24, the resolution was 0.625 or 0.630 mm, and the section thickness was 10 mm. Figure 3.81 shows an example of initial boundaries of the deformable model and its segmentation results after minimizing Eq. 3.21, in which the final state of the deformable model succeeded in capturing the true boundaries of the kidney. When we used the internal energy, the JI between an extracted kidney region and a true one increased by 0.028 on average from 33 cases compared with the results of the same cases without the internal energy, resulting in a mean of 0.865.

Computer-Aided Diagnosis of the Kidneys The number of kidney cancers in Japan is increasing every year [55]. It is desirable to detect the disease at the earliest possible stage using state-of-the-art imaging modalities, such as MDCT, to increase the likelihood of successful diagnosis and treatment.

Some studies described algorithms to measure clinically useful information, e.g., volume of a kidney tumor, in which region-growing- and thresholding-based segmentation algorithms were used to segment tumors [155, 270]. A semiautomated tumor segmentation algorithm from multiphase CT volumes was proposed [177] where a level set-based segmentation algorithm was presented followed by experimental results using 12 tumors from ten cases. The Dice coefficient between a true tumor and an extracted one was 0.80 on average. Figure 3.82 shows an example of extracted tumors [177]. In the segmented lesions, the histograms of curvature-related features were computed to quantify and classify the lesion types [177, 178].

3.9.5 *Digestive Tract Segmentation*

Intestinal Region: Segmentation The digestive tract includes the oropharynx, esophagus, stomach, small intestine, colon, rectum, and anus. The system comprises a wall surrounding a lumen. These regions are natural orphans and contain some airs flown from the mouth.

3.9.5.1 Stomach

The stomach, and, to a certain extent, the esophagus, can be distended with extra liquid contrast agent or with gas from a gas-forming agent administered just before imaging. If the lumen is well distended, the luminal regions of the stomach can be easily segmented by a simple thresholding process followed by the connected component selection process or the region-growing process, the seed point of which is set inside an area of gas density of the stomach. Figure 3.83 shows examples of segmentation results of the luminal regions of the stomach.

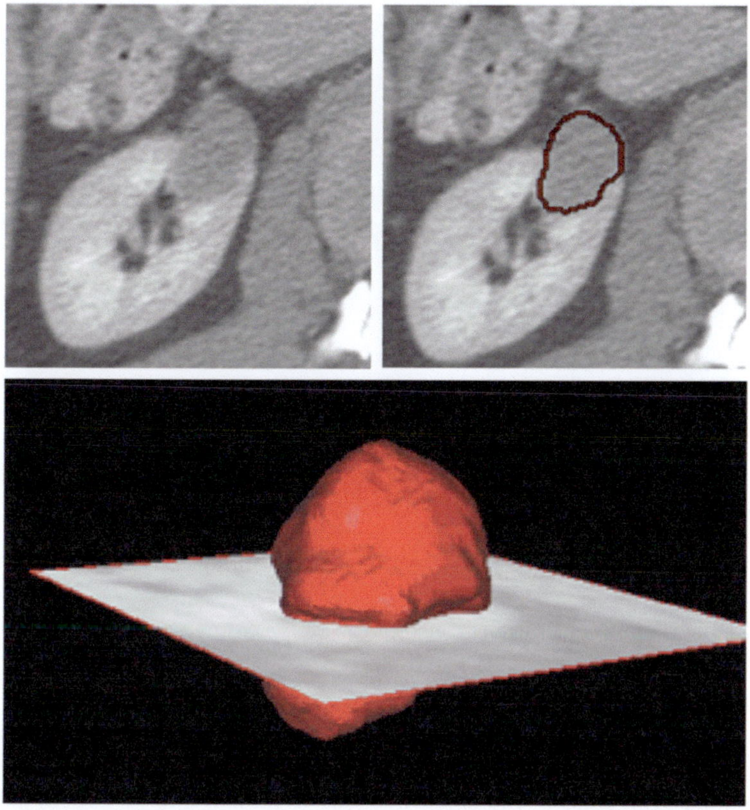

Fig. 3.82 An example of extracted kidney tumors; original CT slice (*left upper*), original CT slice with extracted boundary (*right upper*), and 3D display of the extracted region (*lower*) (Figure 8 of Ref. [177])

3.9.5.2 Intestine

Overview Segmentation information from the small intestine and colon can be used for finding suspicious regions on CT or MR images. The regional segmentation process is difficult if the segment in question is not filled with gas or radiopaque contrast (CT) or sorbitol solution or other hyperosmotic contrast (MRI). It is possible to instill contrast into the small intestine via nasojejunal tube (enteroclysis), but the process is difficult and uncomfortable for the patient, so enterography with oral contrast is preferred. It is a rule of thumb in medical imaging that luminal distension is optimized if the lumen is temporarily occluded at both ends; the pylorus and duodenal bulb are not completely competent valves, and the ileocecal valve is often incompetent, making it difficult to totally distend the small bowel. Because it is difficult to expand the entire small intestine with gas or liquid contrast, and peristalsis empties short sections, it is more difficult for radiologists to follow

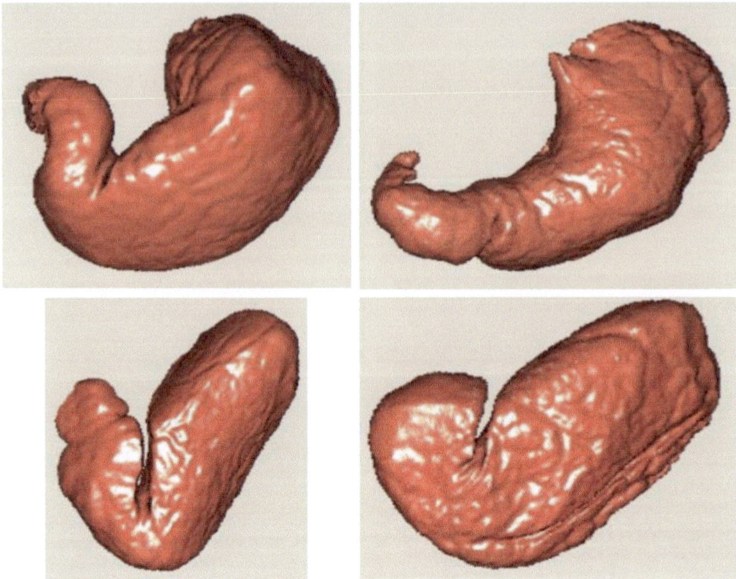

Fig. 3.83 Example of luminal regions of the stomach. Outside views of luminal regions are rendered

the entire course of the long–small intestine, as opposed to the shorter colon, on a CT scan. Typically, the number of "fragments" (opacified segments separated by peristaltic segments) of the small intestine is 20 and that of the colon is five. The large intestine usually runs almost in a simple shape like a U shape. The small intestines run in an unpredictable shape. Fragment connection is much more important in the small intestine extraction from CT images than that of the colon. Oda et al. (2014) tried to solve this issue by enumerating all possibilities of fragment connection [220]. This method first obtains centerlines of the fragmented luminal regions (Fig. 3.84). The i-th centerline is expressed as L_n $(n = 1 \cdots N)$. A terminal point and the other terminal point of the centerline L_n are expressed as v_{2n} and v_{2n+1}, respectively (Fig. 3.85). For one ordered sequence of L, we compute

$$E = w_1 D + w_2 R, \tag{3.22}$$

where D is a distance term and R is a direction term. D is described as

$$D = \frac{1}{N} \sum\nolimits_{j=0}^{N} \frac{d_j}{\widehat{d}_j}. \tag{3.23}$$

Here N is the number of the centerlines of the fragmented luminal regions. Here we introduce C_j $(j = 0 \cdots N - 2)$ that expresses the connection of the centerlines. $\widehat{d}_j$ is

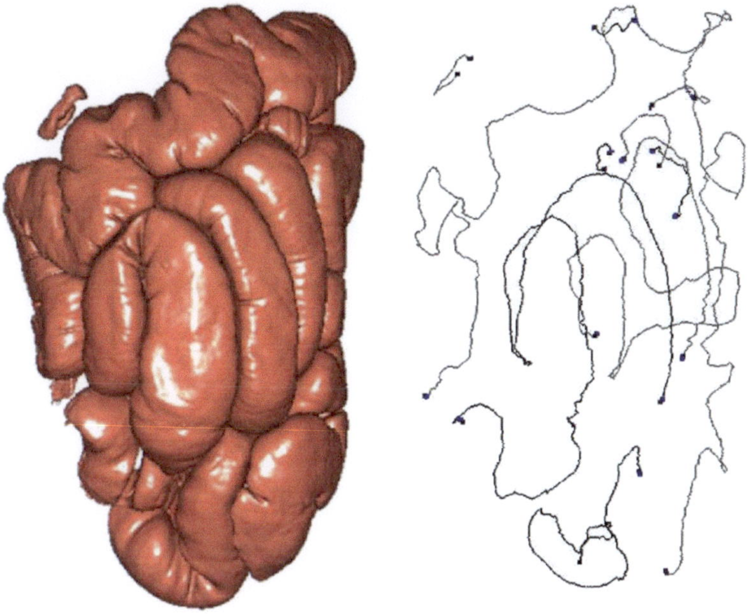

Fig. 3.84 Example of intestinal luminal regions and their centerlines

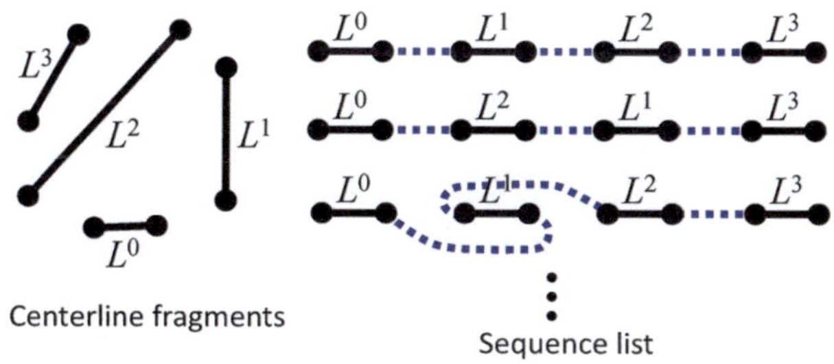

Fig. 3.85 Concept of fragmented region connection

the maximum distance in the connections. Direction term R is defined as

$$R = \frac{1}{N}\sum_{0}^{N-2} \frac{r_{j,1} \bullet r_{j,2} + 1}{2} \tag{3.24}$$

where $r_{j,1}$ and $r_{j,2}$ are running directions of the centerlines connected by C_j. Fragments connection can be obtained by finding a set of connection minimizing Eq. 3.22. Figure 3.86 shows an example where fragmented regions are connected correctly by the above process.

Colon lumen segmentation is easier, given the facts that the colon is shorter and can be opacified with contrast or distended with air in a retrograde fashion if necessary, with the ileocecal valve at the beginning of the ascending colon

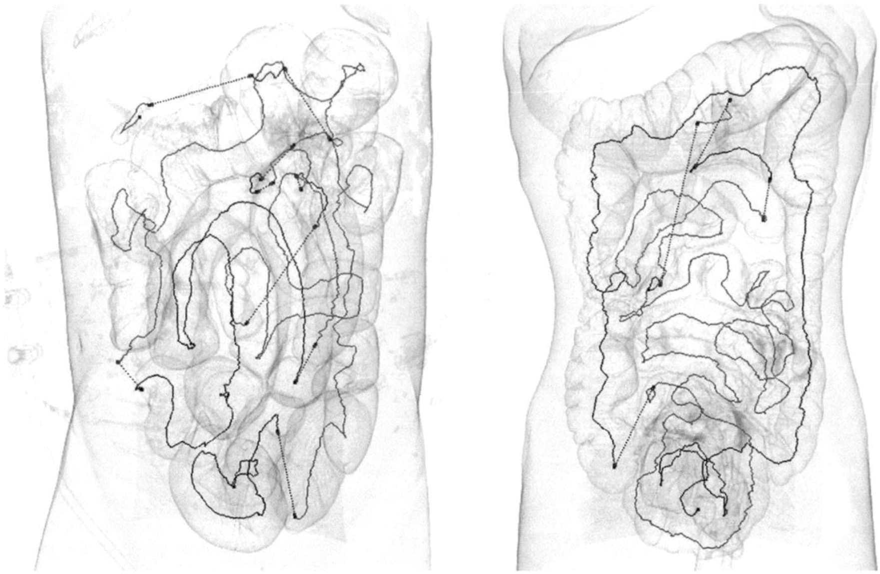

Fig. 3.86 Example of fragmented region connection

and a rectal balloon at the end of the colon acting as plugs to allow distension. Segmentation data on the colonic lumen can be used for eliminating processing regions in automated polyp detection. They can also be used for computing the centerlines of luminal segments, enabling automated fly-through inside the lumen. Such segmentation includes the lumen and the intestinal wall. Because peristalsis changes the luminal shape, it is difficult to employ statistical methods for segmenting intestinal organs. It is important to connect luminal fragments. A typical method uses the distances between the segments. Two segments are connected if the distance between these segments is lower than a certain threshold value.

Large Intestine Threshold value t for segmenting the luminal regions is usually set as

$$t = \frac{f_w + f_a}{2} \tag{3.25}$$

where f_w means a CT value of the large intestine wall and f_a shows a CT value of air. Typically f_w and f_a are −90 H.U. and −1000 H.U., respectively.

Figure 3.87 shows an example of luminal region extraction of the colon.

1. *Intestinal wall segmentation*

Intestinal wall extraction from CT images is a difficult problem when we compare intestinal wall segmentation with intestinal lumen segmentation. This is

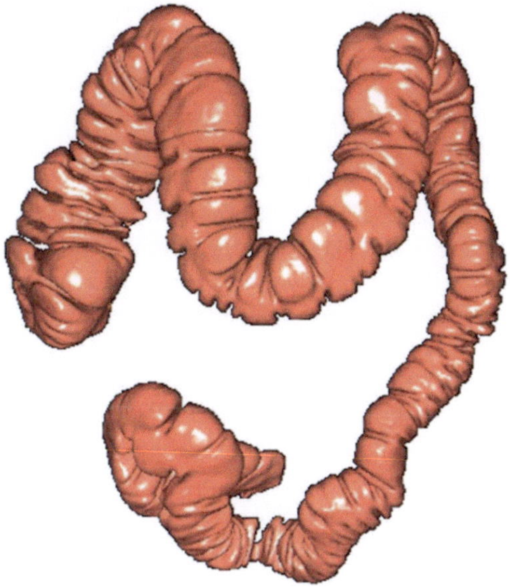

Fig. 3.87 Example of luminal regions of the colon. Outside views of luminal regions are rendered

because normal intestinal wall in a distended segment is almost invisible on CT. Masutani et al. tried to segment the colonic wall regions from abdominal CT images. This method extracts the colonic wall regions in combination of thresholding and binary image manipulations [187].

3.9.6 Multiple Abdominal Organs

Introduction In the human body, various anatomical structures are interrelated in a complex manner. The abdomen is the most appropriate domain to address the problem of representing multi-organ interrelationships. The pioneering work on abdominal multi-organ segmentation by Shimizu et al. was published in 2007 [256]. Since then, several methods have been proposed [180, 223–225, 299]. In this subsection, statistical modeling of interrelated organs is addressed, and one of the latest works is described [225]. The organs to be segmented are the liver, spleen, left and right kidneys, pancreas, gallbladder, aorta, and inferior vena cava.

Basic Unit for Modeling Multiple Organs: Prediction-Based CA Models We begin with two interrelated organs before addressing multiple organs generally. As a typical example of two organs, we first consider the liver and gallbladder. These two organs are closely situated, and the gallbladder handles the bile secreted by the liver. It is desirable to represent interrelations between them in addition to the two separate CA models. To represent two organs statistically, two approaches will be considered, that is, joint probability representation and conditional representation. If we consider joint representation, one possible method is to regard the two organs as one object. One SSM of the two organs can be represented in a hierarchical manner similar to H-SSMs described in the previous section. With conditional representation, the liver can be segmented in a sufficiently accurate manner by a single-organ segmentation method, while the error in gallbladder segmentation is typically much larger than the liver due to its small size and locational variability. Therefore, it would be useful for CA models of the gallbladder to be conditionally modeled under the assumption that the segmentation result of the liver is given. Figure 3.88a shows a conventional as well as a conditional probabilistic atlas of the gallbladder when the liver shape is given, which represents the ambiguity of its shape and location remaining after their prediction from the segmented liver shape. The prediction scheme is formulated using partial least squares regression (PLSR) of the target organ (gallbladder) from the predictor organ, the liver. In this form, the conditional probabilistic atlas is considered to represent the prediction error of PLSR. We call it a *prediction-based probabilistic atlas*. The prediction error is also represented as an SSM, which is a different form of conditional modeling and called a *prediction-based SSM*. This form is applied to various interrelated organs. Figure 3.88b shows the prediction-based probabilistic atlas of the pancreas given the liver and spleen shapes, in comparison with a conventional probabilistic atlas.

Multi-organ Computational Anatomy Modeling: Organ Correlation Graph (OCG) To represent interrelations among multiple organs more generally, a graph representation combining the basic units described in the above paragraph is formulated. Figure 3.89 shows a graph representation called the organ correlation graph (OCG). In the OCG, the basic unit is represented by nodes and directed arcs of a graph, in which the nodes correspond to organs and the directed arcs connect the predictor to target organs. One problem is how to find the arcs between

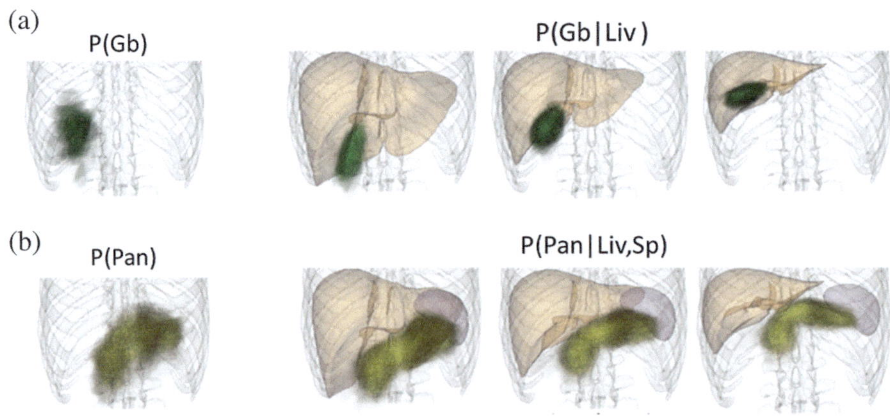

Fig. 3.88 Conventional and prediction-based probabilistic atlas of (**a**) gallbladder and (**b**) pancreas. *Left*: Single-organ modeling. *Right*: Multi-organ modeling incorporating interrelation with the liver and spleen. *P (Pan)* denotes the probabilistic atlas of the pancreas while *P (Pan|Liv, Sp)* the probabilistic atlas of the pancreas under the condition that the regions of the liver and spleen are known

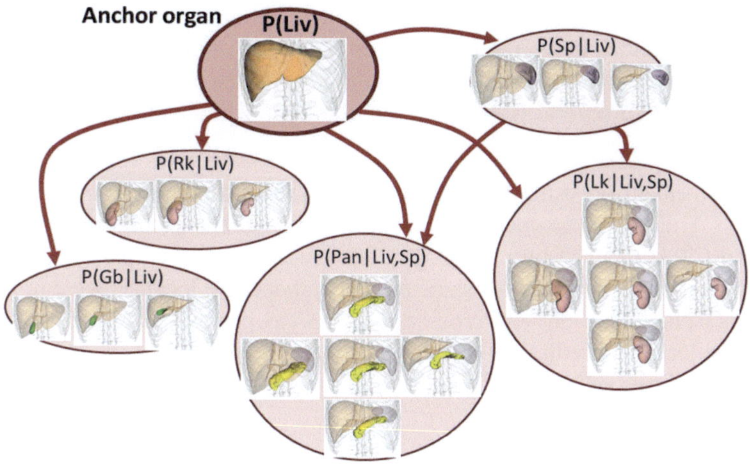

Fig. 3.89 Organ correlation graph (OCG) of the upper abdominal organs

nodes. Basically, the arcs are determined so as to minimize prediction error. For example, the predictor organs predicting the gallbladder are selected so that the prediction error is minimized, and the arcs are connected from the selected predictor organ nodes to the gallbladder node. Manually defined constraints can also be incorporated. For example, the liver is regarded as an anchor organ, and its region is assumed to be segmented beforehand using the methods as described in the previous section. That is, the liver can only be a predictor organ node and any arcs do not direct to it. Conversely, the gallbladder can only be a target organ node because its segmentation may not be accurate enough to use as one of the predictor organs. Once the set of organ nodes and the abovementioned manually defined constraints on them are given, the directed arcs in the OCG are determined automatically by finding the incoming arcs to each target organ node, which minimize the prediction error subject to satisfying these manually defined constraints. By using the OCG, the prediction-based CA models (i.e., prediction-based probabilistic atlas and SSM) can be generated once the regions of the predictor organs are segmented.

Multi-organ Segmentation The OCG is applied to automated multi-organ segmentation. The OCG is used for generating a procedure for automated multi-organ segmentation from abdominal CT data. The basic assumptions are as follows: (1) The field of view (FOV) of input CT data includes the whole liver, which is the anchor organ. (2) The intensity models of each organ and its background, which are the probability distributions of the CT values inside and outside the organ region, respectively, are available (or a procedure for estimating the intensity models by using the OCG and input CT data is available). The segmentation method at each node of the OCG is basically the same as that which was applied to liver segmentation (described in the previous section), but the difference is that the conventional probabilistic atlas and SSM are replaced by the prediction-based ones. Each organ node is prepared to start the segmentation procedure when the segmentation results at the predictor organ nodes are obtained. The segmentation procedures are executed at all the prepared nodes in a synchronized manner in the OCG and repeated several times to obtain the final segmentation results at all the nodes.

The abovementioned multi-organ segmentation method was tested using more than 100 CT datasets obtained under four different imaging conditions in contrast agent and CT scanner at two hospitals. The intensity models were constructed for each imaging protocol while the same priors on shape and location were utilized for all the datasets. Leave-one-out cross validation was performed. Figure 3.90 shows typical results. The prediction-based priors were effective in these results. JI was used for accuracy evaluation.

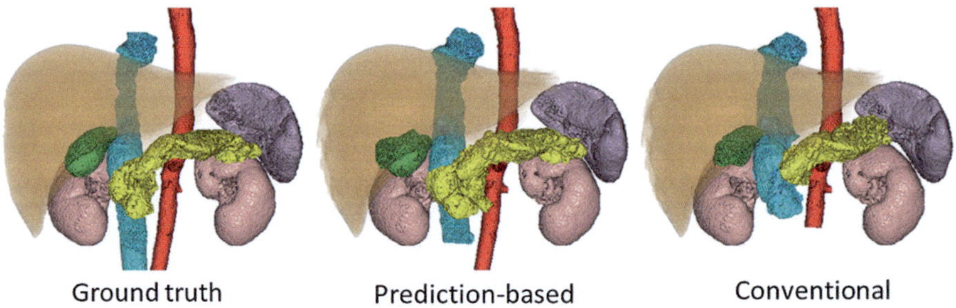

JI	LIV	SP	RK	LK	PAN	GB	AO	IVC
Prediction-based	0.916	0.941	0.980	0.963	0.747	0.543	0.935	0.681
Conventional		0.936	0.985	0.964	0.430	0.591	0.833	0.467

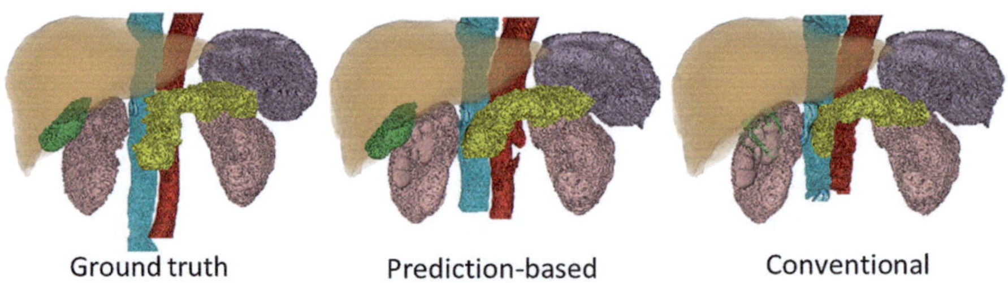

JI	LIV	SP	RK	LK	PAN	GB	AO	IVC
Prediction-based	0.916	0.894	0.864	0.915	0.607	0.829	0.687	0.645
Conventional		0.884	0.852	0.909	0.278	0.000	0.231	0.337

Fig. 3.90 Results of abdominal multi-organ segmentation from CT data. Table shows JI of results of multi-organ (prediction-based) and single-organ (conventional) methods. The improvement of segmentation accuracy was notable in the organs surrounded by red rectangles in the Table

3.9.7 Abdominal Aorta Segmentation

Abdominal Aorta and Anatomical Models

The abdominal aorta is usually located to the left of the center of the body and runs along the spine toward the lower abdomen, where it bifurcates. Anatomically important arteries, such as the celiac and superior mesenteric arteries arise from the abdominal aorta. Therefore, abdominal aortic segmentation is essential to recognize the network of abdominal arteries.

It is easy to segment the abdominal aorta region from a contrast-enhanced CT volume of a patient who does not suffer from aortic diseases such an aneurysm, since the contour of the aorta is very clear. A simple region-growing method can achieve sufficient segmentation. However, for CT volumes from a patient suffering from an abdominal aortic aneurysm, or non-contrast-enhanced CT volumes, the segmentation of the aorta focusing only on the intensity distribution does not work well. It causes the lack of the aneurysm area or over-segmentation of other surrounding organs. Model-based segmentation methods have been reported [46, 73, 157, 227, 334] to cope with these difficulties.

Since the aorta is a tubular organ, the aorta model consists of a tube surface. A centerline of the aorta is also modeled together with a tube surface. The tube surface is constructed by parametric surfaces such as B-spline. A surface model may also have the appearance of the aortic contour which represents the intensity or edge distribution at the vicinity of the contour. The distribution information is analyzed using PCA, and the appearance is often represented as eigenvectors.

Abdominal Aorta Segmentation The region-growing approach to segmentation becomes inaccurate in the face of abnormalities such as aortic aneurysms. There are several approaches to maintain segmentation accuracy.

The Centerline Model and Hough Transformation First, the centerline model of the aorta is fitted to an input CT volume. One can make a likelihood map of the aortic centerline by detecting edges of the aorta and performing the distance transformation to the edge voxels. The central parts of the aorta tend to have larger distance, and the distance information is used as likelihoods of the aortic centerline. A model fitting technique is applied which maximizes likelihoods on the centerline model. Then, the aortic surface is recovered by performing the reverse distance transformation using distance values on the centerline. In the recovery step, the Hough transformation may be adopted for modification of the distance value. If false edges are detected in the construction of a likelihood map, the distance value on the centerline becomes smaller. The Hough transformation is thus applied to the edge voxels, and the center points of the aorta are modified to the location which has the maximum votes.

The Tube Surface Model The tube surface model is used to segment the aortic surface more accurately. After the centerline model of the aorta is fitted, the surface model is deformed so as to fit it to the contour of the aorta. The likelihood map can be used as the deforming energy. The appearance of the aortic contour can also be added to the energy.

3.9.8 Abdominal Blood Vessel Segmentation

Abdominal blood vessel segmentation is an important task in CA. The basic framework of the segmentation process can be summarized as follows: (a) simple thresholding, (b) region growing, and (c) employment of a vessel enhancement filter. Because the density values of abdominal blood vessels and other organs are similar, intravenous contrast is used to enhance the vessels. Relatively large blood vessels have very high image contrast and higher CT values; segmentation of these blood vessels can be performed using simple thresholding techniques with some connected components analysis.

Thresholding-Based Vessel Extraction This method extracts abdominal blood vessel regions from 3D abdominal CT images. As stated before, contrast-enhanced CT acquisition is performed to achieve higher density in the blood vessels.

Figure 3.91 shows an example of abdominal contrast-enhanced CT images, which were taken in the arterial phase. Because abdominal regions are highly contrasted areas, it is easy to segment these regions using a simple thresholding technique followed by connected components analysis, which removes some bone areas.

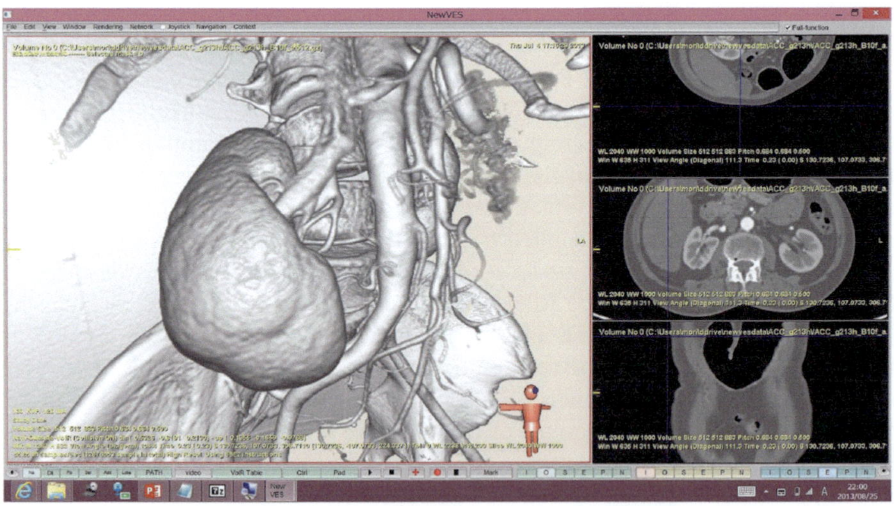

Fig. 3.91 Example of 3D-rendered view of abdominal contrast-enhanced CT image

Region Growing The process extracts abdominal blood vessels by tracing high-intensity regions on contrast-enhanced CT images. The first step, that of continuous growing of regions from a given starting point, often uses spherical structure elements. Typical growing conditions are based on intensity values. Because abdominal blood vessels have a branching structure, growing conditions considering such branching are also developed.

Blood Vessel Enhancement Filter The diameters of abdominal blood vessels visible on CT vary from 30 mm in size to submillimeter. Because of the partial volume effect, the intensity of small blood vessels is lower. Also, absolute intensity values of small blood vessels become smaller than those of large blood vessels. Thresholding or region-growing methods may fail to detect abdominal blood vessels because of such phenomena. This is the reason for using a blood vessel enhancement filter. Second-order differential intensity analysis is a popular technique in enhancing blood vessels on 3D CT images. Hessian-based analysis is widely used in the field of medical image analysis [81, 243]. Blood vessel enhancement filtering can be performed by the method shown in Sect. 3.4.2.2.

First a Hessian matrix H is obtained at a target point $\mathbf{p}$ and eigenvalues $\lambda_1, \lambda_2, \lambda_3$ $(0 \geq \lambda_1 \geq \lambda_2 \geq \lambda_3)$ are computed for a predefined scale. The blood vessel regions are extracted as a set of voxels that satisfy

$$\lambda_3 \approx \lambda_2 \ll \lambda_3 \approx 0. \tag{3.26}$$

There is always an array of diameters of vessels as they branch. Because the Hessian-based approach of blood vessel region extraction is quite sensitive to the diameter of a blood vessel, a multi-scale approach changing the σ in Hessian matrix computation is performed to enable appropriate extraction. Figure 3.92 shows an example of blood vessel regions extracted by Hessian-based analysis.

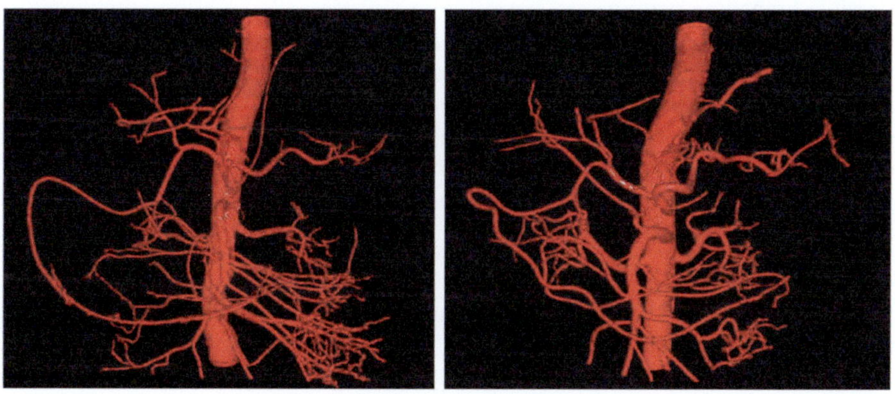

Fig. 3.92 Example of blood vessel extraction based on Hessian analysis

3.9.9 Anatomical Labeling of Abdominal Blood Vessels

Definition In CA, anatomical labeling is important. Abdominal blood vessel labeling can be understood as the procedure to give labels to each branch (edge) of a graph structure representing the abdominal blood vessel network of a subject.

Let b_i be a branch of the graph showing an abdominal blood vessel network. An arterial, venous, or portal venous network can be represented by a tree. Here a set of anatomical label classes are written as

$$\mathbf{C} = \{c_i\}. \tag{3.27}$$

Also a set of branches to be labeled as

$$\mathbf{B} = \{\boldsymbol{b}_i\}. \tag{3.28}$$

Machine Learning-Based Approach One approach for anatomical labeling is to employ machine learning [48, 190, 200].This process can be formulated as the process that finds the suitable category of a branch $\boldsymbol{b}_i$ from feature values f_i of $\boldsymbol{b}_i$. It can be written as

$$L_i = H(f_i). \tag{3.29}$$

Of course, this approach assigns anatomical labels to each branch based on features computed locally. Global information is not considered. A process to correct labeling results will be necessary.

Practical Example of Anatomical Labeling As a given blood vessel region is expressed as a tree structure, likelihoods of candidate anatomical names for each branch in the tree structure are computed using a machine learning-based method. Possible branching patterns are expressed in a graph called a bifurcation graph, a graph structure that is different from the tree structure used to represent the blood vessel region. Each node of the bifurcation graph expresses each anatomical name, and each edge expresses a possible bifurcation. The edges are assigned weights based on the likelihoods of the branches in the tree structure. The directed spanning tree of the bifurcation graph represents a branching pattern. The optimum branching pattern is obtained by computing the directed maximum spanning tree. Each branch in the tree structure is labeled based on the branching pattern.

This algorithm was evaluated using 50 sets of abdominal CT volume data. The recall and precision rates of abdominal arteries were 89.3% and 92.9%, and the recall and precision rates of the hepatic portal system were 86.0% and 86.3%, respectively. Examples of results of automated anatomical labeling are shown in Fig. 3.93. In this experiment, 80.8% of the branching patterns of major blood vessels that have branching variations were obtained correctly.

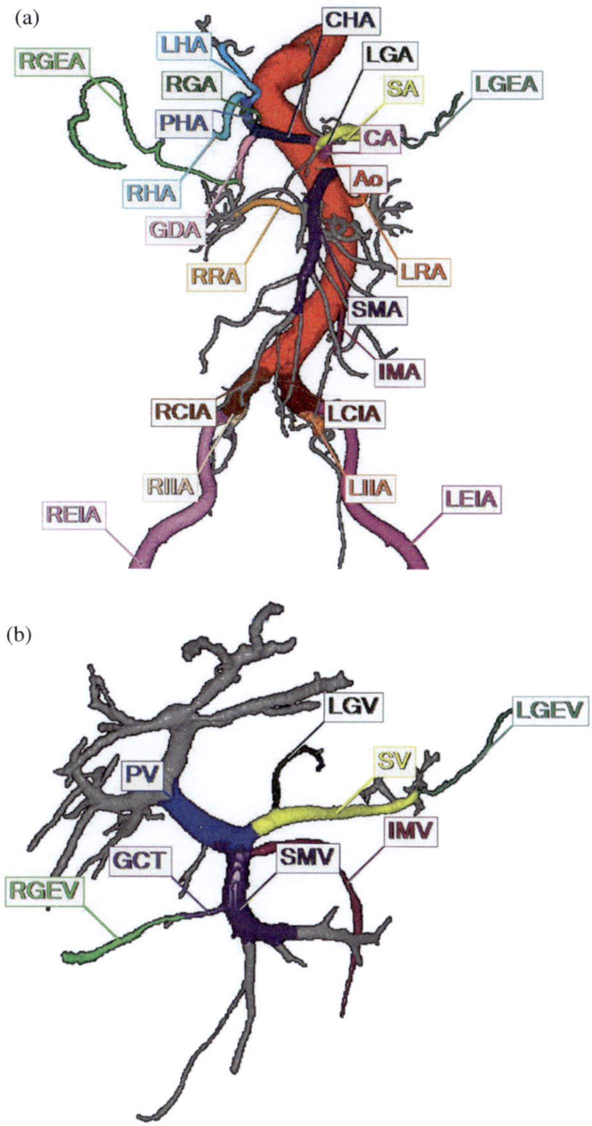

Fig. 3.93 Examples of results of automated anatomical labeling. (**a**) Abdominal arteries. (**b**) Hepatic portal system

References

1. Abe O, Aoki S, Hayashi N et al (2002) Normal aging in central nervous system: quantitative MR diffusion tensor analysis. Neurobiol Aging 23:433–441
2. Agarwala M, Hendriksb EA, Stoela BC et al (2012) Local SIMPLE multi atlas-based segmentation applied to lung lobe detection on chest CT. In: Proceedings of SPIE Medical Imaging. 8314: 831410-1–7
3. Allen PD, Graham J, Farnell DJJ et al (2007) Detecting reduced bone mineral density from dental radiographs using statistical shape models. IEEE Trans Info Technol Biomed 11:601–610
4. Amores J, Sebe N, Radeva P et al (2004) Boosting contextual information in content-based image retrieval. In: Proceedings of the 6th ACM SIGMM international workshop on Multimedia information retrieval, pp 31–38
5. Anderson SE, Lahm R, Iaisso PA (2010) The coronary vascular system and associated medical devices. In: Handbook of cardiac anatomy, physiology, and devices. Springer, Heidelberg
6. Andrews S, Hamarneh G, Yazdanpanah A, HajGhanbari B, Reid WD (2011) Probabilistic multi-shape segmentation of knee extensor and flexor muscles. Proc MICCAI 3:651–658
7. Antiga L, Ene-lordache B, Caverni L et al (2002) Geometric reconstruction for computational mesh generation of arterial bifurcations from CT angiography. Comput Med Imaging Graph 26:227–235
8. Antiga L, Ene-lordache B, Remuzzi A (2003) Computational geometry for patient specific reconstruction and meshing of blood vessels from MR and CT angiography. IEEE Trans Med Imaging 22:674–684
9. Antila K, Lilja M, Kalke M, Lotjonen J (2008) Automatic extraction of mandibular bone geometry for anatomy-based synthetization of radiographs. IEEE Eng Med Bio Soc 2008:490–493
10. Arad Y, Goodman KJ, Roth M et al (2005) Coronary calcification, coronary disease risk factors, C-reactive protein, and atherosclerotic cardiovascular disease events: the St Francis heart study. J Am Coll Cardiol 46:158–165
11. Armato SG, Sensakovic WF (2004) Automated lung segmentation for thoracic CT. Acad Radiol 11:1011–1021
12. Ashburner J, Friston KJ (1999) Nonlinear spatial normalization using basis functions. Hum Brain Mapp 7:254–266
13. Ashburner J, Friston KJ (2000) Voxel-based morphometry-The methods. NeuroImage 11:805–821
14. Ashburner J, Friston KJ (2001) Why voxel based morphometry should be used. NeuroImage 14:1238–1243
15. Ashburner J, Friston KJ (2005) Unified segmentation. NeuroImage 26:839–851
16. Ashburner J, Neelin P, Collins DL, Evans A, Friston K (1997) Incorporating prior knowledge into image registration. NeuroImage 6:344–352
17. Ashburner J, Hutton C, Frackowiak RSJ, Johnsrude I, Price C, Friston KJ (1998) Identifying global anatomical differences: deformation-based morphometry. Hum Brain Mapp 6:348–357
18. Ashihara T, Haraguchi R, Nakazawa K, Namba T, Ikeda T, Nakazawa Y, Ozawa T, Ito M, Horie M, Trayanva NA (2012) The role of fibroblasts in complex fractionated electrograms during persistent/permanent atrial fibrillation: implications for electrogram-based catheter ablation. Circ Res 110:275–284
19. Assaf Y, Basser PJ (2005) Composite hindered and restricted model of diffusion (CHARMED) MR imaging of the human brain. NeuroImage 27(1):48–58
20. Assaf Y, Blumenfeld-Katzir T, Yovel Y, Basser PJ (2008) AxCaliber: a method for measuring axon diameter distribution from diffusion MRI. Magn Reson Med 59(6):1347–1354 . 3. 5.3. Brain CT

21. Avidan S. (2006) SpatialBoost: adding spatial reasoning to AdaBoost. In: Proceedings of European conference on computer vision, pp 386–396
22. Awaya H, Mitchell DG, Kamishima T, Holland G, Ito K, Matsumoto T (2002) Cirrhosis: modified caudate–right lobe ratio 1. Radiology 224(3):769–774
23. Aykac D, Hoffman EA, McLennan G, Reinhardt JM (2003) Segmentation and analysis of the human airway tree from 3D X-ray CT images. In: IEEE transactions of medical imaging, pp 940–950
24. Bae KT, Giger ML, Chen C-T, Kahn CE Jr (1993) Automatic segmentation of liver structure in CT images. Med Phys 20(1):71–78
25. Bagci AM, Shahidi M, Ansari R et al (2008) Thickness profiles of retinal layers by optical coherence tomography image segmentation. Am J Ophthalmol 146:679–687
26. Bakic PR, Albert M, Brzakovic D, Maidment AD (2003) Mammogram synthesis using a three-dimensional simulation. III. Modeling and evaluation of the beast ductal network. Med Phys 30:1914–1925
27. Barbu A, Suehling M, Xu X, Liu D, Zhou SK, Comaniciu D (2012) Automatic detection and segmentation of lymph nodes from CT data. IEEE Trans Med Imaging 31(2):240–250. doi:10.1109/TMI.2011.2168234
28. Baron JC, Chételet G, Desgranges, Perchey G, Landeau B, de la Sayette V, F E (2001) In vivo mapping of gray matter loss with voxel-based morphometry in mild Alzheimer's disease. NeuroImage 14:298–309
29. Barratt DC, Chan CS, Edwards PJ, Penney GP, Slomczykowski M, Carter TJ, Hawkes DJ (2008) Instantiation and registration of statistical shape models of the femur and pelvis using 3D ultrasound imaging. Med Image Anal 12(3):358–374
30. Baskin KM, Stanford W, Thompson BH et al (1995) Comparison of electron beam and helical computed tomography in assessment of coronary artery calcification. Circulation 92:I-651
31. Basser PJ, Mattiello J, Le Bihan D (1994) Estimation of the effective self-diffusion tensor from NMR spin echo. J Magn Reson 103:247–254
32. Basser PJ, Pajevic S, Pierpaoli C, Duda J, Aldroubi A (2000) In vivo fiber-tractography in human brain using diffusion tensor MRI (DT-MRI) data. Magn Reson Med 44(4):625–632
33. Baudin P-Y, Azzabou N, Carlier PG, Paragios N (2012) Prior knowledge, Random walks and human skeletal muscle segmentation. Proc MICCAI 1:495–503
34. Baudin PY, Azzabou N, Carlier PG, Paragios N (2012) Prior knowledge, random walks and human skeletal muscle segmentation. In: Medical Image Computing and Computer-Assisted Intervention–MICCAI 2012. Springer, Berlin/Heidelberg, pp 569–576
35. Behrens TE, MW W, Jenkinson M, Johansen-Berg H, RG N, Clare S, PM M, JM B, SM S (2003) Characterization and propagation of uncertainty in diffusion-weighted MR imaging. Magn Reson Med 50:1077–1088
36. Beichel R, Schmalstieg D (2006) Liver surgery planning using virtual reality
37. Bild DE, Detrano R, Peterson D et al (2005) Ethnic differences in coronary calcification: the Multi-Ethnic Study of Atherosclerosis (MESA). Circulation 111:1313–1320
38. Bishop CM (2010) Pattern recognition and machine learning; 2nd printing, Published by Springer, New York, ISBN 10: 0387310738
39. Blemker SS, Delp SL (2005) Three-dimensional representation of complex muscle architectures and geometries. Ann Biomed Eng 33(5):661–673
40. Bookstein FL (2001) "Voxel-based morphometry" should not be used with imperfectly registered images. NeuroImage 14:1454–1462
41. Bosch JG, Mitchell SC, Lelieveldt BPF, Nijland F, Kamp O, Sonka M, Reiber JHC (2002) Automatic segmentation of echocardiographic sequences by active appearance motion models. IEEE Trans Med Imaging 21:1374–1383
42. Boskamp T, Rinck D, Link F et al (2004) New vessel analysis tool for morphometric quantification and visualization of vessels in CT and MR imaging data sets. Radiographics 24:287–297
43. Bouraoui B, Ronse C, Baruthio J et al (2010) 3D segmentation of coronary arteries based on advanced mathematical morphology techniques. Comput Med Imaging Graph 34:377–387

44. Boyer KL, Herzog A, Roberts C (2006) Automatic recovery of the optic nerve head geometry in optical coherence tomography. IEEE Trans Med Imaging 25:553–570
45. Brown MS, McNitt-Gray MF, Mankovich NJ et al (1997) Method for segmenting chest CT image data using an anatomical model: preliminary results. IEEE Trans Med Imaging 16:828–839
46. Bruijne M, Ginneken B, Viergever M, Niessen WJ (2004/06) Interactive segmentation of abdominal aortic aneurysms in CTA images, Med Image Anal 8(2): 127–138
47. Buerger C, Peters J, Waechter-Stehle I, Weber F, Klinder T, Renisch S (2013) Multi-modal vertebra segmentation from MR Dixon in hybrid whole-body PET/MR. In: MICCAI workshop on computational methods and clinical applications for spine imaging
48. Bui HH, Oda M, Jiang Z, Kitasaka T, Misawa K, Fujiwara M, Mori K (2011/02) A study on automated anatomical labeling to arteries concerning with colon from 3D abdominal CT images. In: Proceedings of SPIE, vol 7962, pp 79623R-1-9
49. Butts K, de Crespigny A, Pauly JM, Moseley M (1996) Diffusion-weighted interleaved echo-planar imaging with a pair of orthogonal navigator echoes. Magn Reson Med 35:763–770
50. Cabrera Fernández D, Salinas HM, Puliafito CA (2005) Automated detection of retinal layer structures on optical coherence tomography images. Opt Express 13:10200–10216
51. Cademartiri F, Grutta LL, Palumbo A et al (2007) Non-invasive visualization of coronary atherosclerosis: state-of-art. J Cardiovasc Med 8:129–137
52. Carballido-Gamio J, Belongie SJ, Majumdar S (2004) Normalized cuts in 3-D for spinal MRI segmentation. IEEE Trans Med Imaging 23:36–44
53. Cardenas VA, Studholme C, Gazdzinski S, Durazzo TC, Meyerhoff DJ (2007) Deformation-based morphometry of brain changes in alcohol dependence and abstinence. NeuroImage 34:879–887
54. Carrino JA, Campbell PD Jr, Lin DC, Morrison WB, Schweitzer ME, Flanders AE, Eng J, Vaccaro AR (2011) Effect of spinal segment variants on numbering vertebral levels at lumbar MR imaging. Radiology 259:196–202
55. Center for Cancer Control and Information Services, National Cancer Center, Japan Vital Statistics Japan (Ministry of Health, Labour and Welfare)
56. Chaudhiri S, Chatterjee S, Katz N et al (1989) Detection of blood vessels in retinal images using two-dimensional matched filter. IEEE Trans Med Imaging 8:263–269
57. Chen Z, Molloi S (2003) Automatic 3D vascular tree construction in CT angiography. Comput Med Imaging Graph 27:469–479
58. Chen D-R, Chang RF, Chen CJ, Ho MF, Kuo SJ, Chen ST, Hung SJ, Moon WK (2005) Classification of breast ultrasound images using fractal feature. Clin Imaging 29:235–245
59. Conturo TE, Lori NF, Cull TS, Akbudak E, Snyder AZ, Shimony JS, McKinstry RC, Burton H, Raichle ME (1999) Tracking neuronal fiber pathways in the living human brain. Proc Natl Acad Sci U S A 96(18):10422–10427
60. Cootes TF, Taylor CJ, Cooper DH, Graham J (1995) Active shape models – their training and application. Comput Vis Image Underst 61:38–59
61. Cootes TF, Edwards GJ, Taylor CJ (2001) Active appearance models. IEEE Trans Pattern Anal Mach Intel 23:681–685
62. Criqui MH, Denenberg JO, Ix JH et al (2014) Calcium density of coronary artery plaque and risk of incident cardiovascular events. JAMA 311:271–278
63. Cristinacce D, Cootes T (2008) Automatic feature localization with constrained local models. Pattern Recogn 41:3054–3067
64. Davatzikos C, Genc A, Xu D, Resnick SM (2001) Voxel-based morphometry using the RAVENS map: methods and validation using simulated longitudinal atrophy. NeuroImage 14:1361–1369
65. Davis SM, Donnan GA, Parsons MW et al (2008) Effects of alteplase beyond 3 h after stroke in the Echoplanar Imaging Thrombolytic Evaluation Trial (EPITHET): a placebo-controlled randomised trial. Lancet Neurol 7:299–309
66. De Bruijne M, Lund MT, Tankó LB, Pettersen PC, Nielsen M (2007) Quantitative vertebral morphometry using neighbor-conditional shape models. Med Image Anal 11(5):503–512

67. DeBuc DC (2011) A review of algorithms for segmentation of retinal image data using optical coherence tomography. In: Ho P-G (ed) Image segmentation. InTech, Rijeka
68. Delmarcelle T, Hesselink L (1993) Visualizing second-order tensor fields with hyperstreamlines. IEEE Comput Graph Appl 13(4):25–33
69. Devlin H, Honer K (2002) Mandibular radiomorphometric indices in the diagnosis of reduced skeletal bone mineral density. Osteoporos Int 13:373–378
70. Drexler W, Fujimoto JG (2008) Optical coherence tomography; Technology and applications. Springer, Berlin/Heidelberg
71. Tuch DS (2002) Diffusion MRI of complex tissue structure, PhD. thesis, Harvard University
72. Dufour PA, Abdillahi H, Ceklic L et al (2012) Pathology hinting as the combination of automatic segmentation with a statistical shape model. In: Proceedings of Medical Image Computing and Computer-Assisted Intervention – MICCAI 2012, pp 599–606
73. Duquette AA, Jodoin P, Bouchot O, Lalande A (2012) 3D segmentation of abdominal aorta from CT-scan and MR images. Comput Med Imaging Graph 36(4):294–303
74. Fabritius T, Makita S, Miura M et al (2009) Automated segmentation of the macula by optical coherence tomography. Opt Express 17:15659–15669
75. Farsiu S, Chiu SJ, Izatt JA et al (2008) Fast detection and segmentation of drusen in retinal optical coherence tomography images. In: Proceedings of SPIE 6844, Ophthalmic Technologies XVIII, 68440D; doi:10.1117/12.768624
76. Feng SSJ, Patel B, Sechopoulos I (2013) Objective models of compressed breast shapes undergoing mammography. Med Phys 40:031902
77. Feuerstein M, Glocker B, Kitasaka T, Nakamura Y, Iwano S, Mori K (2012) Mediastinal atlas creation from {3-D} chest computed tomography images: application to automated detection and station mapping of lymph nodes. Med Image Anal 16(1):63–74. doi:10.1016/j.media.2011.05.005
78. Fioranelli M, Gonnella C, Tonioni S (2009) Clinical anatomy of the coronary circulation. In: CT evaluation of coronary artery disease. Springer, Heidelberg
79. Florin C, Paragios N, Willams J (2005) Particle filters, a quasi-Monte Carlo solution for segmentation of coronaries. MICCAI 3749:246–253
80. Frangi AF, Niessen WJ, Vincken KL, Viergever MA (1998) Multiscale vessel enhancement filtering, LNCS, 1496
81. Frangi AF, Wiro J. Niessen, Vincken KL, Viergever MA (1998) Multiscale vessel enhancement filtering, LNCS, 1496
82. Fraz MM, Remagnino P, Hoppe A et al (2013) Quantification of blood vessel caliber in retinal images of multi-ethnic school children using a model based approach. Comput Med Imaging Graph 37:48–60
83. Freeborough PA, Fox P (1998) Modeling brain deformations in Alzheimer disease by fluid registration of serial MR images. J Comput Assist Tomogr 22:838–843
84. Freund Y, Schapire R (1997) A decision-theoretic generalization of on-line learning and an application to boosting. J Comput Syst Sci 55:119–139
85. Friedman JH (2001) Greedy function approximation: a gradient boosting machine. Ann Stat 29:1189–1232
86. Friedman SL (2008) Hepatic fibrosis—overview. Toxicology 254(3):120–129
87. Friedman J, Hastie T, Tibshirani R (2008) Sparse inverse covariance estimation with the graphical lasso. Biostatistics 9:432–441
88. Friese U, Meindl T, Herpertz SC, Reiser MF, Hampel H, Teipel SJ (2010) Diagnostic utility of novel MRI-based biomarkers for Alzheimer's disease: diffusion tensor imaging and deformation-based morphometry. J Alzheimers Dis 20:477–490
89. Friston KJ, Frith CD, Liddle PF, Frackowiak RSJ (1991) Comparing functional (PET) images: the assessment of significant change. J Cereb Blood Flow Metab 11:690–699
90. Friston KJ, Worsley KJ, Frackowiak RSJ, Mazziotta JC, Evans AC (1993) Assessing the significance of focal activations using their spatial extent. Hum Brain Mapp 1:214–220

91. Friston KJ, Holmes AP, Worsley KJ, Poline J-B, Frith CD, Frackowiak RSJ (1995) Statistical parametric maps in functional imaging: a general linear approach. Hum Brain Mapp 2:189–210
92. Friston KJ, Holmes A, Poline J-B, Price CJ, Frith CD (1996) Detecting activations in PET and fMRI: levels of inference and power. NeuroImage 4:223–235
93. Fujita H, Hara T, Zhou X et al (2013) A01–3 model construction for computational anatomy: progress overview FY2012. In: Proceedings of the fourth international symposium on the project "Computational Anatomy", pp 22–42
94. Fukuda H, Taki Y, Wu K, Sato K, Goto R, Inoue K, Okada K, Kawashima R (2012) Development and aging of the human brain studied with brain magnetic resonance image. In: Nano-Biomedical Engineering, Proceedings of the Tohoku University Global Centre of Excellence Programme, Imperial College Press, pp 179–190
95. Yokota F, Takaya M, Okada T, Takao M, Sugano N, Tada Y, Tomiyama N, Sato Y (2012) Automated muscle segmentation from 3D CT data of the hip using hierarchical multi-atlas method. In: 12th annual meeting of CAOS-international proceedings, pp 30–32
96. Gallagher FA, Tay KY, Vowler SL et al (2011) Comparing the accuracy of initial head CT reporting by radiologists, radiology trainees, neuroradiographers and emergency doctors. Br J Radiol 84(1007):1040–1045
97. Gang L, Chutatape O, Krishnan S (2002) Detection and measurement of retinal vessels in fundus images using amplitude modified second-order Gaussian filter. IEEE Trans Biomed Eng 49:168–172
98. Gao L, Heath DG, Kuszyk BS, Fishman EK (1996) Automatic liver segmentation technique for three-dimensional visualization of CT data. Radiology 201(2):359–364
99. Gao X, Bharath A, Stanton A et al (2000) Quantification and characterization of arteries in retinal images. Comput Methods Prog Biomed 63:133–146
100. Garra BS, Krasner BH, Horii SC, Ascher S, Mun SK, Zeman RK (1993) Improving the distinction between benign and malignant breast lesions: the value of sonographic texture analysis. Ultrason Imaging 15:267–285
101. Garvin MK, Abramoff MD, Kardon R et al (2008) Intraretinal layer segmentation of macular optical coherence tomography images using optimal 3-D graph search. IEEE Trans Med Imaging 27:1495–1505
102. Gaser C, Nenadic I, Buchsbaum BR, Hazlett EA, Buchsbaum MS (2001) Deformation-based morphometry and its relation to conventional volumetry of brain lateral ventricles in MRI. NeuroImage 13:1140–1145
103. Gee JC, Bajcsy RK (1999) Brain warping, chapter 11, pp 183–198, Academic Press
104. Genovese CR, Lazar NA, Nichols T (2002) Thresholding of statistical maps in functional neuroimaging using the false discovery rate. NeuroImage 15:870–878
105. Gilboa G, Sochen N, Zeevi YY (2004) Image enhancement and denoising by complex diffusion process. IEEE Trans Pattern Anal Mach Intell 25:1020–1036
106. Gilles B, Magnenat-Thalmann N (2010) Musculoskeletal MRI segmentation using multi-resolution simplex meshes with medial representations. Med Image Anal 14(3):291–302
107. Gilles B, Moccozet L, Magnenat-Thalmann N (2006) Anatomical modelling of the musculoskeletal system from MRI. In: Medical Image Computing and Computer-Assisted Intervention–MICCAI 2006. Springer, Berlin/Heidelberg, pp 289–296
108. Good CD, Scahill RI, Ashburner J, Henson RN, Friston KJ, Frackowiak RSJ (2001) A voxel-based morphometric study of ageing in 465 normal adult human brains. NeuroImage 14:21–36
109. Gregori G, Knighton RW, Jiao S et al (2005) 3–D OCT maps of retinal pathologies. Invest Ophthalmol Vis Sci. 46: E-Abstract 1055
110. Gubern-Mérida A, Kallenberg M, Martí R, Karssemeijer N (2012) Segmentation of the pectoral muscle in breast MRI using atlas-based approaches. Proc MICCAI 2:371–378
111. Gubern-Mérida A, Kallenberg M, Martí R, Karssemeijer N (2012) Segmentation of the pectoral muscle in breast MRI using atlas-based approaches. In: Proceedings of MICCAI, pp 371–378

112. Hanaoka S, Fritscher K, Welk M, Nemoto M, Masutani Y, Hayashi N et al (2011) 3-d graph cut segmentation with Riemannian metrics to avoid the shrinking problem. In: Medical Image Computing and Computer-Assisted Intervention–MICCAI 2011. Springer, Berlin/Heidelberg, pp 554–561
113. Hanaoka S, Masutani Y, Nemoto M, Nomura Y, Yoshikawa T, Hayashi N, Ohtomo K (2011) Probabilistic modeling of landmark distances and structure for anomaly-proof landmark detection. In: Proceedings of the third international workshop on mathematical foundations of computational anatomy-geometrical and statistical methods for modelling biological shape variability, pp 159–169
114. Hara T, Matoba N, Zhou X, et al (2007) Automated detection of extradural and subdural hematoma for contrast-enhanced CT images in emergency medical care. In: Proceedings of SPIE medical imaging 2007: Computer-aided diagnosis. 6514: 651432-1–651432-4
115. Haraguchi R, Nakao M et al (2014) 3D modeling system for easily displaying complex congenital heart conditions. DIGINFO TV. http://www.diginfo.tv/v/13-0094-r-en.php Last viewed 17 Jan 2014
116. Hayashi T, Chen H, Miyamoto K, Zhou X, Hara T, Yokoyama R, Kanematsu M, Hoshi H, Fujita H (2011) A computerized scheme for localization of vertebral bodies on body CT scans. In: SPIE medical imaging. International Society for Optics and Photonics, pp 796238–796238
117. Hee MR (1997) Optical coherence tomography of the eye. PhD thesis. Massachusetts Institute of Technology
118. Heimann T, Van Ginneken B, Styner MA, Arzhaeva Y, Aurich V, Bauer C, Beck A et al (2009) Comparison and evaluation of methods for liver segmentation from CT datasets. IEEE Trans Med Imaging 28(8):1251–1265
119. Herzog A, Boyer KL, Roberts C (2004) Robust extraction of the optic nerve head in optical coherence tomography. CVAMIA-MMBIA LNCS 3117:395–407
120. Hori M, Okada T, Higashiura K, Sato Y, Chen Y-W, Kim T, Onishi H et al. (2014) Quantitative imaging: quantification of liver shape on CT using the statistical shape model to evaluate hepatic fibrosis. Academic radiology. Academic radiology 22(3):303–309
121. Horsch K, Giger ML, Venta LA, Vyborny CJ (2002) Computerized diagnosis of breast lesions on ultrasound. Med Phys 29:157–164
122. Hu S, Hoffman EA, Reinhardt JM (2001) Automatic lung segmentation for accurate quantitation of volumetric X-ray CT images. IEEE Trans Med Imaging 20:490–498
123. Hua P, Song Q, Sonka M et al (2011) Segmentation of pathological and diseased lung tissue in CT images using a graph-search algorithm. Biomedical imaging: from nano to macro 2011 IEEE international symposium on IEEE, pp 2072–2075
124. Huang SH, Chu YH, Lai SH, Novak CL (2009) Learning-based vertebra detection and iterative normalized-cut segmentation for spinal MRI. IEEE Trans Med Imaging 28:1595–1605
125. Huppi PS, Maier SE, Peled S et al (1998) Microstructural development of human newborn cerebral white matter assessed in vivo by diffusion tensor magnetic resonance imaging. Pediatr Res 44:584–590
126. Ibragimov B, Likar B, Pernus F, Vrtovec T (2013) Segmentation of vertebrae from 3D spine images by applying concepts from transportation and game theories. In: Yao et al (eds), Proceeding of MICCAI 2013 workshop on computational methods and Clinical Applications for Spine Imaging – CSI 2013, Sep 26, Nagoya, pp 1–12
127. ICBM web site, http://www.loni.usc.edu/ICBM/
128. ICBM partner in Asia (Sendai), http://www.idac.tohoku.ac.jp/bir/index.en.html
129. Ijiri T, Ashihara T, Umetani N, Igarashi T, Haraguchi R, Yokota H, Nakazawa K (2012) A kinematic approach for efficient and robust simulation of the cardiac beating motion. PLoS One 7:e36706
130. Ikeda M, Shigeki I, Ishigami T et al (1997) Evaluation of a neural network classifier for pancreatic masses based on CT findings. Comput Med Imaging Graph 21:175–183

131. Ikedo Y, Fukuoka D, Hara T, Fujita H, Takada E, Endo T, Morita T (2007) Computerized mass detection in whole breast ultrasound images: reduction of false positives using bilateral subtraction technique. In: Proceedings of SPIE medical imaging 2007: Computer-Aided Diagnosis. 2007; 6514: 65141 T-1-65141 T-10
132. Isgum I, Rutten A, Prokop M et al (2007) Detection of coronary calcifications from computed tomography scans for automated risk assessment of coronary artery disease. Med Phys 34:1450–1461
133. Ishikawa I, Stein DM, Wollstein G et al (2005) Macular segmentation with optical coherence tomography. Invest Ophthalmol Vis Sci 46:2012–2017
134. Izard C, Jedynak B, Stark CEL (2006) Spline-based probabilistic model for anatomical landmark detection. In: Proceedings MICCAI, pp 849–856
135. Jain A, Zongker D (1997) Feature selection: evaluation, application, and small sample performance. IEEE Trans Pattern Anal 19:153–158
136. Jensen JH, Helpern JA, Ramani A et al (2005) Diffusional kurtosis imaging: the quantification of non-gaussian water diffusion by means of magnetic resonance imaging. Magn Reson Med 53:1432–1440
137. Joshi S, Pizer S, Fletcher PT et al (2002) Multiscale deformable model segmentation and statistical shape analysis using medial descriptions. IEEE Trans Med Imaging 21:538–550
138. Kadoury S, Labelle H, Paragios N (2013) Spine segmentation in medical images using manifold embeddings and higher-order MRFs. IEEE Trans Med Imaging 32:1227–1238
139. Kainmueller D, Lamecker H, Zachow S, Heller M, Hege H C (2008) Multi-object segmentation with coupled deformable models. In: Proceedings of medical image understanding and analysis, pp 34–38
140. Kainmueller D, Lamecker H, Zachow S, Hege H-C (2009) An articulated statistical shape model for accurate hip joint segmentation. In: Engineering in Medicine and Biology Society, 2009. EMBC 2009. Annual international conference of the IEEE, pp 6345–6351
141. Kainmüller D, Lange T, Lamecker H (2007) Shape constrained automatic segmentation of the liver based on a heuristic intensity model. In: Proceedings of MICCAI workshop 3D segmentation in the clinic: a grand challenge, pp 109–116
142. Kajic V, Povazay B, Hermann B et al (2010) Robust segmentation of intraretinal layers in the normal human fovea using a novel statistical model based on texture and shape analysis. Opt Express 18:14730–14744
143. Kajic V, Esmaeelpour M, Povazay B et al (2012) Automated choroidal segmentation of 1060 nm OCT in healthy and pathologic eyes using a statistical model. Biomed Opt Express 3:86–103
144. Kallergi M, Hersh M, Manohar A (2004) Automated segmentation of pancreatic tumors in computed tomography. In: Suri W, Laxminarayan (eds) Handbook of medical image analysis, vol I Segmentation and registration models, IEEE Press Series, pp 176–226
145. Kamiya N, Zhou X, Chen H, Muramatsu C, Hara T, Yokoyama R, Kanematsu M, Hoshi H, Fujita H (2012) Automated segmentation of psoas major muscle in X-ray CT images by use of a shape model: preliminary study. Radiol Phys Technol 5:5–14
146. Kamiya N, Zhou X, Chen H, Muramatsu C, Hara T, Fujita H (2013) Model-based approach to recognize the rectus abdominis muscle in CT images. IEICE Trans Inform Syst E-96-D:869–871
147. Kanazawa K, Kawata Y, Niki N et al (1998) Computer-aided diagnosis for pulmonary nodules based on helical CT images. Comput Med Imaging Graph 22:157–167
148. Kaneko M, Eguchi K, Ohmatsu H et al (1996) Peripheral lung cancer: screening and detection with low-dose spiral CT versus radiography. Radiology 201:798–802
149. Kang D, Woo J, Slomka PJ, Dey D, Germano G, Kuo CCJ (2012) Heart chambers and whole heart segmentation techniques: review. J Electron Imaging 21:010901
150. Karas GB, Burton EJ, Rombouts SA, van Schijndel RA, O'Brien JT, Scheltens P, McKeith IG, Williams D, Ballard C, Barkhof F (2003) A comprehensive study of gray matter loss in patients with Alzheimer's disease using optimized voxel based morphometry. NeuroImage 18:895–907

151. Kass M, Witkin A, Terzopoulos D (1988) Snakes: active contour models. Int J Comput Vis 1:321–331
152. Khalvati F, Martel A (2013) Atlas-based segmentation of breast MR images. In: Proceedings of workshop on breast image analysis in conjunction with MICCAI, pp 74–81
153. Kido S, Tsunomori A (2009) Automated extraction of pleural effusion in three-dimensional thoracic CT images. Proceedings of SPIE Medical Imaging. 7260: 72600D–72600D8
154. Kim Y, Kim D (2009) A fully automatic vertebra segmentation method using 3D deformable fences. Comput Med Imaging Graph 33:343–352
155. Kim DY, Park JW (2004) Computer-aided detection of kidney tumor on abdominal computer tomography scans. Acta Radiol 45:791–795
156. Kitasaka T, Mori K, Hasegawa J-i, Toriwaki J-i (2002/12) A method for extraction of bronchus regions from 3D chest X-ray CT images by analyzing structural features of the bronchus, FORMA, 17(4): 321–338
157. Kitasaka T, Mori K, Hasegawa J, Toriwaki J Katada K (2002/08) A method for automated extraction of aorta and pulmonary artery using line models from 3D chest X-ray CT images with contrast medium. In: 16th international conference on pattern recognition 2002 (ICPR2002), vol 3, pp 273–276
158. Kitasaka T, Sakashita M, Mori K et al (2008) A method for extracting pancreas regions from four-phase contrasted 3D abdominal CT images. Int J Comput Assist Radiol Surg 3(Suppl 1):S40
159. Klemetti E, Kolmakov S, Kroger H (1994) Pantomography in assessment of the osteoporosis risk group. Scand J Dent Res 102:68–72
160. Klinder T, Lorenz C, von Berg J, Dries SP, Bülow T, Ostermann J (2007) Automated model-based rib cage segmentation and labeling in CT images. Med Image Comput Assist Interv 10:195–202
161. Klinder T, Ostermann J, Ehm M, Franz A, Kneser R, Lorenz C (2009) Automated model-based vertebra detection, identification, and segmentation in CT images. Med Image Anal 13:471–482
162. Ko TH, Fujimoto JG, Schuman JS et al (1922) Comparison of ultrahigh- and standard-resolution optical coherence tomography for imaging macular pathology. Ophthalmology 2005(112):e1–e15
163. Kobashi M, Shapiro LG (1995) Knowledge-based organ identification from CT images. Pattern Recogn 28:475–491
164. Kobashi S, Fujimoto S, Nishiyama T, Kanzaki N, Fujishiro T, Shibanuma N et al (2010) Robust pelvic coordinate system determination for pose changes in multidetector-row computed tomography images. Int J Fuzzy Logic Intel Syst 10(1):65–72
165. Koozekanani D, Boyer KL, Roberts C (2001) Retinal thickness measurements in optical coherence tomography using a Markov boundary model. IEEE Trans Med Imaging 20:900–916
166. Kubo M, Niki N, Nakagawa S et al (1999) Extraction algorithm of pulmonary fissures from thin-section CT images based on linear feature detector method. IEEE Trans Nucl Sci 46:2128–2133
167. Lamecker H, Seebass M, Hege HC, Deuflhard P (2004) A 3D statistical shape model of the pelvic bone for segmentation. In: Medical Imaging 2004. International Society for Optics and Photonics, pp 1341–1351
168. Lamecker H, Lange T, Seebass M (2004) Segmentation of the liver using a 3D statistical shape model. Konrad-Zuse-Zentrum für Informationstechnik
169. Lassen B, van Rikxoort EM, Schmidt M et al (2012) Automatic segmentation of the pulmonary lobes from chest CT scans based on fissures, vessels, and bronchi. IEEE Trans Med Imaging 32:210–222
170. Le Bihan D, Jones DK (eds) (2010) Diffusion MRI: theory, methods, and applications. Oxford University Press, Oxford

171. Leber AW, Becker A, Knez A et al (2006) Accuracy of 64-slice computed tomography to classify and quantify plaque volumes in the proximal coronary system: a comparative study using intravascular ultrasound. J Am Coll Cardiol 47:672–677
172. Lee J, Beighley P, Ritmana E et al (2007) Automatic segmentation of 3D micro-CT coronary vascular images. Med Image Anal 11:630–647
173. Lee K, Abramoff MD, Niemeijer M et al (2010) 3-D segmentation of retinal blood vessels in spectral-domain OCT volumes of the optic nerve head. In: Proceedings of SPIE medical imaging: biomedical applications in molecular, structural, and functional imaging. 7626: 76260 V
174. Li H, Chutatape O (2004) Automated feature extraction in color retinal images by a model based approach. IEEE Trans Biomed Eng 51:246–254
175. Li H, Hsu W, Lee ML, Wang H (2003) A piecewise Gaussian model for profiling and differentiating retinal vessels. In: Proceedings of international conference on image processing, pp 1069–1072
176. Lin D-T, Lei C-C, Hung S-W (2006) Computer-aided kidney segmentation on abdominal CT images. IEEE Trans Inform Technol Biomed 10:59–65
177. Linguraru MG, Yao J, Gautam R et al (2009) Renal tumor quantification and classification in contrast-enhanced abdominal CT. Pattern Recogn 42:1149–1161
178. Linguraru MG, Wang S, Shah F et al (2009) Computer-aided renal cancer quantification and classification from contrast-enhanced CT via histograms of curvature-related features. In: Conference of proceedings IEEE Eng Medical Biology Society 2009: 6679–6682
179. Linguraru MG, Richbourg WJ, Liu J, Watt JM, Pamulapati V, Wang S, Summers RM (2012) Tumor burden analysis on computed tomography by automated liver and tumor segmentation. IEEE Trans Med Imaging 31(10):1965–1976
180. Linguraru MG, Pura JA, Pamulapati V, Summers RM (2012) Statistical 4D graphs for multi-organ abdominal segmentation from multiphase CT. Med Image Anal 16(4):904–914
181. Lo P, van Ginneken B, Reinhardt JM, Yavarna T, de Jong PA, Irving B, Fetita C, Ortner M, Pinho R, Sijbers J, Feuerstein M, Fabijanska A, Bauer C, Beichel R, Mendoza CS, Wiemker R, Lee J, Reeves AP, Born S, Weinheimer O, van Rikxoort EM, Tschirren J, Mori K, Odry B, Naidich DP, Hartmann I, Hoffman EA, Prokop M, Pedersen JH, de Bruijne M (2012/11) Extraction of airways from CT (EXACT'09), IEEE Trans Med Imag 31(11): 2093–2107
182. Lombaert H, Peyrat JM, Croisille P, Rapacchi S, Fanton L, Clarysse P, Delingette H, Ayache N (2011) Statistical analysis of the human cardiac fiber architecture from DT-MRI. In: Metaxas DN, Axel L (eds) Functional imaging and modeling of the heart. Springer, Berlin/Heidelberg, Berlin
183. Lowell J, Hunter A, Steel D (2004) Measurement of retinal vessel widths from fundus images based on 2-D modeling. IEEE Trans Med Imaging 23:1196–1204
184. Ma J, Lu L (2013) Hierarchical segmentation and identification of thoracic vertebra using learning-based edge detection and coarse-to-fine deformable model. Med Image Comput Comput Assist Interv
185. Malcolm DTK, Goh CM, Balarenda Gamage TP, Hou BS, Nielsen PMF, Nash MP (2013) Modelling breast deformation using partial least-square regression. In: Proceedings of workshop on breast image analysis in conjunction with MICCAI, pp 9–16
186. Masumoto J, Hori M, Sato Y, Murakami T, Johkoh T, Nakamura H, Tamura S (2001) Automated liver segmentation using multislice CT images. IEICE Trans DII-84(9):2150–2161 (in Japanese)
187. Masutani Y, Yoshida H, MacEneaney PM, Dachman AH (2001) Automated segmentation of colonic walls for computerized detection of polyps in CT colonography. J Comput Assist Tomogr 25(4):629–638
188. Masutani Y, Aoki A, Abe O, Hayashi N, Otomo K (2003) MR diffusion tensor imaging: recent advance and new techniques for diffusion tensor visualization. Eur J Radiol 46(1):53–66
189. Matsuhiro M, Suzuki H, Kawata Y et al (2013) Extraction method of interlobar fissure based on multi-slice CT images. IEICE Trans Inf Syst J96D:834–843 (in Japanese)

190. Matsuzaki T, Oda M, Kitasaka T, Hayashi Y, Misawa K, Mori K (2013/02) A method for automated anatomical labeling of abdominal veins extracted from 3D CT images. In: Proceedings of SPIE, vol 8669, pp 86691Y-1-9
191. McIntosh AR, Gonzalez-Lima F (1994) Structural equation modeling and its application to network analysis in functional brain imaging. Hum Brain Mapp 2:2–22
192. Melhem ER, Hoon AH, Ferrucci JT III et al (2000) Brain MR imaging in periventricular leukomalacia: relationship between lateral ventricular volume and severity of cognitive and motor impairment. Radiology 214:199–204
193. Michael Kelm B, Wels M, Kevin Zhou S, Seifert S, Suehling M, Zheng Y, Comaniciu D (2013) Spine detection in CT and MR using iterated marginal space learning. Med Image Anal 17:1283–1292
194. Mirzaalian H, Wels M, Heimann T, Kelm BM, Suehling M (2013) Fast and robust 3D vertebra segmentation using statistical shape models. In: Conference of Proceedings on IEEE Eng Medical and Biology Society, pp 3379–3382
195. Mori K, Hasegawa J-i, Toriwaki J-i, Anno H, Katada K (1996/08) Recognition of bronchus in three dimensional X-ray CT images with applications to virtualized bronchoscopy system. In: Proceedings of 13th international conference of pattern recognition, pp 528–532
196. Mori S, Crain BJ, Chacko VP, van Zijl PC (1999) Three-dimensional tracking of axonal projections in the brain by magnetic resonance imaging. Ann Neurol 45(2):265–269
197. Mori K, Hasegawa J-i, Suenaga Y, Toriwaki J-i (2000/2002) Automated anatomical labeling of the bronchial branch and its application to the virtual bronchoscopy system, IEEE Trans Med Imag 19(2): 103–114
198. Mori K, Ema S, Kitasaka T, Mekada Y, Ide I, Murase H, Suenaga Y, Takabatake H, Mori M, Natori H (2005/10) Automated nomenclature of bronchial branches extracted from CT images and its application to biopsy path planning in virtual bronchoscopy. In: 8th international conference on Medical Image Computing and Computer Assisted Intervention (MICCAI 2005), Palm Springs, October 26–30, 2005, Proceedings, Part II, LNCS 3750, James S. Duncan, Guido Gering (eds), pp 854–861
199. Mori K, Ota S, Deguchi D, Kitasaka T, Suenaga Y, Iwano S, Hasegawa Y, Takabatake H, Mori M, Natori H (2009/09) Automated anatomical labeling of bronchial branches extracted from CT datasets based on machine learning and combination optimization and its application to bronchoscope guidance, Lecture Notes in Computer Science, vol 5762, pp 707–714
200. Mori K, Bui HH, Matsuzaki T, Oda M, Kitasaka T, Fujiwara M, Misawa K (2012/06) Automated anatomical labeling of abdominal arteries extracted from CT images based on machine learning, Int J Comput Assist Radiol Surg 7(1):S46–S47
201. Morra JH, Tu Z, Apostolova LG et al (2008) Automatic subcortical segmentation using a contextual model. Med Image Comput Comput Assist Interv 11:194–201
202. Moseley ME, Butts K, Yenari MA, Marks M, de Crespigny A (1995) Clinical aspects of DWI. NMR Biomed 8:87–96
203. Mowatt G, Cumminus F, Waugh N et al (2008) Systematic review of the clinical effectiveness and cost-effectiveness of 64-slice or higher computed tomography angiography as an alternative to invasive coronary angiography in the investigation of coronary artery disease. Health Technol Assess 12:iii–iiv ix-143
204. Mujat M, Chan R, Cense B et al (2005) Retinal nerve fiber layer thickness map determined from optical coherence tomography images. Opt Express 13:9480–9491
205. Muller D, Maeder A (2008) Robust semi-automated path extraction for visualizing stenosis of the coronary arteries. Comput Med Imaging Graph 32:463–475
206. Muramatsu C, Matsumoto T, Hayashi T et al (2013) Automated measurement of mandibular cortical width on dental panoramic radiographs. Int J Computer Assisted Radiology and Surgery 8:877–885
207. Murgio A, Patrick PD, Andrade FA, Boetto S, Leung KM, Munoz Sanchez MA (2001) International study of emergency department care for pediatric traumatic brain injury and the role of CT scanning. Childs Nerv Syst 17(4–5):257–262

208. Naegel B (2007) Using mathematical morphology for the anatomical labeling of vertebrae from 3D CT-scan images. Comput Med Imaging Graph 31:141–156
209. Nagashima H, Harakawa T (2008) Computer-aided diagnostic scheme for detection of acute cerebral infarctions on brain CT images. J Signal Process 12:73–80
210. Nakagomi K, Shimizu A, Kobatake H, Yakami M, Fujimoto K, Togashi K (2013) Multi-shape graph cuts with neighbor prior constraints and its application to lung segmentation from a chest CT volume. Med Image Anal 17:62–77
211. Nakaguchi T, Okui M, Tsumura N et al (2004) Pancreas extraction using a deformable model on abdominal CT image. In: International workshop on nonlinear circuits and signal processing, pp 387–390, Waikiki Beach Marriott Resort, Honolulu
212. Nakamura Y, Nimura Y, Kitasaka T et al (2013) Automatic abdominal lymph node detection method based on local intensity structure analysis from 3-D X-ray CT images. Proc SPIE 8670:86701 K-1–7
213. Nakao M, Maeda K, Haraguchi R, Kurosaki K, Kagisaki K, Shiraishi I, Nakazawa K, Minato K (2012) Cardiovascular modeling of congenital heart disease based on neonatal echocardiographic images. IEEE Trans Inf Technol Biomed 16:70–79
214. Narasimha-lyer H, Mahadevan V, Beach JM, Roysam B (2008) Improved detection of the central reflex in retinal vessels using a generalized dual-Gaussian model and robust hypothesis testing. IEEE Trans Inf Technol Biomed 12:406–410
215. Neil JJ, Shiran SI, McKinstry RC et al (1998) Normal brain in human newborns: apparent diffusion and diffusion anisotropy measured by using diffusion tensor MR imaging. Radiology 209:57–66
216. Nemoto M, Masutani Y, Hanaoka S et al (2011) A unified framework for concurrent detection of anatomical landmarks for medical image understanding. Proc SPIE 7962:79623E
217. Netter FH (2010) Atlas of human anatomy: with student consult access, 5e (Netter Basic Science), Saunders (ISBN 1416059512)
218. Nishii R, Eguchi S (2005) Supervised image classification by contextual AdaBoost based on posteriors in neighborhoods. IEEE Trans Geosci Remote Sens 43:2547–2554
219. Nowinski WL, Qian G, Kirgaval BP et al (2006) Analysis of ischemic stroke MR images by means of brain atlases of anatomy and blood supply territories. Acad Radiol 13:1025–1034
220. Oda M, Kitasaka T, Furukawa K, Watanabe O, Ando T, Hirooka Y, Goto H, Mori K (2014/06) An automated method for connecting separated luminal regions of intestine segmented from CT volumes, Int J Comp Assis Radiol Surg 9(1): S223–S224
221. Okada T, Shimada R, Sato Y, Hori M, Yokota K, Nakamoto M, Chen Y-W, Nakamura H, Tamura S (2007) Automated segmentation of the liver from 3D CT images using probabilistic atlas and multi-level statistical shape model. In: Medical Image Computing and Computer-Assisted Intervention–MICCAI 2007. Springer, Berlin/Heidelberg, pp 86–93
222. Okada T, Shimada R, Hori M, Nakamoto M, Chen Y-W, Nakamura H, Sato Y (2008) Automated segmentation of the liver from 3D CT images using probabilistic atlas and multilevel statistical shape model. Acad Radiol 15(11):1390–1403
223. Okada T, Yokota K, Hori M, Nakamoto M, Nakamura H, Sato Y (2008) Construction of hierarchical multi-organ statistical atlases and their application to multi-organ segmentation from CT images. In: Medical Image Computing and Computer-Assisted Intervention–MICCAI 2008. Springer, Berlin/Heidelberg, pp 502–509
224. Okada T, Linguraru MG, Yoshida Y, Hori M, Summers RM, Chen Y-W, Tomiyama N, Sato Y (2012) Abdominal multi-organ segmentation of CT images based on hierarchical spatial modeling of organ interrelations. In: Abdominal imaging. Computational and clinical applications. Springer, Berlin/Heidelberg, pp 173–180
225. Okada T, Linguraru MG, Hori M, Summers RM, Tomiyama N, Sato Y (2013) Abdominal multi-organ CT segmentation using organ correlation graph and prediction-based shape and location priors. In: Medical Image Computing and Computer-Assisted Intervention–MICCAI 2013. Springer, Berlin/Heidelberg, pp 275–282
226. Olabarriaga S, Breeuwer M, Niessen W et al (2003) Minimum cost path algorithm for coronary artery central axis tracking in CT images. MICCAI 2879:687–694

227. Olabarriaga SD, Rouet J, Fradkin M, Breeuwer M, Niessen WJ (2005/04) Segmentation of thrombus in abdominal aortic aneurysms from CTA with nonparametric statistical grey level appearance modeling, IEEE Trans Med Imaging 24(4)
228. Pamulapati V, Venkatesan A, Wood BJ, Linguraru MG (2012) Liver segmental anatomy and analysis from vessel and tumor segmentation via optimized graph cuts. In: Abdominal imaging. Computational and clinical applications. Springer, Berlin/Heidelberg, pp 189–197
229. Park H, Bland PH, Meyer CR (2003) Construction of an abdominal probabilistic atlas and its application in segmentation. IEEE Trans Med Imaging 22(4):483–492
230. Pohle R, Toennies KD (2001) Segmentation of medical images using adaptive region growing. Proc Int Soc Opt Eng 4322:1337–1346
231. Pu J, Leader JK, Zheng B et al (2009) A computational geometry approach to automated pulmonary fissure segmentation in CT examinations. IEEE Trans Med Imaging 28:710–719
232. Pu J, Fuhrman C, Duric J et al (2010) Computerized assessment pulmonary fissure integrity using high resolution CT. Med Phys 37:4661–4672
233. Quellec G, Lee K, Dolejsi M et al (2010) Three dimensional analysis of retinal layer texture: identification of fluid-filled regions in SD-OCT of the macula. IEEE Trans Med Imaging 29:1321–1330
234. Rajamani KT, Styner MA, Talib H, Zheng G, Nolte LP, Ballester MAG (2007) Statistical deformable bone models for robust 3D surface extrapolation from sparse data. Med Image Anal 11(2):99–109
235. Rangayyan RM, Deglint HJ, Boag GS (2006) (2006). Method for the automatic detection and segmentation of the spinal canal in computed tomographic images. J Electron Imaging 15:033007–033007
236. Rathke F, Schmidt S, Schnorr C (2011) Order preserving and shape prior constrained intra-retinal layer segmentation in optical coherence tomography. Proceedings of Medical Image Computing and Computer-Assisted Intervention – MICCAI 2011, pp 370–377
237. Reisert M, Mader I, Anastasopoulos C, Weigel M, Schnell S, Kiselev V (2011) Global fiber reconstruction becomes practical. NeuroImage 54(2):955–962
238. Roger VL, Go AS, Loyd-Jones DM et al (2012) Heart disease and stroke statistics—2012 update: a report from the American heart association. Circulation 125:e2–e220
239. Rosen HJ, Gorno-Tempini MI, Goldman WP, Perry RJ, Schuff N, Weiner M, Feiwell R, Kramer JH, Miller BL (2002) Patterns of brain atrophy in frontotemporal dementia and sematic dementia. Neurology 58:198–208
240. Rueda S, Gil JA, Pichery R, Alcaniz M (2006) Automatic segmentation of jaw tissues in CT using active appearance models and semi-automatic landmarking. In: Larsen R, Nielsen M, Sporring J (eds) MICCAI 2006; LNCS 4190, pp 167–174
241. Saita S, Kubo M, Kawata Y et al (2006) An algorithm for the extraction of pulmonary fissures from low-dose multislice CT image. Syst Comput Jpn 37:63–76
242. Salmond CH, Ashburner J, Vargha-Khadem F, Connelly A, Gadian DG, Friston KJ (2002) Distributional assumptions in voxel-based morphometry. NeuroImage 17:1027–1030
243. Sato Y, Nakajima S, Shiraga N (1998) Three-dimensional multi-scale line filter for segmentation and visualization of curvilinear structures in medical images. Med Image Anal 2(2):143–168
244. Sato K, Taki Y, Fukuda H, Kawashima R (2003) Neuroanatomical database of normal Japanese brains. Neural Netw 16:1301–1310
245. Sato K, Kinomura S, Taki Y, Goto R, Inoue K, Okada K, Kawashima R, Fukuda H (2009) Inclusion of affine transformation into no-linear deformation matrix in brain image analysis. NeuroImage 47(Supplement 1):S59
246. Schaap M, Metz CT, van Walsum T et al (2009) Standardized evaluation methodology and reference database for evaluating coronary artery centerline extraction algorithms. Med Image Anal 13:701–714
247. Schaap M, van Walsum T, Neefies L et al (2011) Robust shape regression for supervised vessel segmentation and its application to coronary segmentation in CTA. IEEE Trans Med Imaging 30:1974–1986

248. Schmid J, Magnenat-Thalmann N (2008) MRI bone segmentation using deformable models and shape priors. In Medical Image Computing and Computer-Assisted Intervention–MICCAI 2008. Springer, Berlin/Heidelberg, pp 119–126
249. Schmid J, Kim J, Magnenat-Thalmann N (2011) Robust statistical shape models for MRI bone segmentation in presence of small field of view. Med Image Anal 15(1):155–168
250. Schormann T, Henn S, Zilles K (1996) A new approach to fast elastic alignment with application to human brain. Lect Notes Comput Sci 1131:337–342
251. Seim H, Kainmueller D, Heller M, Lamecker H, Zachow S, Hege HC (2008) Automatic segmentation of the pelvic bones from CT data based on a statistical shape model. In: Proceedings of the first eurographics conference on visual computing for biomedicine, pp 93–100
252. Selle D, Preim B, Schenk A, Peitgen H-O (2002) Analysis of vasculature for liver surgical planning. IEEE Trans Med Imaging 21(11):1344–1357
253. Shankar PM, Dumane VA, Reid JM, Genis V, Forsberg F, Piccoli CW, Goldberg BB (2000) Use of the K-distribution for classification of breast masses. Ultrasound Med Biol 26:1503–1510
254. Shankar PM, Dumane VA, Reid JM, Genis V, Forsberg F, Piccoli CW, Goldberg BB (2001) Classification of ultrasonic B-mode images of breast masses using Nakagami distribution. IEEE Trans Ultrason Ferroelectr Freq Control 48:569–580
255. Shimizu A, Ohno R, Ikegami T et al (2007) Segmentation of multiple organs in non-contrast 3D abdominal CT images. Int J Comput Assist Radiol Surg 2:135–142
256. Shimizu A, Ohno R, Ikegami T, Kobatake H, Nawano S, Smutek D (2007) Segmentation of multiple organs in non-contrast 3D abdominal CT images. Int J Comput Assist Radiol Surg 2(3–4):135–142
257. Shimizu A, Narihira T, Furukawa D, Kobatake H, Nawano S, Shinozaki K (2008) Ensemble segmentation using AdaBoost with application to liver lesion extraction from a CT volume. In: Proceedings of MICCAI workshop on 3D segmentation in the clinic: a grand challenge, USA
258. Shimizu A, Kimoto T, Kobatake H, Nawano S, Shinozaki K (2010) Automated pancreas segmentation from three-dimensional contrast enhanced computed tomography. Int J Comput Assist Radiol Surg 5:85–98
259. Shimizu A, Kitasaka T, Nawano S, Shinozaki K, Tateno Y (2011) 3D medical image processing algorithm competition in Japan. In: Proceedings of International Forum on Medical Imaging in Asia (IFMIA) 2011: S1–2
260. Shimizu A, Narihira T, Kobatake H, Furukawa D, Nawano S, Shinozaki K (2013) Ensemble learning based segmentation of metastatic liver tumours in contrast-enhanced computed tomography. IEICE Trans Inf Syst 96(4):864–868
261. Shimizu A, Shindo K, Kobatake H, Nawano S, Shinozaki K (2013) Proposal of a novel boosting algorithm regularized by a statistical shape feature and its application to organ segmentation. Med Imag Technol 31:121–131 (in Japanese)
262. Sluimer IC, Niemeijer M, van Ginneken B et al (2005) Toward automated segmentation of the pathological lung in CT. IEEE Trans Med Imaging 24:1025–1038
263. Sluimer IC, Schilham A, Prokop M et al (2006) Computer analysis of computed tomography scans of the lung: a survey. IEEE Trans Med Imaging 25:385–405
264. Soler L, Delingette H, Malandain G, Montagnat J, Ayache N, Koehl C, Dourthe O et al (2001) Fully automatic anatomical, pathological, and functional segmentation from CT scans for hepatic surgery. Comput Aided Surg 6(3):131–142
265. SPM. http://www.fil.ion.ucl.ac.uk/spm/
266. Sporns O, Tononi G, Kötter R (2005) The human connectome: a structural description of the human brain. PLoS Comput Biol 1(4):e42
267. Staal J, van Ginneken B, Viergever MA (2007) Automatic rib segmentation and labeling in computed tomography scans using a general framework for detection, recognition and segmentation of objects in volumetric data. Med Image Anal 11:35–46

268. Štern D, Likar B, Pernuš F, Vrtovec T (2011) Parametric modelling and segmentation of vertebral bodies in 3D CT and MR spine images. Phys Med Biol 56:7505
269. Styner M, Lieberman JA, McClure RK, Weinberger DR, Jones DW, Gerig G (2005) Morphometric analysis of lateral ventricles in schizophrenia and healthy controls regarding genetic and disease-specific factors. Proc Natl Acad Sci U S A 102(13):4872–4877
270. Summers RM, Agcaoili CML, McAuliffe MJ et al (2001) Helical CT of von Hippel–Lindau: semi-automated segmentation of renal lesions. In: Proceeding of international conference on image processing, pp 93–296
271. Suzuki Y, Iwase A, Araie M, Yamamoto T, Abe H, Shirato S, Kuwayama Y, Mishima HK, Shimizu H, Tomita G, Inoue Y, Kitazawa Y, Tajimi Study Group (2006) Risk factors for open-angle glaucoma in a Japanese population, The Tajimi study. Ophthalmology 113:1613–1617
272. Szymczak A, Stillman A, Tannenbaum A et al (2006) Coronary vessel trees from 3D imagery: a topological approach. Med Image Anal 10:548–559
273. Tabuchi A, Katsuda T, Eguchi M, Gotanda R, Gotanda T, Mitani M, Takeda Y (2009) Functional residual capacity breath hold for subtraction image of dynamic liver MRI. Eur J Radiol 71:506–512
274. Tabuchi A, Katsuda T, Gotanda R, Gotanda T, Mitani M, Takeda Y (2009) High resolution T2 weighted liver MR imaging using functional residual capacity breath-hold with a 1.0-Tesla scanner. Eur J Radiol 72:300–305
275. Takada E, Ikedo Y, Fukuoka D, Hara T, Fujita H, Endo T, Morita T (2007) Semi-automatic ultrasonic full-breast scanner and computer assisted detection system for breast cancer mass screening. In: Proceedings of SPIE medical imaging 2007: Computer-Aided Diagnosis. 2007; 6514: 65141 T-1–65141 T-10
276. Takahashi N, Lee Y, Tsai DY et al (2007) Novel noise reduction filter for improving visibility of early computed tomography signs of hyperacute stroke: evaluation of the filter's performance–preliminary clinical experience. Radiat Med 25:247–254
277. Takahashi N, Lee Y, Tsai DY et al (2008) Improvement of detection of hypoattenuation in acute ischemic stroke in unenhanced computed tomography using an adaptive smoothing filter. Acta Radiol 49:816–826
278. Takemura A, Shimizu A, Hamamoto K (2010) Discrimination of breast tumors in ultrasonic images using an ensemble classifier based on the AdaBoost algorithm with feature selection. IEEE Trans Med Imaging 29:598–609
279. Takemura A, Shimizu A et al (2010) A cost-sensitive extension of AdaBoost with Markov random field priors for automated segmentation of breast tumors in ultrasonic images. Int J Comput Assist Radiol Surg 5:537–547
280. Taki Y, Goto R, Evans A, Zijdenbos A, Neelin P, Lerch J, Sato K, Ono S, Kinomura S, Nakagawa M, Sugiura M, Watanabe J, Kawashima R, Fukuda H (2004) Voxel-based morphometry of human brain with age and cerebrovascular risk factors. Neurobiol Aging 25:455–463
281. Taki Y, Kinomura S, Awata S, Inoue K, Sato K, Ito H, Goto R, Uchida S, Tsuji I, Arai H, Kawashima R, Fukuda H (2005) Male elderly subthreshold depression patients have smaller volume of medial part of prefrontal cortex and precentral gyrus compared with age-matched normal subjects: a voxel-based morphometry. J Affect Disord 88:313–320
282. Taki Y, Thyeau B, Kinomura S, Sato K, Goto R, Kawashima R, Fukuda H (2011) Correlations among brain gray matter volumes, age, gender, and hemisphere in healthy individuals. PLoS One 6:1–13
283. Taki Y, Kinomura S, Sato K, Goto R, Kawashima R, Fukuda H (2011) A longitudinal study of gray matter volume decline with age and modifying factors. Neurobiol Aging 32:907–915
284. Tavakoli V, Amini AA (2013) A survey of shaped-based registration and segmentation techniques for cardiac images. Comput Vis Image Underst 117:966–989
285. The National Institute of Neurological Disorders and Stroke rt-PA Stroke Study Group (1995) Tissue plasminogen activator for acute ischemic stroke. N Engl J Med 333:1581–1587
286. The National Lung Screening Trial Research Team (2011) Reduced lung-cancer mortality with low-dose computed tomographic screening. N Engl J Med 365:395–409

287. Tomoshige S, Oost E, Shimizu A, Watanabe H, Nawano S (2014) A conditional statistical shape model with integrated error estimation of the conditions: application to liver segmentation in non-contrast CT images. Med Image Anal 18(1):130–143
288. Tomoshige S, Oost E, Shimizu A, Watanabe H, Nawano S (2014) A conditional statistical shape model with integrated error estimation of the conditions; application to liver segmentation in non-contrast CT images. Med Image Anal 18:130–143
289. Tsagaan B, Shimizu A, Kobatake H et al (2002) An automated segmentation method of kidney using statistical information. Med Image Comput Comput Assist Interv 1:556–563
290. Tschirren J, McLennan G, Palagyi K, Hoffman EA, Sonka M (2005) Matching and anatomical labeling of human airway tree. In: IEEE Transactions of Medical Imaging, pp 1540–1547
291. Tsui P-H, Liao Y-Y, Chang C-C, Kuo W-H, Chang K-J, Yeh C-K (2010) Classification of benign and malignant breast tumors by 2-d analysis based on contour description and scatterer characterization. IEEE Trans Med Imaging 29:513–522
292. Tsui P-H, Yeh C-K, Liao Y-Y, Chang C-C, Kuo W-H, Chang K-J, Chen C-N (2010) Ultrasonic Nakagami imaging: a strategy to visualize the scatterer properties of benign and malignant breast tumors. Ultrasound Med Biol 36:209–217
293. Tuch DS (2004) Q-ball imaging. Magn Reson Med 52(6):1358–1372
294. Ukai Y, Niki N, Satoh H et al (1997) A coronary calcification diagnosis system based on helical CT images. IEEE Trans Nucl Sci 45:3083–3088
295. Ukil S, Reinhardt JM (2009) Anatomy-guided lung lobe segmentation in X-ray CT images. IEEE Trans Med Imaging 28:202–214
296. Vadakkumpadan F, Arevalo H, Prass AJ, Chen J, Kickinger F, Kohl P, Plank G, Trayanova NA (2010) Image-based models of cardiac structure in health and disease. Wiley Interdiscip Rev Syst Biol Med 2:489–506
297. van Rikxoor EM, van Ginneken B, Klik M et al (2008) Supervised enhancement filters: application to fissure detection in chest CT scans. IEEE Trans Med Imaging 27:1–10
298. Vital Statistics of Japan (2013) Ministry of Health, Labour and Welfare, Japan http://www.mhlw.go.jp/toukei/saikin/hw/jinkou/kakutei13/dl/10_h6.pdf
299. Robin W, Chu C, Kazunari M, Kensaku M, Daniel R (2012) Multi-organ abdominal CT segmentation using hierarchically weighted subject-specific atlases. In: Medical Image Computing and Computer-Assisted Intervention–MICCAI 2012. Springer Berlin/Heidelberg, pp 10–17
300. Wang C, Smedby O (2007) Coronary artery segmentation and skeletonization based on competing fuzzy connectedness tree. MICCAI 4791:311–318
301. Wang H, Yushkevich PA (2013) Multi-atlas segmentation with joint label fusion and corrective learning—an open source implementation. Front Neuroinform 7
302. Wang L, Bhalerao A, Wilson R (2007) Analysis of retinal vasculature using a multiresolution Hermite-Gaussian model. IEEE Trans Med Imaging 26:137–152
303. Wedeen VJ, Wang RP, Schmahmann JD, Benner T, Tseng WYI, Dai G, Pandya DN, Hagmann P et al (2008) Diffusion spectrum magnetic resonance imaging (DSI) tractography of crossing fibers. NeuroImage 41(4):1267–1277
304. Wehbe H, Ruggeri M, Jiao S et al (2007) Automatic retinal blood flow calculation using spectral domain optical coherence tomography. Opt Express 15:15193–15206
305. Weickert J (1999) Coherence-enhancing diffusion filtering. Int J Comput Vis 31:111–127
306. Wenk JF, Klepach D, Lee LC, Zhang Z, Ge L, Tseng E, Martin A, Kozerke S, Gorman JH III, Gorman RC, Guccione JM (2012) First evidence of depressed contractility in the border zone of a human myocardial infarction. Ann Thorac Surg 93:1188–1193
307. Wenkel E, Heckmann M, Heinrich M, Schwab SA, Uder M, Schulz-Wendtland R, Bautz WA, Janka R (2008) Automated breast ultrasound: lesion detection and BI-RADS classification-a pilot study. Röfo 180:804–808
308. Whitmarsh T, Treece GM, Poole KES (2013) Automatic segmentation and discrimination of connected joint bones from CT by multi-atlas registration. In: MICCAI workshop on computational methods and clinical applications for spine imaging

309. Whitwell JL (2009) Voxel-based morphometry: an automated technique for assessing structural changes in the brain. J Neurosci 29:9661–9664
310. Whitwell JL, Josephs KA (2007) Voxel-based morphometry and its application to movement disorders. Parkinsonism Relat Disord 13(supple 3):s406–s416
311. Whitwell JL, Jack CR Jr, Boeve BF, Senjem ML, Baker M, Rademakers R, Ivnik RJ, Knopman DS, Wsdolek ZK, Peterson RC, Josephs KA (2009) Voxel-based morphometry patterns of atrophy in FTLD with mutations in MAPT or PGRN. Neurology 72:813–820
312. Wiegell MR, Larsson Henrik BW, Wedeen Van J (2000) Fiber crossing in human brain depicted with diffusion tensor MR imaging1. Radiology 217:897–903
313. Wolz R, Chu C, Misawa K, Fujiwara M, Mori K, Rueckert D (2013) Automated abdominal multi-organ segmentation with subject-specific atlas generation. IEEE Trans Med Imaging 32(9):1723–1730
314. Woods RP, Grafton ST, Holmes CJ, Cherry SR, Mazziotta JC (1998) Automated image registration: I. General methods and intrasubject, intramodality validation. J Comput Assist Tomogr 22:139–152
315. Wright IC, McGuire PK, Poline JB, Travere JM, Murray RM, Frith CD, Frackowiak RS, Friston KJ (1995) A voxel-based method for the statistical analysis of gray and white matter density applied to schizophrenia. NeuroImage 2:244–252
316. Wu S, Weinstein SP, Conant EF, Kontos D (2013) Automated fibroglandular tissue segmentation and volumetric density estimation in breast MRI using an atlas-aided fuzzy C-means method. Med Phys 40:122302
317. Xu Y, Liang G, Hu G et al (2012) Quantification of coronary arterial stenosis in CTA using fuzzy distance transform. Comput Med Imaging Graph 36:11–24
318. Yang Y, Tannenbaum A, Giddens D et al (2007) Automatic segmentation of coronary arteries using Bayesian driven implicit surfaces. In: Proceedings of 4th IEEE ISBI, pp 189–192
319. Yang Q, Reisman CA, Wang Z et al (2010) Automated layer segmentation of macular OCT images using dual-scale gradient information. Opt Express 18:21294–21307
320. Yao J, Taylor R (2003) Assessing accuracy factors in deformable 2D/3D medical image registration using a statistical pelvis model. In: Ninth IEEE international conference on Computer Vision, 2003. IEEE proceedings, pp 1329–1334
321. Yao J, O'Connor SD, Summers RM (2006). Automated spinal column extraction and partitioning. In: Biomedical imaging: nano to macro. 3rd IEEE International Symposium, pp 390–393. IEEE
322. Yazdanpanah A, Hamarneh G, Smith B et al (2009) Intra-retinal layer segmentation in optical coherence tomography using an active contour approach. Med Image Comput Comput Assist Interv Pt 2:649–656
323. Yokota F, Okada T, Takao M, Sugano N, Tada Y, Sato Y (2009) Automated segmentation of the femur and pelvis from 3D CT data of diseased hip using hierarchical statistical shape model of joint structure. In: Medical Image Computing and Computer-Assisted Intervention–MICCAI 2009. Springer, Berlin/Heidelberg, pp 811–818
324. Yokota F, Okada T, Takao M, Sugano N, Tomiyama N, Sato Y, Tada Y (2012) Automated localization of pelvic anatomical coordinate system from 3D CT data of the hip using statistical atlas. Med Imaging Technol 30(1):43–52 (in Japanese)
325. Yokota F, Okada T, Takao M, Sugano N, Tada Y, Tomiyama N, Sato Y (2013) Automated CT segmentation of diseased hip using hierarchical and conditional statistical shape models. In: Medical Image Computing and Computer-Assisted Intervention–MICCAI 2013. Springer, Berlin/Heidelberg, pp 190–197
326. You JS, Kim SW, Lee HS, Chung SP (2010) Use of diffusion-weighted MRI in the emergency department for unconscious trauma patients with negative brain CT. Emerg Med J 27(2):131–132
327. Zhan Y, Maneesh D, Harder M, Zhou XS (2012) Robust MR spine detection using hierarchical learning and local articulated model. In: Medical Image Computing and Computer-Assisted Intervention–MICCAI 2012. Springer, Berlin/Heidelberg, pp 141–148

328. Zhang L, Hoffman EA, Reinhardt JM (2006) Atlas-driven lung lobe segmentation in volumetric X-ray CT images. IEEE Trans Med Imaging 25:1–16
329. Zheng B, Leader JK, Maitz GS et al (2003) A simple method for automated lung segmentation in X-ray CT images. Proc SPIE Med Imaging 5032:1455–1463
330. Zheng G, Gollmer S, Schumann S, Dong X, Feilkas T, González Ballester MA (2009) A 2D/3D correspondence building method for reconstruction of a patient-specific 3D bone surface model using point distribution models and calibrated X-ray images. Med Image Anal 13(6):883–899
331. Zhou X, Kitagawa T, Hara T, Fujita H, Zhang X, Yokoyama R, Kondo H, Kanematsu M, Hoshi H (2006) Constructing a probabilistic model for automated liver region segmentation using non-contrast X-ray torso CT images. In: Medical Image Computing and Computer-Assisted Intervention–MICCAI 2006. Springer, Berlin/Heidelberg, pp 856–863
332. Zhou X, Hayashi T, Han M, Chen H, Hara T, Fujita H, Yokoyama R, Kanematsu M, Hoshi H (2009) Automated segmentation and recognition of the bone structure in non-contrast torso CT images using implicit anatomical knowledge. Proc SPIE 7259:72593S. doi:10.1117/12.812945
333. Zhou C, Chan HP, Chughtai A et al (2012) Automated coronary artery tree extraction in coronary CT angiography using a multiscale enhancement and dynamic balloon tracking (MSCAR-DBT) method. Comput Med Imaging Graph 36:1–10
334. Zhuge F, Rubin GD, Sun S, Napel S (2006) An abdominal aortic aneurysm segmentation method: level set with region and statistical information. Med Phys 33:1440

Chapter 4
Applied Technologies and Systems

Kensaku Mori, Noboru Niki, Yoshiki Kawata, Hiroshi Fujita, Masahiro Oda, Hyoungseop Kim, Hidetaka Arimura, Akinobu Shimizu, Sakon Noriki, Kunihiro Inai, and Hirohiko Kimura

K. Mori (✉)
Information and Communications, Nagoya University, Furo-cho, Chikusa-ku, 464-8601, Nagoya, Japan
e-mail: kensaku@is.nagoya-u.ac.jp

N. Niki • Y. Kawata
Graduate School of Science and Technology, Tokushima University, 2-1 Minamijyousanjima-cho, 770-8506, Tokushima, Japan
e-mail: niki@tokushima-u.ac.jp; kawata@tokushima-u.ac.jp

H. Fujita
Department of Intelligent Image Information, Graduate School of Medicine, Gifu University, 1-1 Yanagido, 501-1194, Gifu, Japan
e-mail: fujita@fjt.info.gifu-u.ac.jp

M. Oda
Department of Media Science, Graduate School of Information Science, Nagoya University, Furo-cho,Chikusa-ku, 464-8601, Nagoya, Japan
e-mail: moda@is.nagoya-u.ac.jp

H. Kim
Department of Mechanical and Control Engineering, Graduate School of Engineering, Kyushu Institute of Technology, 1-1 Sensui-cho, 804-0015, Kitakyushu, Tobata-ku, Japan
e-mail: kimhs@cntl.kyutech.ac.jp

H. Arimura
Faculty of Medical Sciences, Division of Medical Quantum Science, Department of Health Sciences, Kyushu University, 3-1-1, Maidashi, 812-8582, Fukuoka, Higashi-ku, Japan
e-mail: arimurah@med.kyushu-u.ac.jp

A. Shimizu
Institute of Engineering, Tokyo University of Agriculture and Technology, 184-8588, Tokyo, Japan
e-mail: simiz@cc.tuat.ac.jp

S. Noriki
Division of Tumor Pathology, Department of Pathological Sciences, Faculty of Medical Sciences, University of Fukui, 23-3 Matsuoka-Shimoaizuki, Eiheiji, Fukui 910-1193, Fukui, Japan
e-mail: noriki@u-fukui.ac.jp

K. Inai
Division of Molecular Pathology, Department of Pathological Sciences, Faculty of Medical Sciences, University of Fukui, 23-3 Matsuoka-Shimoaizuki, Eiheiji, Fukui 910-1193, Fukui, Japan
e-mail: kinai@u-fukui.ac.jp

H. Kobatake, Y. Masutani (eds.), *Computational Anatomy Based on Whole Body Imaging*, DOI 10.1007/978-4-431-55976-4_4

Abstract This chapter shows applied technologies using computational anatomy (CA) models. CA systems based on clinical images assist physicians by providing useful information related to diagnostic and therapeutic procedures. Such systems include computer-aided diagnosis and computer-assisted surgery systems. A thorough understanding of anatomy is essential when designing these systems. It is important to understand how anatomical information extracted by a computer is used. In this chapter, we introduce applications of CA in three categories: (a) computer-aided diagnosis, (b) computer-assisted therapy and intervention, and (c) computer-assisted autopsy imaging. The technical details of these applications are discussed.

Keywords Computer-aided diagnosis • Computer-assisted surgery medical image processing applications

4.1 Application and Systematization of CA

Computational anatomy (CA) has the potential to change the world of medical imaging techniques. Its detailed analysis of anatomy holds the promise of improved quality and precision of diagnostic imaging and therapy.

Well-designed CA-assisted medical imaging and surgical systems are beginning to play a role in many centers. These systems are mainly classified into three categories: (a) computer-aided diagnosis (CAD), (b) surgical assistance systems, and (c) fusion-aid system of diagnosis and surgery.

CA aims to systematize whole-body anatomy based on medical images with the aim of providing supporting technologies for medical image interpretation and surgery. For example, in automated detection of lymph nodes from computed tomographic (CT) images, anatomical structure information may reduce false-positive regions from candidate sets of lymph nodes. The Hessian-based approach for lymph node extraction from volumetric CT images detects many false-positive foci in regions such as the small or large intestine. Such false-positive regions can be easily removed if we could understand patient anatomy from CT images.

Another example is utilization of patient anatomy data from medical images in computer-assisted intervention. Surgical navigation assistance is used widely for efficiency and achieving optimal outcomes. A typical surgical navigation system can show the location of a forceps in real time on three-dimensional (3D) rendered views of a surgical patient. A surgical navigation system provides 3D rendering of

H. Kimura
Division of Radiology, Department of Radiology and Laboratory Medicine, Faculty of Medical Sciences, University of Fukui, 23-3 Matsuoka-Shimoaizuki, Eiheiji, Fukui 910-1193, Fukui, Japan
e-mail: kimura@u-fukui.ac.jp

patient anatomy, which is often called the "virtual endoscopic view," synchronized with endoscopic motion. If these views are color-coded to anatomical structures, the surgical navigation view becomes much informative. Such views may streamline surgeons' intraoperative decisions. This review will explore new horizons of surgical assistance.

4.2 Computer-Assisted Diagnosis

CAD has been rapidly developing over the past three decades. Using dedicated computer systems, CAD interprets medical images and provides a "second opinion." The final medical decision, however, is ultimately made by a physician [1–3]. Studies on CAD systems and technology reveal that CAD improves diagnostic accuracy of physicians and lightens the burden of increasing workload. Moreover, CAD reduces the number of lesions overlooked because of fatigue or high volumes of data and improves inter- and intrareader variability. CAD has been successfully applied in a variety of clinical areas such as mammography, chest radiography and CT, and CT colonography. All of these technologies are approved by the US Food and Drug Administration (FDA) and are commercially available in the USA and some other countries. These CAD systems are classified as computer-aided detection (CADe) systems and can be differentiated from so-called computer-aided diagnosis (CADx) systems, which evaluate conspicuous structures and determine whether they are benign or malignant.

Enlisting the assistance of a computer to analyze medical images is not a new idea. In fact, in 1963, Lodwick et al. investigated using a computer to diagnose bone tumors [4]. In 1964, Meyers et al. proposed a system to automatically distinguish normal chest radiographs from abnormal ones by measuring the cardiothoracic ratio [5]. In 1967, Winsberg et al. employed optical scanning to detect radiographic abnormalities in mammograms [6]. Although these earlier systems differ from CAD as we know it today, most of these researchers were well on their way to designing automated diagnostic systems. One of the most important years in the history of CAD was 1998. This year marked the transition of CAD technologies from the research phase to clinical practice with the success of ImageChecker (R2 Technology, Inc., Sunnyvale CA, USA; later acquired by Hologic, Inc., Bedford MA, USA in 2006), which obtained FDA approval. ImageChecker is a computer system intended to mark regions of interest (ROIs) on routine screening mammograms. Today, it is estimated that more than 10,000 mammography CAD systems are used by hospitals, clinics, and screening centers across the USA.

CAD systems target lesions in every human organ and tissue and work with every imaging modality currently in clinical use. Many of the CAD technologies developed to date rely on brute force. For CADe, this means systematically locating regions displaying lesion-like features. With breast mass detection, for example, CADe systems search for regions having the potential characteristics of a tumor by looking for oval shapes with a certain level of contrast. Similarly, CADx systems measure and analyze malignancy characteristics of candidate regions (e.g., the state of the edges) using feature values. Incorporating the new concept of CA into existing CAD systems is very rare [7]. Regardless, new techniques are being developed and applied, to a limited extent, to basic research in image analysis and as a component of CAD systems.

While a number of CAD systems have been implemented for clinical use, several technical problems exist that must be addressed. In CADe systems, false-positive (FP) cases, where the computer erroneously detects a lesion, always exist. The FP rates in current CADe systems are known to be five to ten times greater than those of physicians. To further reduce the FP rate and enhance CADe performance, new techniques must be incorporated that consider anatomical structural information in regions surrounding candidate lesions. For example, if a CADe system can accurately determine whether a feature found in a thoracic CT is a nodule or part of a vessel, the FP rate can potentially be greatly reduced. For this to be possible, a technique based on anatomical information that accurately differentiates between the two types of vessels (arteries and veins) is required.

A more sophisticated CAD system would be one with the capability of identifying lesions across multiple organs. The system would first need to determine where the target organ or organs are located within the image. In the case of multiple organs, information regarding their location in relation to each other would also be required. Techniques that can address issues such as automatic landmark detection, a probabilistic atlas construction method, a statistical shape model (SSM) construction method, and utilization of an anatomical knowledge database are explained in Chap. 2. These techniques must be expressed mathematically to achieve these goals.

It is expected that techniques based on CA will lead to the development of next-generation CAD systems that are even more sophisticated, with the capability to handle multiple organs and multiple diseases.

4.2.1 Detection

4.2.1.1 Lung Nodule Detection

Introduction

This section discusses CAD of lung cancer on CT images. A promising application of computational anatomical models is the automatic detection of pulmonary nodules as lung cancer candidates. Another application is to suggest malignancy or benignity of nodules to support radiologists/physicians in diagnosing the disease. Other applications, including characterization of time-interval changes of pulmonary nodules using follow-up CT images, and prognosis generation, are important topics in the implementation of CAD in lung cancer management.

Lung cancer is the most common cause of cancer death worldwide [8, 9] . CT is the modality of choice for lung imaging. There is abundant evidence that CT screening with a low-dose imaging protocol improves sensitivity for identification of pulmonary nodules compared with plain radiography [10–12]. Research results from the National Lung Screening Trial (NLST) revealed that screening for lung nodules with low-dose CT (LDCT) reduced lung cancer mortality in heavy smokers by 20% compared with plain chest radiography [13].

Recently, results of two lung cancer screening studies showed the ability of CT to differentiate malignant from benign nodules before invasive biopsy procedures are considered [14, 15]. In a study by Aberle et al., an analysis of NLST results found that by the third annual screening, radiologist can distinguish malignant from benign nodules detected on earlier screening rounds based on change over time [14]. Since even benign nodules can increase in size, the quantitative tools in CAD system might assist radiologists to assess the growth ratio of nodules in annual screening. McWilliams et al. found that analyzing patient and nodule characteristics can be used to estimate the malignant potential of pulmonary nodules detected on baseline LDCT screening [15]. The authors created two prediction models to determine whether a nodule detected on the first CT scan was cancerous or not. One model consisted of the limited malignancy predictors that were significant, and the other included additional variables thought to be associated with a higher risk of malignancy. Analysis of nodule characteristics included nodule type (solid, nonsolid, or part solid), whether nodule margins were smooth or spiculated, nodule location in the lung, and presence of visible emphysema. Analyses of these nodule characteristics were carried out by visual assessment. They concluded that predictors in the full model included older age, female sex, family history of lung cancer, emphysema, larger nodule size, location of the nodule in the upper lobe, part-solid nodule type, lower nodule count, and spiculation. The readers who are interested in the management of solitary and multiple pulmonary nodules can refer the statement from the Fleischner Society [16]. Subjective radiologic descriptors of pulmonary nodules are replaced by quantitative metrics that enable statistical comparisons between features and clinical outcomes; computer-aided diagnosis for lung nodules (CADx) has become one of the most active research areas as well as CADe [17–25].

Generic Scheme in CADe Systems for Lung Cancer

A generic scheme in CADe systems of LDCT screening for lung cancer usually consists of the following major steps: LDCT acquisition, preprocessing, segmentation of pulmonary structures, initial candidate detection, reduction of false-positive (FP) detections, and lung nodule detection.

(a) **LDCT acquisition:** The scan and reconstruction parameters of CT are important factors in the performance of CADe. Narrow collimation with reconstructions of thin sections is recommended to improve the detection of nodules [26–28]. Radiation dose is a key concern [29, 30]. With the advance of iterative reconstruction techniques, ultralow-dose chest CT with a radiation dose comparable to that of chest radiography might be considered when designing future screening protocols [30]. To develop, train, and validate CADe systems, public databases are available. The Lung Image Database Consortium (LIDC) offers annotated chest CT scans [31]. To compare CADe systems for nodule detection, the ANODE09 dataset is another publicly available database in which all data were provided by the University Medical Center Utrecht and originate from the NELSON study, the largest CT lung cancer screening trial in Europe [32].To benchmark the performance of developed CAD systems, the publicly available CT databases become more important [33, 34].

(b) **Preprocessing:** Before beginning with detection steps, some initial processing is performed on the original CT images to remove defects caused by the image acquisition process such as noise and to enhance the characteristics of lung nodule candidates [25, 35–37].

(c) **Segmentation of pulmonary structures:** Figure 4.1 shows anatomical structures on an axial chest CT section. The segmentation of the left and right lungs from chest CT images is performed to restrict the nodule detection to the lung volumes. The major portion of the lungs comprises lung parenchyma that is

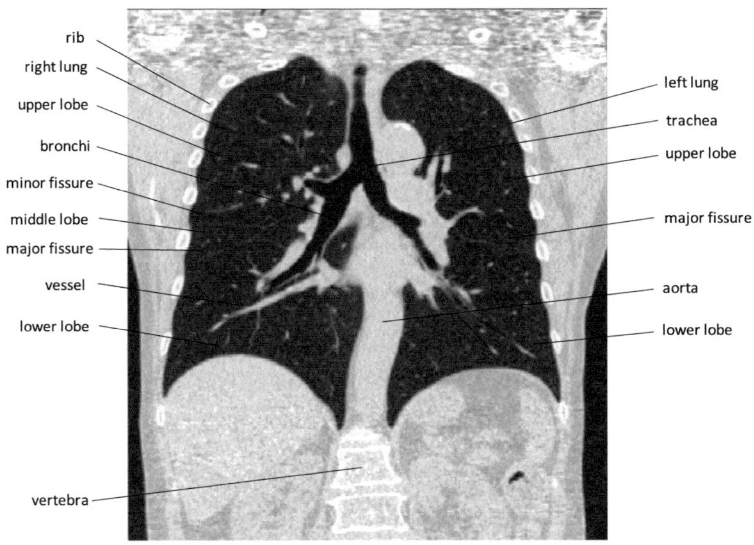

Fig. 4.1 Pulmonary structures on CT images. Within the thorax, the ribs enclose the lungs, and the diaphragm lies beneath the bases of the lungs, separating the thoracic and abdominal cavities. The mediastinum between the two lungs consists of the heart, major blood vessels, the esophagus, and the trachea. The pulmonary arteries enter the lungs; the pulmonary veins exit the lungs. The blood vessels, airways, and lymphatics at the root of each lung collect in the hilum and enter the mediastinum. The lungs consist of airways, vessels, and the lung parenchyma. The left and right lungs are usually subdivided into two lobes (the upper and lower lobes) and three lobes (the upper, middle, and lower lobes), respectively. The five lobes are separated by fissures of varying completeness, which are potential spaces lined by the visceral pleura. The visceral pleura covers all the lung parenchyma, and a second layer of parietal pleura is attached to the chest wall and the mediastinum. The lobes are further subdivided into segments, which are defined by bronchial supply

involved in gas exchange. Because the lung parenchyma has a lower density, around −900 HU, than the surrounding tissue in chest CT images of healthy subjects, many lung segmentation algorithms are based on a thresholding approach. The threshold-based methods consist of three major steps: (1) extraction of the preliminary lung regions using thresholding, (2) identification of lungs and separation between left and right lungs, and (3) refinement of the lung shapes to smooth the borders and include vessels in the segmentation result [18, 23, 24]. When the higher densities of the abnormalities are included in the lungs compared with the density of normal lung parenchyma, the conventional threshold-based lung segmentation methods result in segmentation errors [24, 38]. To handle this situation, atlas-based segmentation of pathological lungs [39], hybrid lung segmentation in which a conventional threshold-based method and a multi-atlas-based algorithm using nonrigid registration are combined [40], and a graph cut-based segmentation in which multiple possible shapes of lungs can be taken into account [41] have been proposed. Segmentation methods of vessels, airways, pleurae, lobes, segments, and ribs have been studied [18, 24, 42]. Figure 4.2 presents some segmentation results of pulmonary structures [42–45].

(d) **Initial candidate detection:** After preprocessing, initial candidate detection is employed to locate potential lung nodules. There are many strategies to detect nodule candidates [18–25, 32]: multiple gray-level thresholding [36, 46, 47], fuzzy clustering and surface curvature [45], template matching [48], a model-based image understanding technique [49], a mathematic model of anatomic structures [50, 51], mathematical morphology [52, 53], a convergence index filter [54], Gaussian curve fitting [55], shape-based genetic algorithm [56], geometric model based on the analysis of the signed distance field [57], shape index [35, 58], intensity structure enhancement [37, 59, 60], and gradient analysis [37, 61, 62].

(e) **Reduction of FP detections and lung nodule detection:** The pattern features of lesion candidates such as gray-level-based features, texture features, and morphological features are extracted. Once a candidate's characteristics are obtained, the step tries to remove FPs and retain potential lung nodules. In this procedure, classifiers are widely used [18–25, 32]. The role of the classifiers is to determine optimal boundaries between lung nodules and non-lesions in the multidimensional feature space, which is generated by the input features of the candidates. There are a number of classification techniques: linear discriminant analysis [36, 46, 54, 60–62], rule-based classifier [45, 47, 48, 53, 56, 59], 3D MRF models [50], neural network [37, 52, 55, 63], Bayesian classifier [51], fuzzy logic [49], shape similarity [57], support vector machine [58], and k-nearest neighbor [35].

Table 4.1 summarizes the principal methods of lung nodule detection that are covered in this section, looking at sensitivity, FP rate, number of nodules used in validation, and size of nodules. The performances given are the best performances when performances based on several conditions are included in the study

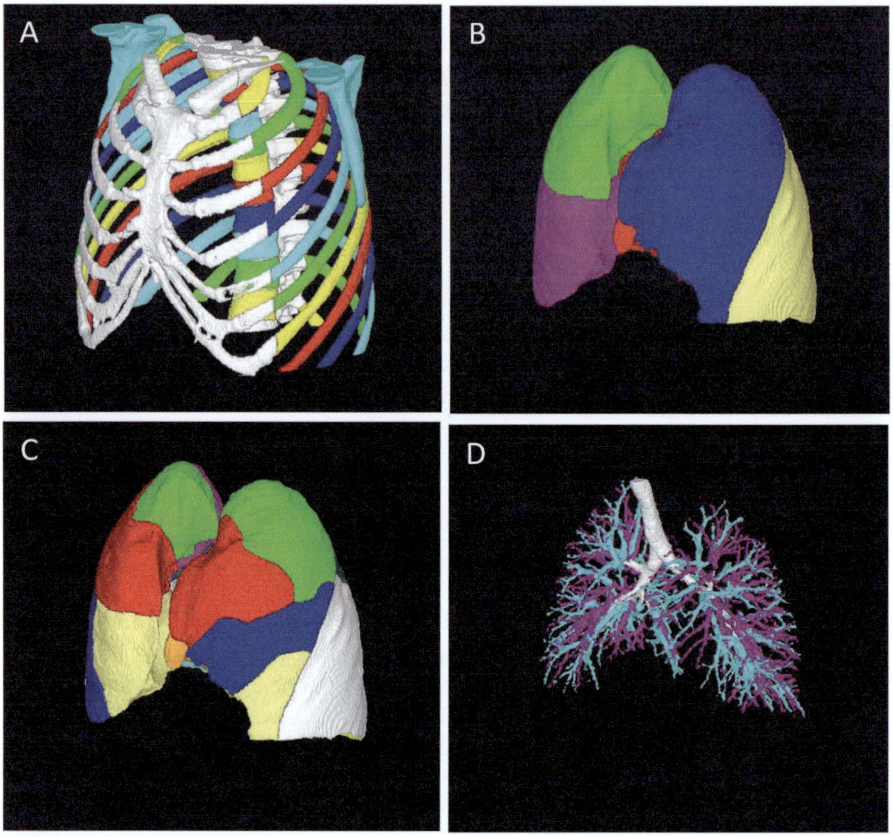

Fig. 4.2 Examples of the segmentation of pulmonary structures. (**a**) Bones including ribs and vertebrae. Color codes represent the classifications of the ribs and vertebrae. (**b**) Lung lobes. *Green*, *pink*, and *red* represent the upper, middle, lower lobes of the right lung. *Blue* and *yellow* represent the upper and lower lobes of the left lung. (**c**) Lung segments. Color codes represent the separation of each lobe into lung segments. (**d**) Pulmonary vessels, trachea, and bronchi. *White*: trachea and bronchi. *Blue* and *pink* represent the segmentation result of pulmonary vein and artery

description. The comparison of the results of CADe performance using different datasets, the different natures and characteristics of the nodules, and various evaluation methods [20] is of limited value. Furthermore, though the reference standard for positive cases is thought of as a gold standard (ground truth), the determination of a perfect gold standard is not an easy task. Because the reference standard is usually determined by an expert panel, substantial variability has been reported in the definition of a gold standard for identifying nodules on CT images [31]. In the ANODE09 study, a comparative study on the same dataset was performed [32]. The performances of six CADe systems (five from academic groups and one commercially available system) for lung nodule detection were evaluated using a database of 55 scans and demonstrated that combining the outputs of the CADe

Table 4.1 Summary of the reported best performance of CADe systems for detection lung nodules

Author	Year	Detection schemes initial candidate detection/FP reduction and lung nodule detection	No. of nodules	CAD sensitivity (%)	FP rate (%)	Data (No. of patients, section thickness, radiation dose)	Nodule size
Armato et al. [46]	1999	Multiple gray-level thresholding/linear discriminant analysis	187	70	3/section	17 patients, 10 mm,	3.1–27.8 mm
Kanazawa et al. [45]	1998	Fuzzy clustering and surface curvature/rule-based classifier	230	90	2.8/scan	450 patients, 10 mm, low dose	–
Lee et al. [48]	2001	Template-matching/rule-based classifier	98	72	1.1/section	20 patients, 10 mm, low dose	5–30 mm
Brown et al. [49]	2001	Model-based image understanding technique/fuzzy logic	36	86	11/scan	17 patients, 5–10 mm, normal dose	5–30 mm
Armato et al. [47]	2002	Multiple gray-level thresholding/rule-based classifier	50	80	1/section	31 patients, 10 mm, low dose	5–25 mm
Suzuki et al. [63]	2003	Supervised lesion enhancement filter based on a massive-training ANN (MTANN)	71	80.3	4.8/scan	71 patients, 10 mm, low dose	Mean, 13.5 mm
McCulloch et al. [51]	2004	Mathematic model of anatomic structures/Baysian classifier	43	69.8	8.3/scan	50 patients, 2.5 mm, low dose	5–17.1 mm
Awai et al. [52]	2004	Mathematical morphology/neural network	78	80	0.87/section	82 patients, 7.5 mm, normal dose	3–30 mm
Ge et al. [61]	2005	Gradient analysis/linear discriminant analysis	116	87.9	0.5/section	56 patients, 1.0–2.5 mm,	3–30.6 mm
Bae et al. [53]	2005	Mathematical morphology/rule-based classifier	107	97.2	4/scan	20 patients, 1 mm, normal dose	>3 mm

(continued)

Table 4.1 (continued)

Author	Year	Detection schemes initial candidate detection/FP reduction and lung nodule detection	No. of nodules	CAD sensitivity (%)	FP rate (%)	Data (No. of patients, section thickness, radiation dose)	Nodule size
Kim et al. [55]	2005	Gaussian curve fitting/neural network	297	94.3	0.89/section	14 patients, 1.0–5.0 mm, normal dose	5–28 mm
Roy et al. [62]	2006	Gradient analysis/linear discriminant analysis	82	70	0.28/section	38 patients, 7 mm, normal dose	3–30 mm
Yuan et al. [69]	2006	ImageCheker CT (R2 Technology)	628	73	3.19/scan	150 patients, 1.25 mm, normal dose	>4 mm
Matsumoto et al. [54]	2006	Convergence index filter/linear discriminant analysis	50	90	1.67/section	5 patients, 5.0–7.0 mm, normal dose	3–12 mm
Das et al. [64]	2006	ImageCheker CT (R2 Technology)	116	73	6/scan	25 patients, 2 mm,	Mean, 3.4 mm
		Nodule enhanced viewing (NEV) (Siemens Medical Solutions)	116	75	8/scan	25 patients, 2 mm,	Mean, 3.4 mm
Dehmeshki et al. [56]	2007	Shape-based genetic algorithm/rule-based classifier	178	90	14.6/scan	70 patients, 0.5–1.25 mm, normal dose	3–20 mm
Li et al. [59]	2008	Intensity structure enhancement/rule-based classifier	153	86	6.6/scan	117 patients, 1.25–5.0 mm, low/normal dose	4–28 mm
Pu et al. [57]	2008	Geometric model based on the analysis of the signed distance field/scoring based on shape similarity	184	81.5	6.5/scan	52 patients, 2.5 mm, low dose	3–28.9 mm
Ye et al. [58]	2009	Shape index and dot features/rule-based filtering and support vector machine	220	90.2	8.2/scan	108 patients, 0.5–2.0 mm, low/normal dose	–

Murphy et al. [35]	2009	Shape index and curvedness/K-nearest-neighbor	1525	80	4.2/scan	813 patients, 1 mm, low dose	>3 mm
Yanagawa et al. [65]	2009	Lung VCAR (GE Healthcare)	229	40	5.7/scan	48 patients, 0.625 mm	>4 mm
Messay et al. [36]	2010	Multiple gray-level thresholding/linear discriminant analysis	143	82.7	3/scan	84 patients, 1.3–3.0 mm,	3–30 mm
Tan et al. [37]	2011	Intensity structure enhancement/neural network	80	87.5	4/scan	125 patients, 0.75–3.0 mm,	3–30 mm
Guo et al. [60]	2012	Intensity structure enhancement/linear discriminant analysis	111	85	17.3/scan	85 patients, 1.25–3.0 mm, low/normal dose	3–115 mm
Zhao et al. [66]	2012	LungCAD VB10A (Siemens AG Healthcare)	151	96.7	3.7/scan	400 patients, 1 mm, low dose	2.3–6.9 mm
Godoy et al. [75]	2013	VD10A (Siemens Healthcare)	155	79	3/scan	46 patients, 0.67–1.0 mm, normal dose	4–27.5 mm

systems led to performance improvement. In the system combination, the results of multiple nodule CAD systems were combined without their internals, like the feature values of candidates [32]. The combination method used only the findings (coordinates and degree of suspicion for each finding) and performance information of each system.

Figure 4.3 shows a snapshot of CADe output for lung nodule referred as a pure ground-glass nodule (GGN) [42]. According to the Fleischner Society glossary of terms for thoracic imaging, a GGN is defined as "a hazy increased attenuation in the lung that does not obliterate the bronchial and vascular margins" [67]. The term "pure GGN" represents nodules of only ground-glass attenuation on CT, whereas the term "part-solid GGN" refers to nodules comprising both ground-glass and solid attenuation areas [16]. The term "subsolid" nodules includes both pure GGN and part-solid GGN [16, 68]. Subsolid nodules are increasingly being detected on LDCT screening and have a high likelihood of representing adenocarcinomas [16, 68] For example, Henschke and colleagues reported that 34% of detected subsolid nodules proved malignant, while only 7% of solid nodules proved malignant [68]. However, most CADe schemes were focused on detecting solid nodules because detecting subsolid nodules with low attenuation is not an easy task. The currently commercially available CADe systems are designed and optimized for detecting solid nodules and have a low sensitivity for subsolid nodules [65, 69]. Improving detection accuracy of subsolid nodules is one of the important research areas in CADe [20, 42, 55, 75].

Several studies have investigated the effect of CADe on clinicians' interpretation of CT examinations for the detection of lung nodules [52, 64, 65, 70–75]. These studies compared the clinicians' performances without and with CADe systems. The assessment approaches included evaluation with use of sensitivity and FP rate, use of ROC analysis, use of localization ROC (LROC), and free-response ROC (FROC) analysis. The LROC and FROC analyses are categorized as location-specific ROC analyses, which require data such as the identified location of suspected regions and a rating for each region [77]. The LROC analysis constrains a mark-rating pair of the most suspicious region in an image, whereas the FROC analysis can deal with an arbitrary number of mark-rating pairs for each image [76–78]. For FROC data, either an alternative FROC (AFROC) method or Jackknife FROC (JAFROC) method is used for evaluating clinicians' performance in the detection of lung nodules without and with CADe systems [79, 80]. The AFROC method reduces FROC data to pseudo-ROC data that can be analyzed by tools developed for ROC analysis [79]. The AFROC analysis can be used to analyze the data acquired from FOC studies. The AFROC curve is a plot of lesion localization fraction against false-positive fraction, and the area under the AFROC curves is used to define lesion detestability [81]. The JAFROC method provides a figure of merit (FOM) to summarize FROC and statistically compares the FOMs of clinician/system performances [80, 82].

Another growing research area of interest in CADe is nodule follow-up [49, 83–86]. Kubo et al. described a comparative reading system using baseline and follow-up 10-mm section scans of the same patients [85]. Their approach comprised section matching, nodule matching, and quantitative evaluation steps for evaluating

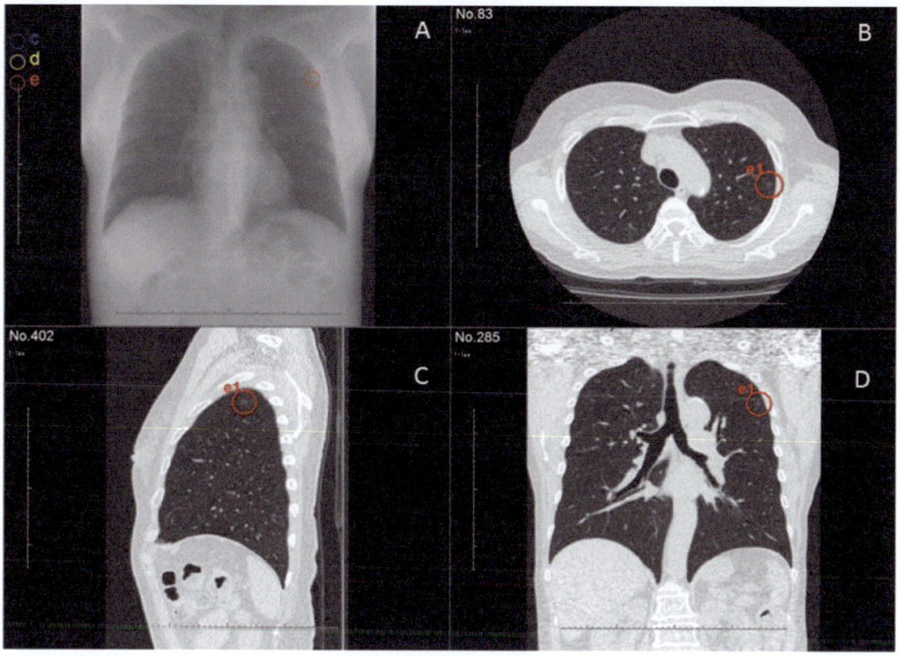

Fig. 4.3 Snapshot of CADe output for lung nodule. Detected pure ground-glass nodule is marked with a *circle*. (**a**) Scout image (A non-tomographic image). (**b**) Transverse image. (**c**) Sagittal image. (**d**) Coronal image

the growth and shrinkage of lung nodules. The section matching rate of 99.8% was obtained through performance evaluation using CT scans for 85 patients with 198 nodules. Brown et al. described a patient-specific model, which was derived from the segmentation results of lung architecture and lung nodules using baseline 10-mm section scans [49]. Their model guided the segmentation of subsequent scans for relocalization of previously detected nodules. Their pilot study reported that the correct relocalization rate was 81% using the follow-up scans of 27 nodules. Ko and Betke described a method to automatically detect lung nodules on a CT scan with 5–10-mm section thicknesses and then relocalize them on follow-up scans [83]. Their method involved global registration of baseline and follow-up scans by translation and rotation to align the centroids of lung architecture landmarks and applied the same transformation to the detected nodules on the baseline scan for relocalizing the nodule on the follow-up scan. The preliminary testing for assessment of nodule change over time was correlated between the radiologist and the computer (Spearman rank correlation coefficient, 0.932). Lee et al. evaluated the performance of automated matching software (LungCARE VB20, Siemens Medical Solutions, Forchheim, Germany) of 30 metastatic patients imaged with two serial CT scans with a 5-mm section thickness [84]. This study included 30 consecutive enrolled patients with metastatic pulmonary nodules from a pulmonary primary tumor (n = 9) or a non-pulmonary primary tumor (n = 21). The overall matching rate of a total of 210 nodules was 66.7%. In a recent study, Aoki et al. introduced a temporal subtraction (TS) method to enhance interval changes in their CADe scheme and assessed the effects of the TS method on radiologist performance in nodule detection on thin-section CT images with 2-mm section thickness [86]. Their observer study reported that the average sensitivity of the eight participants improved from 73.5% to 83.4% with an FP rate of 0.15 per case in the detection of nodules using 30 nodules ranging in size from 5 to 19 mm.

Impact of CADe Systems on Clinicians' Performance for Detecting Lung Nodules on CT Images

The Imaging Physics Committee of the American Association of Physicists in Medicine (AAPM) formed a Computer-Aided Detection in Diagnostic Imaging Subcommittee (CADSC) to develop recommendations on approaches for assessing computer-aided detection and diagnosis (CADe/CADx) system performance [87, 88]. The CADSC reported that these were the following major areas for assessing CADe/CADx systems: training and test datasets, reference standards, mark-labeling criteria, stand-alone performance assessment metrics and methodologies, reader performance assessment metrics and methodologies, and study sample size estimation [87]. In the development of CADe systems for detecting lung nodules on CT images, not only the evaluation of a stand-alone CADe system but also the evaluation of the effects of the CADe system on clinician accuracy is indispensable from the viewpoint of the fundamental role of CADe systems, which is to support a clinician who has the final responsibility to make the decision for each case.

The results of the investigation of the effect of CADe systems on clinicians' interpretation of CT examinations that are covered in this session are briefly summarized as follows: Awai et al. reported that average areas under the AFROC curves without and with their CADe system were 0.64 and 0.67, respectively, using a dataset of 50 CT examinations with ten observers [52]. The difference was statistically significant. Li et al. reported that the average area under the ROC curve improved from 0.76 to 0.85 for 14 radiologists with their CADe system. The LROC curve also showed improvement using a dataset [70]. Brown et al. evaluated the incremental effects of their CADe system using a dataset of eight CT examinations with 202 observers at a national radiology meeting. The JAFROC analysis involving 13 observers who read all cases indicated an improvement of 22% in FOM with the CADe system, which was not statistically significant [71]. Das et al. compared the effectiveness of two commercial CADe systems (ImageCheker CT, R2 Technologies and Nodule Enhanced Viewing (NEV), Siemens Medial Solutions) using a dataset of 116 small nodules (a mean diameter of 3.4 mm) with three radiologists [64]. The sensitivities with the CADe systems were improved for all radiologists. Hirose et al. reported that the average sensitivity improved from 39.5% to 81.0% with a decrease in the average number of FPs from 0.14 to 0.89 per case using a dataset of 21 patients with six radiologists [72]. The average FOM values without and with a commercial CADe (ZIOCAD LE version 1.15, Ziosoft Inc., Tokyo, Japan) were 0.390 and 0.845, respectively, and the difference was statistically significant. White et al. reported a multicenter study involving 109 patients from four sites to evaluate the performance of CADe as a second reader. The average increase in area under the ROC curve for ten radiologists with a CADe system (Philips Extended Brilliance Workspace, Philips Healthcare, Enthoven, the Netherlands) (not commercially available in the USA) was 1.9% for a 95% confidence interval (0.8–8.0%) [73]. Sahiner et al. reported the impact of their CADe system on six radiologists' performance using 85 CT examinations stratified by nodule size [74]. Their study results, which evaluated the sensitivity and FOMs without and with the CADe system, indicated that the CADe system improved radiologists' performance for detecting lung nodules smaller than 5 mm. Yanagawa et al. evaluated the impact of a commercial CADe system (Lung VCAR, GE Healthcare, Milwaukee WI, USA) on radiologists' performance in the detection of lung nodules with or without ground-glass opacity (GGO), using a dataset of 48 patients with three radiologists [65]. They established the reference standard to be used in the evaluation by a consensus panel of the radiologists. The pulmonary nodules in their dataset were categorized into three patterns (GGO, solid, or part solid). The FOM values without and with the CADe system were significantly different for overall patterns and solid pattern of nodules. Godoy et al. evaluated the effect of a commercial prototype CADe system (VD10A, Siemens Healthcare) on the detection of subsolid and solid nodules on thin- and thick-section CT images using a dataset of 46 CT examinations with four radiologists [75]. Their study included 155 nodules (74 solid nodules, 22 part GGNs, and 59 pure GGNs) with a 5.5-mm average diameter and used various combinations of thick (5 mm) and thin (0.67 or 1 mm) sections. Sensitivities for both subsolid and solid nodules were significantly improved with the CADe system.

On the basis of these observations, there are a number of initiatives underway to improve CADe systems for implementation and utilization in clinical practice as a second reader for increasing the detection of lung nodules in LDCT screening.

4.2.1.2 Lesion Detection in the Abdominal Region

Local Intensity Structure Analysis

The local intensity structure analysis technique is widely used for analysis of object structures in images. Applications of this technique to medical images were originally proposed by Sato et al. [89] and Frangi et al. [90]. Local intensity structure analysis technique enhances blob, line, and sheet structures in images. Blob structure enhancement is commonly used for lesion detection in CAD. For instance, polyp detection in the colon [91], ulcer detection in the small and large intestines [92], and enlarged lymph node detection [93] methods were developed using the blob structure enhancement technique. Line and sheet structure enhancements are used in segmentations of blood vessels and lung lobes. This technique utilizes eigenvalues of the Hessian matrix to enhance the target structures.

Enhancement processes based on local intensity structure analysis consist of three steps including (1) local region clipping, (2) a Hessian matrix elements calculation, and (3) an enhancement filter value calculation. Details of each step are described below.

In the local region clipping step, a small local region is clipped from an input image. Local region clipping is performed for a local region centered at each voxel in the image. Local region size is defined based on the size of the detection target. If the detection target is a sphere of diameter 1 cm, a local region will be a square of =1 cm or a sphere of diameter =1 cm. The local region size should be chosen so as to include the target. A large local region size will reduce the detection performance of the enhancement filter.

In the Hessian matrix elements calculation step, the matrix elements are calculated for each local region. The Hessian matrix is written as

$$\mathbf{H} = \begin{bmatrix} \frac{\partial^2 f}{\partial x^2} & \frac{\partial^2 f}{\partial x \partial y} & \frac{\partial^2 f}{\partial x \partial z} \\ \frac{\partial^2 f}{\partial y \partial x} & \frac{\partial^2 f}{\partial y^2} & \frac{\partial^2 f}{\partial y \partial z} \\ \frac{\partial^2 f}{\partial z \partial x} & \frac{\partial^2 f}{\partial z \partial y} & \frac{\partial^2 f}{\partial z^2} \end{bmatrix}. \tag{4.1}$$

The Hessian matrix elements are the second-order partial differential coefficients of a function f. The function f represents image intensity values in the local region. The function is obtained based on the image intensity values. For 3D medical images, a second-order polynomial

$$\begin{aligned} f(x, y, z; \mathbf{a}) = {} & a_1x^2 + a_2y^2 + a_3z^2 + \\ & a_4xy + a_5yz + a_6zx + a_7x + a_8y + a_9z + a_{10}, \end{aligned} \tag{4.2}$$

can be used as the function. $\mathbf{a} = (a_1, \ldots, a_{10})$ is a coefficient vector of the polynomial. The coefficient vector $\mathbf{a}$ is calculated by minimizing squared errors between the image intensity values and the polynomial. Instead of obtaining the function that represents image intensity values, the second-order partial differential coefficients of the function can be directly calculated from image intensity values. The second-order partial differential coefficients of the function are estimated from difference values of the image.

Finally, the enhancement filter value calculation step is performed. In this step, three eigenvalues of the Hessian matrix $\lambda_1, \lambda_2, \lambda_3 (\lambda_1 \geq \lambda_2 \geq \lambda_3)$ and their corresponding eigenvectors $\mathbf{e}_1, \mathbf{e}_2, \mathbf{e}_3$ are calculated. The intensity structure in the local region can be classified into blob, line, or sheet categories by checking the magnitude relationship of the three eigenvalues. If the intensity structure in the local region shows a blob structure, three eigenvalues have nearly equal values and they are smaller than 0. Eigenvalue conditions of three local intensity structures are shown in Table 4.2. A blob structure enhancement filter [89] can be defined that gives high outputs when the eigenvalues satisfy the eigenvalue condition of the blob

Table 4.2 Eigenvalue conditions of three local intensity structures [89]

Local intensity structure	Eigenvalue condition
Blob	$\lambda_3 \simeq \lambda_2 \simeq \lambda_1 \ll 0$
Line	$\lambda_3 \simeq \lambda_2 \ll \lambda_1 \simeq 0$
Sheet	$\lambda_3 \ll \lambda_2 \simeq \lambda_1 \simeq 0$

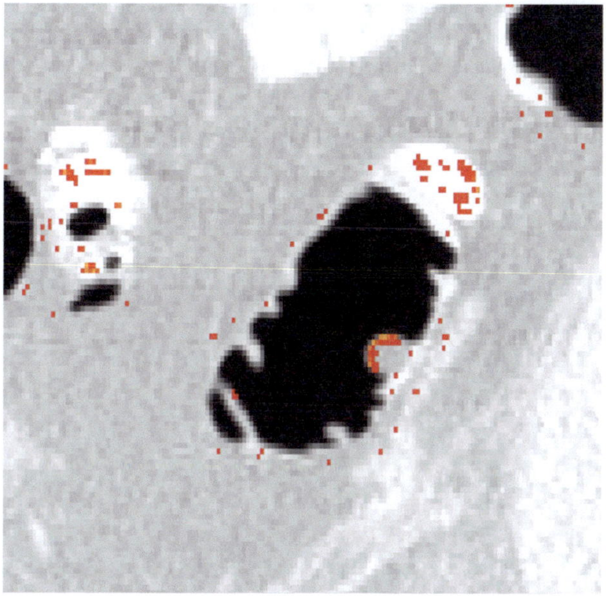

Fig. 4.4 An example of output of the blob structure enhancement filter applied to the colon including a polyp. *Red* and *yellow colors* indicate low and high output values, respectively. High output values are shown in a blob structure

structure. The blob structure enhancement filter is given by

$$S(\lambda_1, \lambda_2, \lambda_3) = \begin{cases} |\lambda_3|\psi(\lambda_2; \lambda_3) \cdot \psi(\lambda_1; \lambda_2) & \textit{if } \lambda_3 \leq \lambda_2 \leq \lambda_1 < 0, \\ 0 & \textit{otherwise}, \end{cases} \tag{4.3}$$

where

$$\psi(\lambda_s; \lambda_t) = \begin{cases} \left(\frac{\lambda_s}{\lambda_t}\right)^\gamma & \textit{if } \lambda_t \leq \lambda_s < 0, \\ 0 & \textit{otherwise}. \end{cases} \tag{4.4}$$

The blob structure enhancement filter outputs high values in blob structures. Figure 4.4 shows an example of blob structure enhancement filter output. The line and sheet structure enhancement filters are defined similarly based on each eigenvalue condition. After applying the blob structure enhancement filter to all voxels in the image, voxels having high output values of the filter can be classified into lesion candidate voxels.

Radial Difference Filter

For detection of small spherical targets, a radial difference filter is used. This filter is utilized in detection of metastatic lymph nodes [94] and small liver tumors [95] from CT images. This filter effectively detects small spherical-shaped targets. However, detection performance of the filter for detection of large targets is not ideal because of shape variations of the targets.

In the detection processes, the value of the radial difference filter is calculated for each voxel in an image. The center voxel in the image is selected. Searching rays are cast from the center voxel. For each searching ray, the different values of intensities of the image are calculated along the searching ray direction. A voxel having a high difference value of the intensity is selected as a point on boundary of the target. If the center voxel is located in a spherical-shaped region, the selected points are located on the border of the spherical-shaped region. The targets including spherical lesions can be detected by checking positions of the selected points.

4.2.2 *Quantification and Classification*

4.2.2.1 Lung Cancer Prognostication Using CT Image-Based Features

Other issues in the implementation of LDCT screening for lung cancer are overdiagnosis and disease recurrence after curative surgery on nodules detected at an early stage. Pulmonary nodules detected on lung cancer screening include not only aggressive lesions but also indolent lesions that may not progress to significant disease. This overdiagnosis is a potential drawback of screening because of unnecessary treatment, additional follow-up cost, and patient anxiety [96–98]. Patz et al. examined data from NLST to estimate the overdiagnosis rate and reported that more than 18% of all lung cancers detected by LDCT appeared to be indolent [98]. The authors emphasize the need for better biomarkers and imaging techniques to determine which lung cancers are more or less aggressive for optimization of patient care and enhancement of the value of screening. LDCT screening for lung cancer increases the rate of detection of early-stage lung cancer [13, 14]. In NLST, 51.8% of lung cancers detected were reportedly at stage IA [13]. However, early detection does not always guarantee cure. A significant number of early-stage lung cancers recur even after complete surgical resection of the primary tumor and pathological confirmation of absence of any regional lymph node metastasis [99]. In a Japanese lung cancer registry study, the five-year survival rate of patients with stage IA lung cancer was under 80% [100]. Recent studies have demonstrated that the use of adjuvant chemotherapy improves the survival of patients with early-stage lung cancers [101]. A proportion of patients with stage I lung cancer have a poorer prognosis and may benefit significantly from adjuvant chemotherapy. Therefore, the identification of patients with early-stage lung cancer who have a higher risk for recurrence and who require more aggressive surveillance, or who may benefit from adjuvant therapy, has been another target of intense investigation [102–104, 116].

In this subsection, we present an example of 3D computerized quantification and show its association with histopathologic findings and postoperative outcomes [102, 105, 106]. The focus of our approach is on lung cancer prognostication on the basis of the analysis of the internal structure of a solitary pulmonary nodule. The relationship between the subjective CT appearance of pulmonary nodules and histopathologic findings, including the Noguchi classification [107] and postoperative outcomes, has been investigated in several studies [108–116].

CT Value Histogram-Based Classification Framework

Volume measurement is one of several approaches to the management of pulmonary nodules detected by CT scanning [117]. This approach often encounters cases where nodules are volumetrically stable in spite of internal CT value variation. These authors have attempted to develop a 3D computerized method for evaluating the volumetric distribution of CT values within pulmonary nodules. We found that the analysis of CT histograms is a potentially useful method for the quantitative classification of pulmonary nodules without requiring measurement of the proportion of nonsolid and solid components [102, 105, 106]. In [102], we developed a five-category classification approach based on the analysis of CT value histograms and investigated the impact of nodule segmentation on classification and the effect of classification on disease-free survival. We also extended the approach to compute a histogram-based score of recurrence risk to track time-interval changes in pulmonary nodules via variational Bayesian mixture modeling for the features obtained from analysis of CT histograms [105, 106]. The key contribution to the computational anatomical models is to represent the internal structure of pulmonary nodules for computing a histogram-based risk score that correlates with prognostic factors. The framework consists of five steps: (1) nodule segmentation, (2) computation of a CT histogram, (3) nodule categorization by applying the variational Bayesian model to cluster CT histograms, (4) computation of the histogram-based risk score by using the combination of the contribution that each category makes to describing the nodule [105, 106], and (5) prognostic prediction using the histogram-based risk score. A schematic overview of the prognostic prediction approach is shown in Fig. 4.5.

Statistical Analysis

We employed Cox regression analysis to assess the effect of quantitative classification on disease-free survival with adjustments for potential confounding factors [118], calculating the hazard ratio (HR) and the 95% confidence interval (CI). To estimate the probabilities of disease-free survival according to the quantitative classification, we generated disease-free survival curves using the Kaplan-Meier

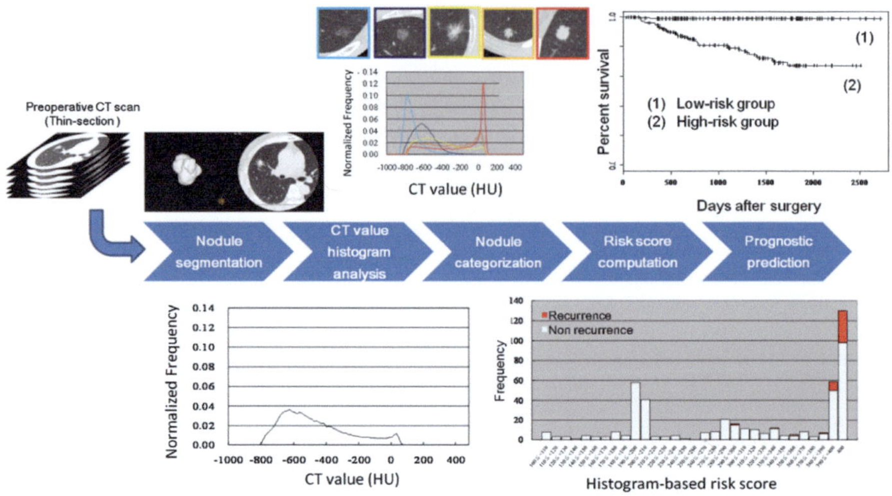

Fig. 4.5 Overview of the classification approach for prognostic prediction of lung cancer

method with a log-rank test to confirm significance [119]. All reported P values were two sided. We defined statistical significance as $p < 0.05$. All statistical analyses were performed using R (a freely available software environment for statistical computing, version 3.0.2).

Experiments and Results

We retrospectively identified data on 454 patients with non-small cell lung cancer (NSCLC). Each patient had undergone a preoperative thin-section CT examination under identical settings. All patients subsequently had histologically or cytologically confirmed NSCLC and had records containing information on the clinical and pathological features and any recurrence of disease. Preoperative CT scans of the entire NSCLC lesion had been acquired using multi-detector row CT scanners (Aquilion; Toshiba Medical Systems, Tochigi, Japan) at the National Cancer Center Hospital East.

We generated CT histograms from segmented nodules using a bin size of 15 HU ranging from −1000 to 500 HU. The frequency value of each histogram was normalized by the nodule volume to allow a comparison of histograms among the nodules. From CT histograms, we extracted ten quantitative features: mean and standard deviation of CT value, skewness, kurtosis, CT value at the peak of the histogram, frequency of the peak of the histogram, and the 10th, 25th, 75th, and 90th percentiles (representing the CT values yielding 10%, 25%, 75%, and 90%, respectively, of the area under the histogram from the minimum CT value). In our study of feature selection [105], we found that the use of two features, frequency of

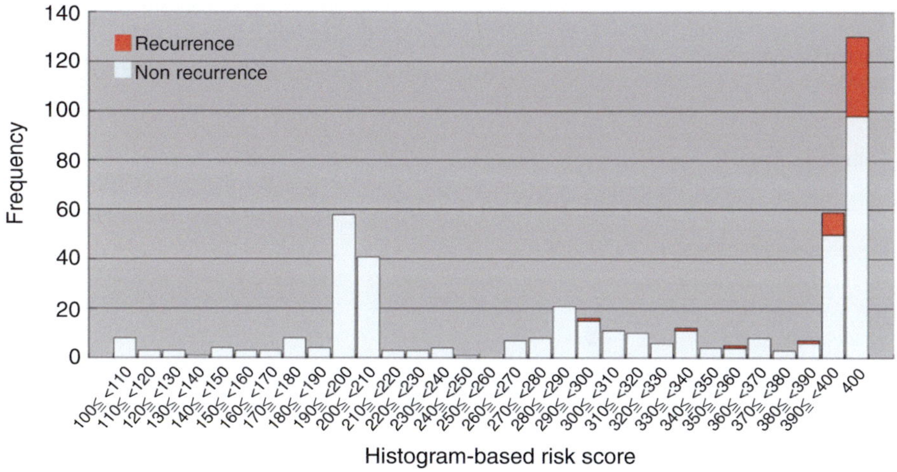

Fig. 4.6 Relationship between CT value histogram-based risk score and recurrence of the lung cancers

the peak of the CT histogram and 90th percentile, was an appropriate combination for representing the histogram pattern.

Through applying the cluster analysis of CT histograms using the selected features, we computed a histogram-based risk score for each lung cancer. Figure 4.6 shows the relationship between histogram-based risk score and tumor recurrence. We selected an appropriate cutoff score for the expression of every pulmonary nodule using X-tile plots based on the association with the patients' recurrence-free survival. X-tile plots provide a method to assess the association between variables and survival [120]. We generated the X-tile plots (X-tile software version 3.6.1. Yale University School of Medicine, New Haven CT, USA). We included those patients with a histogram-based risk score >340 in the group at high risk of disease recurrence (high-risk group) and those with a histogram-based risk score of ≤ 340 in the group at low risk of disease recurrence (low-risk group). Figure 4.7 shows the multivariate Cox regression analysis. This analysis result indicated that the classification (HR: 7.87; 95% CI: 1.75 – 35.37; $P = 0.007$), the pathological stage (HR: 8.39; 95% CI: 4.15 – 16.96; $P < 0.001$), and the lymphatic permeation (HR: 2.02; 95% CI: 1.07 – 3.83; $P = 0.03$) remained significant independent factors in disease-free survival. The disease-free survival curves for the patients with NSCLC according to the histogram-based risk score illustrate that the five-year disease-free survival probability for patients with NSCLC with high-risk scores was 67.3% (95% CI, 58.7 – 77.7). The five-year disease-free survival probability for patients with low-risk scores was 99.1% (95% CI, 97.9 – 100.0). The difference in the disease-free survival rates between the two groups was also found to be significant ($P < 0.001$).

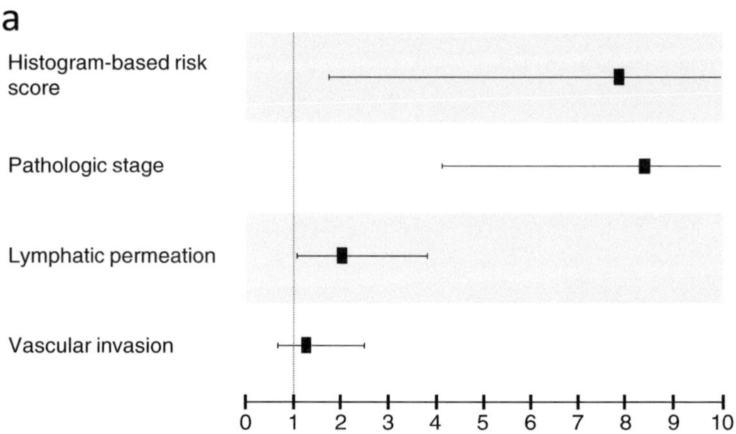

b

Variables	No. of Patients		Multivariate analysis	
	Recurrence	Total	HR (95% CI)	P value
Histogram-based risk score				
≦340 *	2	238	1.00	
> 340	43	216	7.87 (1.75 - 35.37)	0.007
Pathologic stage				
IA*	14	390	1.00	
IB or IIA or IIB or IIIA or IIIB	31	64	8.39 (4.15 - 16.96)	<0.001
Lymphatic permeation				
negative *	21	390	1.00	
positive	24	64	2.02 (1.07 - 3.83)	0.03
Vascular invasion				
negative *	18	351	1.00	
positive	27	103	1.28 (0.66 - 2.49)	0.473

*Reference category

Fig. 4.7 Multivariate analysis of prognostic factors. (**a**): Hazard ratios along with 95% confidence intervals. (**b**): Multivariate analysis

Conclusion

This subsection has described a framework within which CT value histograms are used to represent the internal structure of pulmonary nodules as a computational anatomical model, with the aim of stratifying patients into high- and low-risk groups. The approach has been illustrated using data from preoperative thin-section CT images of lung cancer. This framework provides prognostic value that complements clinicopathological risk factors and more accurately predicts recurrence for patients with early-stage lung cancers. Because the computational anatomical model based on CT value histograms extends naturally to benign cases, the framework described in this chapter might be used for other applications, such as risk stratification of pulmonary nodules detected in LDCT screening.

4.2.3 Miscellaneous

4.2.3.1 Nonrigid Image Registration for Detecting Temporal Changes on Thoracic MDCT Images

Introduction

Image warping is a widely used technique for image registration that deals with geometric transformation techniques in the computer vision and image processing fields. It was introduced for geometric correction applications based on affine transformation, elastic deformation, and optical flow in remote sensing in the mid-1960s. There are many techniques to detect opacities such as solid and/or GGO on thoracic computed tomography (CT) images [121].

The temporal subtraction technique [122] is used in medical imaging to emphasize subtle differences by subtracting a previous image from a later one. This technique can enhance interval changes such as new lesions and/or worsening existing abnormalities by subtracting the two image sets. Some commercial versions have been introduced.

In temporal subtraction, the image warping technique is the most important method for accurately deforming previous images to match current ones. When a misregistration occurs, image artifacts will appear on the subtraction images, often consisting of remaining normal structures such as blood vessels or airways. Ishida et al. [123] proposed a 2D image warping method in chest radiographs for CAD. However, in the temporal subtraction with MDCT, it is necessary to employ a more complex and accurate 3D registration scheme. To overcome this problem, we propose a new technique for automatic image warping to reduce subtraction artifacts on temporal subtraction [124]. In this section, we describe the new temporal subtraction techniques and its application to detect abnormalities such as lung nodules on MDCT images. In the first step for image registration, we use a global image matching technique to correct for the global displacement caused by variations in patient positioning by use of a 3D cross correlation technique. In the second step, a local matching technique and a 3D elastic matching technique are used on the volume of interest (VOI). In the third step, a voxel matching technique is applied to register the two image sets. Finally, temporal subtraction images are generated by subtraction of the previous image from the current image. We have applied our computerized scheme to 31 MDCT image sets, which include examinations performed at two time points.

Methods

The registration problem is one of the most important tasks to find the optimum spatial transformation between two sets of features. To reduce subtraction artifacts, we propose a new temporal subtraction method based on a voxel matching technique by warping the previous image to match the current one [124]. Figure 4.8 shows the overall scheme of our proposed method for temporal subtraction. The main steps are (1) global matching, (2) local matching, (3) 3D nonlinear image warping based on voxel matching technique, and (4) 3D image subtraction steps.

Preprocessing

In general, images generated at different patient visits can vary in voxel size. In this study, we adjusted the voxel size on the X- and Y-axes of a previous CT image to match that of a current CT image using linear interpolation. To segment the lung region as a VOI, we identified a lung region in the current CT image by applying a 3D Gaussian filter before the binarization technique based on the Otsu method [125] and a morphological filter.

Global Matching

A global matching technique was applied for the global displacement caused by positional deviation or respiratory motion. In global matching, we perform a rigid 3D affine transformation. The transformation can be represented as $\mathbf{T}_{global}$:

$$\mathbf{T}_{global} = \begin{pmatrix} \cos\theta_x \cos\theta_z & -\cos\theta_x \sin\theta_z + \sin\theta_x \sin\theta_y \cos\theta_z & \sin\theta_x \sin\theta_z + \cos\theta_x \sin\theta_y \cos\theta_z & t_x \\ \cos\theta_y \sin\theta_z & \cos\theta_x \cos\theta_z + \sin\theta_x \sin\theta_y \sin\theta_z & -\sin\theta_x \cos\theta_z + \cos\theta_x \sin\theta_y \sin\theta_z & t_y \\ -\sin\theta_y & \sin\theta_x \cos\theta_y & \cos\theta_x \cos\theta_y, t_z & \\ 0 & 0 & 0 & 1 \end{pmatrix} \tag{4.5}$$

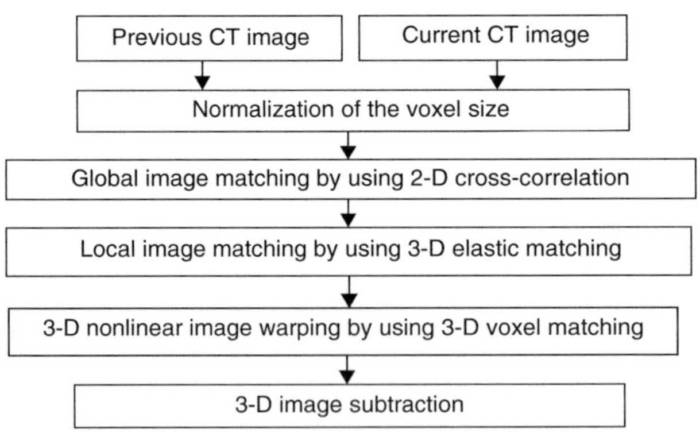

Fig. 4.8 Illustration of the overall scheme of the 3D temporal subtraction method [124]

where θ_x, θ_y, and θ_z are the rotation around x-, y-, and z-axes, respectively, and t_x, t_y, and t_z are the displacement on the x-, y-, and z-axes, respectively. To obtain the voxel pairs between the current image and the transformed previous one, a linear interpolation is adopted. In this global matching procedure, the global shift vector, which is obtained from the template location, with the maximum of the cross correlation value obtained for the similarity of the current and previous images, is shown in Eq. 4.6:

$$CC_{\text{global}} = \frac{\sum_{x,y,z}(I_c(x,y,z) - \bar{I}_c)(I_p(x,y,z) - \bar{I}_p)}{\sqrt{\sum_{x,y,z}(I_c(x,y,z) - \bar{I}_c)^2(I_p(x,y,z) - \bar{I}_p)^2}} \tag{4.6}$$

Here I_c and I_p show the current and previous image, respectively. $\bar{I}_c$ and $\bar{I}_p$ are mean values of the current and previous images, respectively.

Local Matching Based on the 3D Voxel Matching Technique

In our voxel matching technique, an image warping technique is first applied to the current and the previous image to obtain shift vectors which represent the extent of deformation of the previous image relative to the current image [124]. Based on these shift vectors on the current image, the previous image can be warped to produce a temporal subtraction image. However, the temporal subtraction image obtained by the subtraction of the warped previous image from the current image usually contains considerable subtraction artifacts. With the voxel matching technique, for a given location in the current image, we initially identify the corresponding location in the warped previous image. We then search a voxel in the previous image within a small search volume, which is called a kernel, in order to identify the matching voxel which has a value identical or nearly equal to the value of the given voxel in the current image. This search for the matching voxel is repeated for all of the voxels in the current image. The warped previous image is then replaced by the matched voxel, warped previous image in which the voxel values are generally identical to or nearly equal to the voxel values in the current image except for the voxel values in new lesions or changes in existing abnormalities as shown in Fig. 4.9. The voxel matching technique is one of the optimized registration tools for removal of the subtraction artifacts in a temporal subtraction technique by searching the best matched voxels. In order to assess the temporal changes in two images, the voxel matching technique can be applied to 2D images such as chest radiographs by use of a pixel-matching method instead of the voxel matching in 3D images.

Experimental Results

We demonstrate the effectiveness of our image registration algorithms on 31 datasets of chest MDCT images with early and later scans of each subject. The difference in time between the previous images and current images was in the range of 38 months.

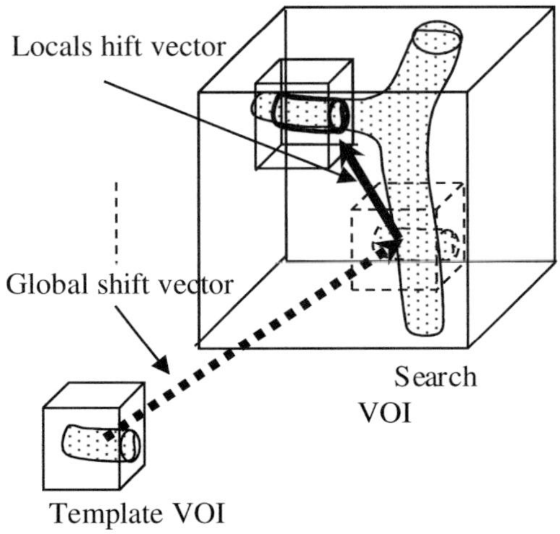

Fig. 4.9 Search for local matching

Figure 4.10 illustrates the results of the use of our temporal subtraction method. Figure 4.10a, b show the previous and current CT images with the lung nodule, respectively. Figure 4.10c–e show the subtraction images based on the global matching technique (using cross correlation value), local image matching (using 3D elastic matching), and with the voxel matching technique as shown in Fig. 4.8. The temporal subtraction enhances the differences between the previous and the current examination, caused by new or changed abnormalities. It is obvious that the majority of the subtraction artifacts in Fig. 4.10c, d are removed in Fig. 4.10e, with a clean and smooth background.

Discussion and Conclusion

In this subsection, we proposed a nonrigid image registration algorithm for temporal subtraction of chest MDCT images using the voxel matching technique. Satisfactory generation of a temporal subtraction image was achieved. To evaluate the usefulness of the voxel matching technique for removal of subtraction artifacts, we performed our new technique on clinical examinations without and with voxel matching. With our new method, subtraction artifacts due to normal structures such as blood vessels were substantially removed on temporal subtraction images. This computerized method can enhance lung nodules on chest MDCT images without the disturbance of misregistration artifacts. Figure 4.11 illustrates an example of generation of a temporal subtraction image. Figure 4.11a shows the results without the voxel matching technique. Figure 4.11b, c show registered images using commercial software and our voxel matching technique, respectively. A brain tumor (arrow) is enhanced on all images. However, subtraction artifacts still remain in (a) and (b). We think this CAD system may be a useful tool in screening examinations.

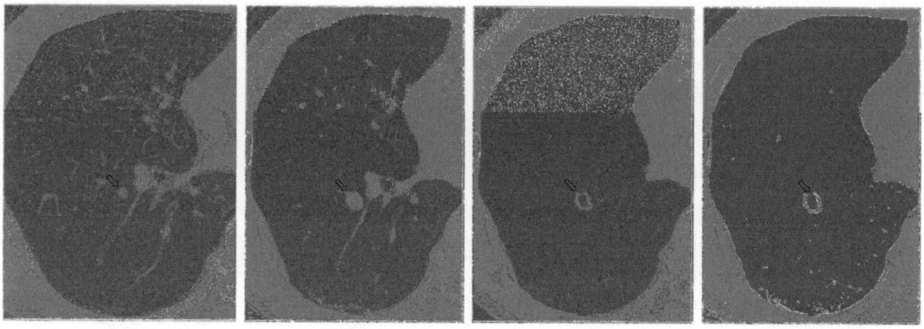

(a) (b) (c) (d)

Fig. 4.10 Experimental results [124]: (**a**) and (**b**) show the original previous and current MDCT images, respectively. (**c**) illustrates the temporal subtraction image based on global image matching based on the 2D cross correlation technique. (**d**) shows the temporal subtraction image based on local image matching using 3D elastic matching. (**e**) shows the temporal subtraction image based on 3D nonlinear image warping using 3D voxel matching. In figure (**e**), most normal structures such as airway and vessels are cleanly removed. The circle in each image shows the lung nodule (Previous; 6.7 mm diameter, Current; 10.8 mm diameter)

(a) (b) (c)

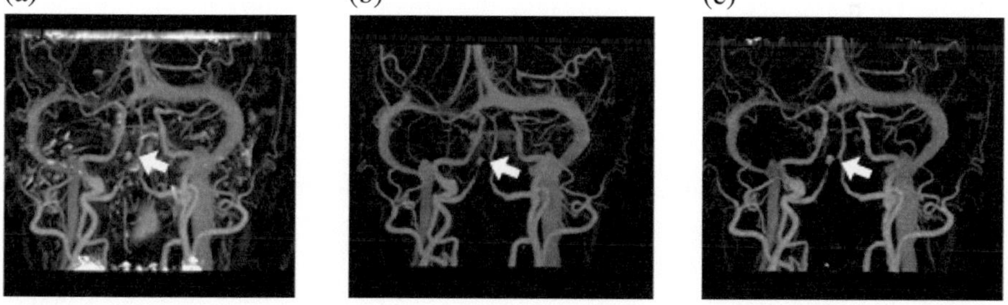

Fig. 4.11 Temporal subtraction images from contrast medium to non-contrast medium head MR image. A brain tumor is enhanced on the subtraction image (*arrow area*). (**a**) Without VM. (**b**) Commercial version. (**c**) Voxel matching

4.2.4 Perspective

CA helps a computer to analyze detailed human anatomy, with the aim of supplementing the work of the radiologist with the advantage of a second opinion via automated diagnostic processes. Critical decision support will someday be implemented by utilizing the power of CA.

4.3 Computer-Aided Surgery and Therapy

4.3.1 Preoperative Support

CA enables us to understand the anatomy of each individual surgery patient. Such understanding can be achieved by detailed segmentation of preoperative CT images, to provide organ shape information that can be used for preoperative simulations such as for procedures such as laparoscopic surgery. In the following subsections, we will show several examples of preoperative simulations based on the CA model.

4.3.1.1 Laparoscopic Surgery Simulation

The main goal of laparoscopic surgery simulation is classified into the following categories: (a) surgical training, (b) surgical rehearsal, and (c) surgical procedure planning. In surgical training, the simulator can train novice surgeons. The main purpose of this system is to impart basic laparoscopic surgery skills such as suturing or membrane peeling. Because this involves learning to manipulate an endoscope or a forceps, detailed representation of patient anatomy is not necessary. Basic representation of an organ shape is sufficient for this kind of simulation. In surgical rehearsal, a great deal of the anatomical information of a specific patient is required. This information can be obtained by processing multi-organ CT segmentation data using surface rendering or volume rendering. Endoscope- or forceps-like user interface devices are used to achieve more realistic reference image generation.

Surgical planning and rehearsal include patient-specific anatomical information and organ deformation simulations using individual patients' organ morphology. CA is crucial to accurate depiction of patient-specific organ morphology.

4.3.1.2 Example of Laparoscopic Surgery Simulation System

This subsection shows examples of laparoscopic simulation systems mainly based on patient-specific organ models generated by CA. In this section, we show an example of laparoscopic gastrectomy.

Organ Segmentation and Labeling

Organ regions are extracted from preoperative CT volume datasets. To enable the algorithm to distinguish between artery and vein regions on CT images, arterial-

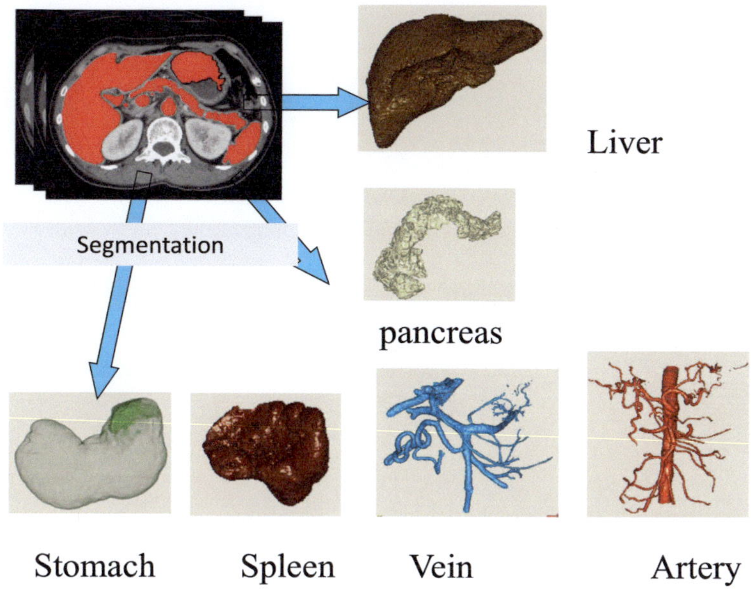

Fig. 4.12 Abdominal organ segmentation for laparoscopic surgery simulation

and venous-phase data are utilized. For a gastrectomy, the following organs are segmented: (a) artery, (b) vein, (c) liver, (d) spleen, (e) kidneys, and (f) pancreas (Fig. 4.12). Segmentation is performed in automated or semiautomated way and utilized for image generation (Fig. 4.13).

Virtual Pneumoperitoneum

In laparoscopic surgery, the abdominal cavity is expanded by inflation with CO_2 gas. In arthroscopic surgery simulation, such demotion should be simulated to generate a more realistic simulation image. The abdominal wall region is extracted from CT data and is separated from the rest of the body. Elastic modes based on a spring-mass model are generated to simulate large deformations of the abdominal wall. By combining $n \times n \times n$ voxels into one deformation element, we allocate springs, dampers, and masses.

Inflation force is added to all nodes located at the boundary of the abdominal wall and the rest of the body. The deformation elements are deformed until deformation converges. Deformation parameters are determined based on the physical properties of a real human body. After the convergence of deformation, we deform an original CT volume to obtain a CT volume with a pneumoperitoneum. This process

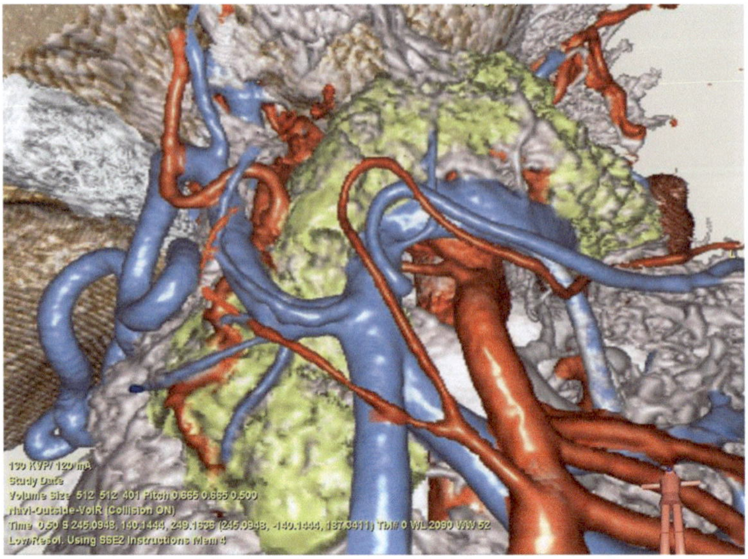

Fig. 4.13 Simulation image generation based on segmentation results

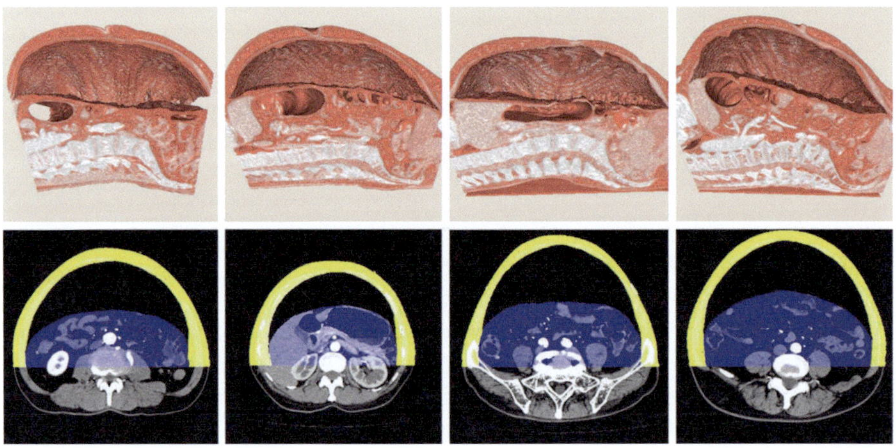

Fig. 4.14 Example of virtual pneumoperitoneum

generates a deformed CT volume reflecting virtual pneumoperitoneum (Fig. 4.14). By rendering deformed volume with multi-organ segmentation results, it is possible to generate a virtual laparoscopic view (Fig. 4.15).

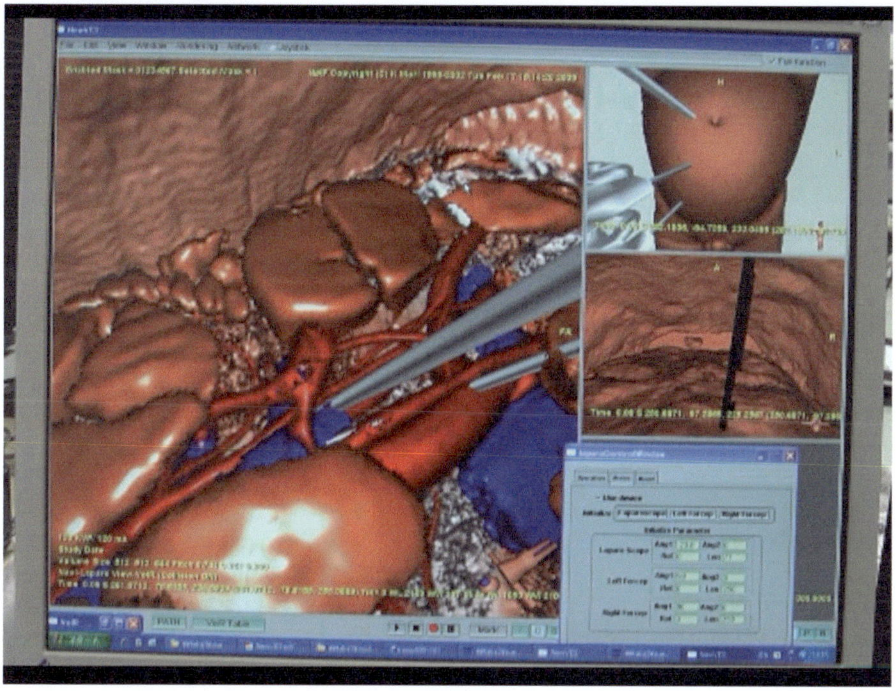

Fig. 4.15 Example of laparoscopic surgery simulation

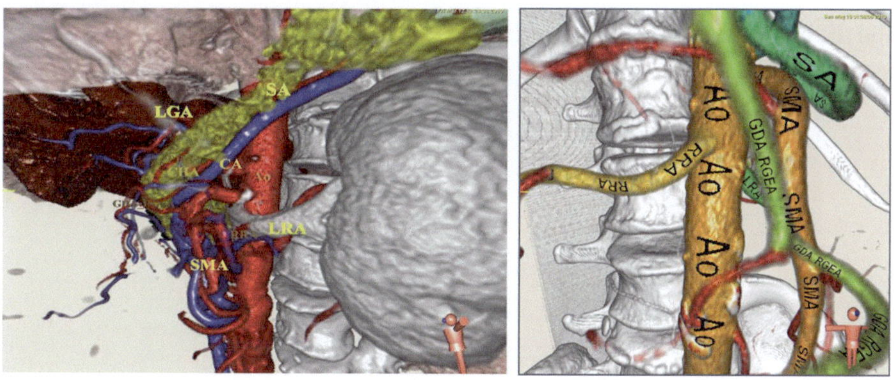

Fig. 4.16 Example of anatomical name display on simulation image

Anatomical Name Display

Anatomical labeling information gives more information for simulation. Anatomical name overlay is also performed in generating a simulation image (Fig. 4.16).

4.3.2 Intraoperative Support

CA also plays an important role in intraoperative assistance. Representative examples of CA in intraoperative assistance, or surgical navigation are provided in this section.

4.3.2.1 Tracking

Tracking of the positions and the orientations of an endoscope and a forceps is vital during interventions. There are two methods for tracking an endoscopic camera: (a) sensor-based tracking and (b) image-based tracking. A positional tracker is mostly used in surgical navigation because it is quite robust. Image-based tracking uses preoperative and intraoperative images to track an endoscope.

Sensor-Based Tracking

An optical or electromagnetic tracking system is typically utilized in surgical navigation (Fig. 4.17). These positional tracking systems give position and orientation in their coordinate system S. To generate a virtual laparoscopic view corresponding to a current real endoscopic view from preoperative CT images, it is necessary to obtain the transformation matrix representing the relationship between two coordinate systems $C^{(s)}$ and $C^{(c)}$, where $C^{(s)}$ and $C^{(c)}$ show the coordinate systems of the sensor and the preoperative coordinate systems, respectively. When the position of known

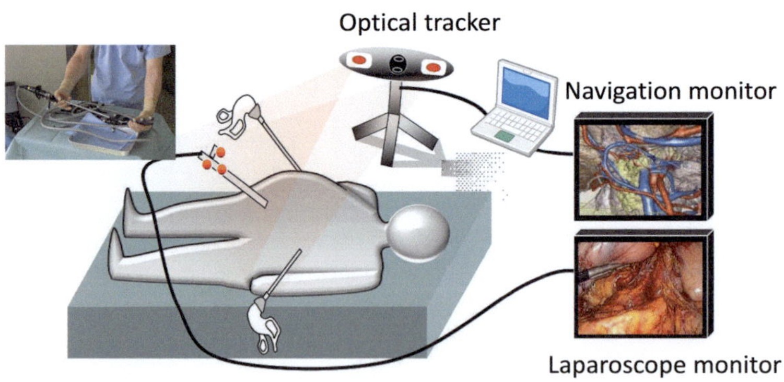

Fig. 4.17 Laparoscopic surgery navigation using an optical positional sensor (Courtesy by Dr. Yuichiro Hayashi)

points $\mathbf{p}_i$ in $C^{(s)}$ and $C^{(c)}$ is measured, the transformation matrix between the two coordinates are obtained by solving the following equation:

$$T^* = arg\max_T \|T\mathbf{p}_i^{(s)} - \mathbf{p}_i^{(c)}\|, \tag{4.7}$$

where $\mathbf{p}_i^{(s)}$ and $\mathbf{p}_i^{(c)}$ are locations of ith known point, which are measured on CT images and on the patient. Anatomical landmarks or fiducial landmarks attached to a patient's body are utilized. To obtain T, we need three points if we consider only rotation and translation. If we obtain T^*, we transform an endoscope position or a forceps position by T^* and generate a surgical assistance image.

Image-Based Tracking

This method estimates endoscopic camera motion based on image matching of the real endoscopic camera image and the virtual endoscopic image (Fig. 4.18). It is possible to generate a virtual view $\mathbf{V}(\mathbf{p}, Q)$ from preoperative CT data by giving a program a viewpoint p and an orientation Q. A similarity-measure function $S(\mathbf{A}, \mathbf{B})$, which measures the similarity of two images of $\mathbf{A}$ and $\mathbf{B}$, can also be generated. Image-based tracking can be formulated as

$$(\mathbf{p}^*, Q^*) = arg\max_{(\mathbf{p}, Q)} S(V(\mathbf{p}, Q), \mathbf{R}), \tag{4.8}$$

where $\mathbf{R}$ means the real endoscopic image. The result of this optimization process $(\mathbf{p}^*, Q^*)$ gives a position and an orientation for the real endoscope camera, which is represented in the CT coordinate system $C^{(c)}$ (Fig. 4.18).

Although this is a simple formulation of image-based endoscope tracking, the performance of this method depends heavily on the accuracy of the initial guess used in Eq. 4.8. There are many methods for guess optimization. Mori et al. used epipolar geometry analysis for obtaining a good initial guess. They recovered endoscopic

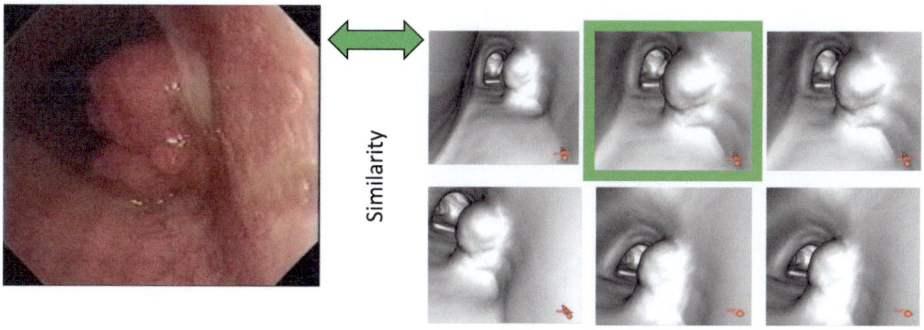

Fig. 4.18 Basic concept of image-based endoscope tracking

motion only from real endoscopic images. Luo et al. introduced a stochastic process for preventing Eq. 4.8 goes to local minima. Combination with physical sensors shown in Sect. 4.3.2.1 is a good way to optimize the initial guess.

4.3.2.2 Intraoperative Assistance Information Image Display

Intraoperative surgical information display is an important function. There are two methodologies: (a) synchronized display and (b) reference display. In synchronized display, virtual views generated from a CA model are generated in synchrony with endoscope/forceps movement. The reference display method shows reference images during surgery. No synchronization with real organs is implemented.

Synchronized Display

If a surgeon moves an endoscope, corresponding views are generated from tracking results. A tip of a forceps is also displayed on a virtual view as a virtual forceps. Anatomical names are also overlain on virtual images for helping a surgeon to navigate patient anatomy. Figure 4.19 shows a situation where synchronized display is performed during gastrectomy. Figure 4.20 shows an example of bronchoscope navigation based on image registration.

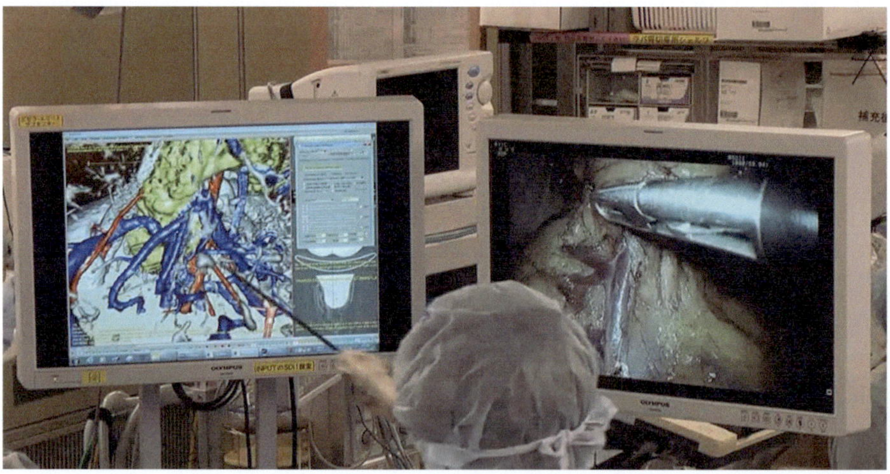

Fig. 4.19 Example of surgical navigation based on synchronized display

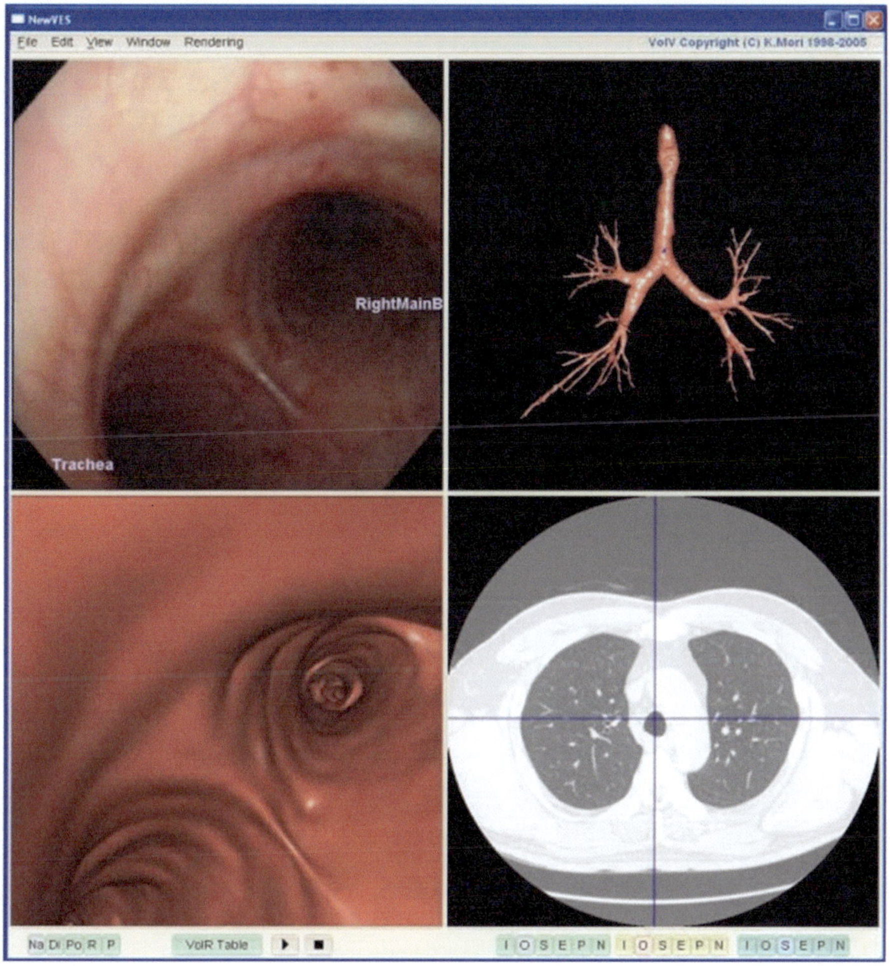

Fig. 4.20 Example of bronchoscope navigation based on synchronized display

Reference Image Display

This display method shows reference images during surgery. Reference images are typically prerendered before surgery and displayed on a monitor or with an interactive display. Tablet devices are often used for reference image display (Fig. 4.21).

Now it is possible to create 3D organ models for image display. Because a 3D print model can be observed during surgery without any devices, it is quite an intuitive display for a surgeon. Translucent models can represent internal structures of an organ. Such models can assist a surgeon in making decisions regarding

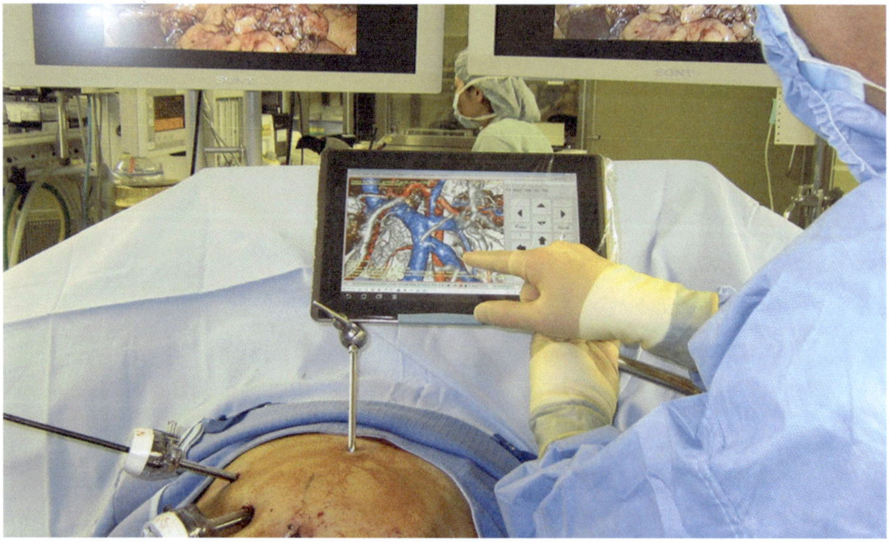

Fig. 4.21 Example of reference image display using tablet device

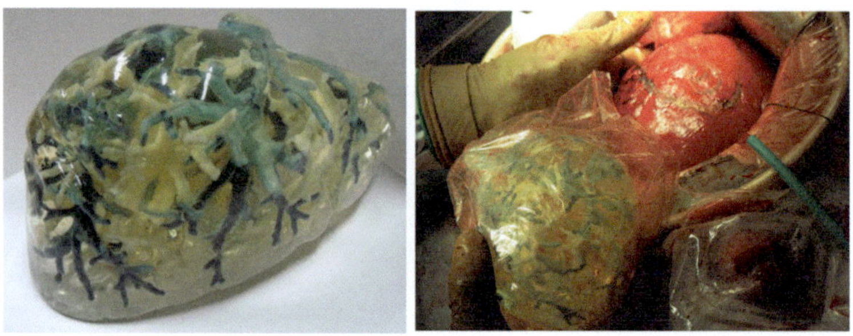

Fig. 4.22 Example of 3D printed organ model and utilization during surgery

resection. It is also useful for improving communications among medical staff. Figure 4.22 shows an example of such 3D print model utilization during surgery.

4.3.3 Respective

There is no doubt that CA enhances interventional procedures by providing useful information before, during, and after intervention, enabling the surgeon to observe important structures that might be missed. One future plan is to connect intraoperative assistance systems based on CA with robotic surgery systems. Robotic surgery enables surgeons to operate in fields where human-operated endoscopic surgery cannot reach. Micro-robotic surgery will allow us to perform sub-sub millimeter procedures. This means much more detailed information about human anatomy is strongly required for future surgery.

4.3.4 Feasibility of Intelligent Image Analysis with CA in High-Precision Radiation Treatment Planning

4.3.4.1 Introduction

The aim of radiation therapy is to deliver as high a dose as possible to a target (i.e., a malignant lesion) to kill as many malignant cells as possible while minimizing the dose to healthy tissues or organs at risk (OAR). This requires exact localization of the target and OARs. Delineation of the contours of the gross tumor volume (GTV), which is defined as the visible tumor region identified in medical images, is key, because patient positioning and the optimization of radiation treatment planning (RTP) are performed based on the planning target volume (PTV) obtained from GTV [126]. The OAR contours are also deeply involved in the optimization of the RTP. To address these requirements, many computational methods based on intelligent image analysis and/or CA have been studied with the aim of refining segmentation of targets and assisting determination of beam directions.

In this section, the feasibility of using intelligent image analysis and/or CA as an aid in high-precision RTP, including particle therapy, is explored.

4.3.4.2 Automated Delineation of Target Regions for Radiation Treatment Planning

The GTV and the clinical target volume (CTV) (larger than the GTV, the CTV defines the volume including gross tumor and areas of likely microscopic involvement) are the starting points in radiation treatment planning as mentioned above. There are two reasons why computer-assisted delineation is important:

(1) The intra- and interobserver variability of GTV delineation by radiation oncologists is plagued by low reproducibility [127–130]. Chao et al. reported that the variation in contours in CTVs delineated by radiation oncologists is a critical issue and that using computer-assisted methods reduced volumetric variation and improved geometric consistency in cases of base of tongue and nasopharyngeal cancer [130].
(2) Manual contouring is tedious and time-consuming. According to one study, the average time saved by using a computer-assisted method is 26% to 29% for experienced physicians and 38% to 47% for less-experienced physicians [130].

Automated contouring based on statistical CTV models is one of the approaches to reduce the inter- and intraobserver variability in GTV or CTV contour generation. Arimura developed a computational method for producing statistical CTV shape models for low-, intermediate-, and high-risk prostate cancers based on a point distribution model, which can be used as a CTV template for automated contouring in prostate cancer radiation treatment planning [131]. First, fifteen radiation oncologists delineated CTV contours for three risk levels. The low-risk, intermediate-risk, and high-risk CTVs included the prostate, the prostate plus the proximal 1 cm of the seminal vesicles, and the prostate plus the proximal 2 cm of the seminal vesicles, respectively. The statistical CTV models for the three risk types were derived based on principal component analysis (PCA), which statistically incorporated the interobserver variability. CTV surfaces were triangulated using a marching cubes algorithm. For matching the number of points on the surfaces of all CTV regions, the number of vertices on each CTV polygonal surface was reduced to 1,000 with quadric error metrics. All CTV regions were registered with a reference CTV using an iterative closest point algorithm for calculation of a covariance matrix to be employed for the PCA-based CTV modeling. CTV models of the three risk types were produced, which consisted of a mean CTV and PCA coefficients multiplied by eigenvectors. Figure 4.23 shows shape variations of statistical CTV models

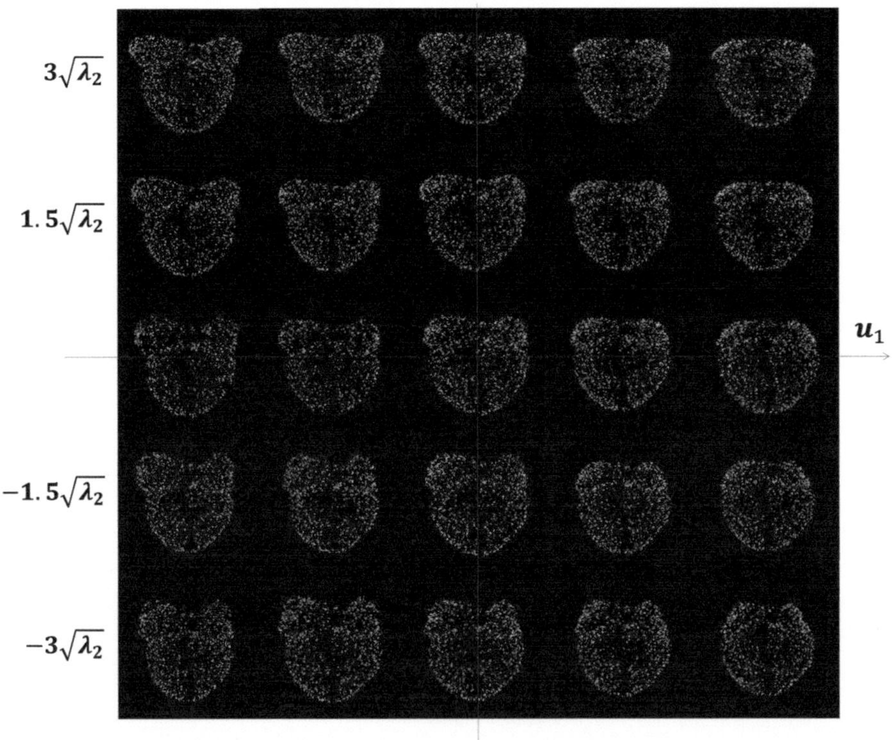

Fig. 4.23 Shape variations of statistical CTV models produced by the first and second largest modes for an intermediate-risk group

produced by the first and second largest modes for an intermediate-risk group. These computational anatomical techniques and mathematical modeling of targets and organs may allow for adaptive target delineation during the course of radiotherapy.

A number of automated delineation methods for determining the GTV or CTV have been proposed to reduce the inter- and intraobserver variability and planning time, as well as to increase the segmentation accuracy of the GTV. Methods using PET images are based on thresholding of the standardized uptake value (SUV) [132, 133]; region-growing methods also use the SUV [134]. Gaussian mixture model-based segmentation [135], gradient-based segmentation methods [136], the fuzzy locally adaptive Bayesian approach [137], the fuzzy c-means algorithm [138], and model-based methods [139] have all been studied. Methods such as MR atlas-based [140] and probabilistic atlas-based [141] segmentation have also been proposed.

Segmentation methods for GTV based on positron-emission tomography/computed tomography (PET/CT) datasets, which include metabolic as well as morphological information, have been assessed. A tumor's higher rate of aerobic glycolysis is directly quantified by ^{1}8F-fluorodeoxyglucose (FDG) PET. El Naqa et al. developed a multimodality segmentation method using a multivalued level set method by combining imaging data obtained from different modalities, including PET/CT [142]. In their study, the level set method was applied to a vector image including CT and PET images so that an energy function could be minimized for determination of CTV regions. As a result, the corresponding Dice similarity coefficient was 0.90 ± 0.02 when CT, MR, and PET images were used. We [143] attempted to incorporate PET biological and CT morphological information on tumor contours determined by radiation oncologists into an optimum contour selection (OCS) framework [144] using a machine learning protocol. We have proposed an automated method for extracting GTVs using a machine learning classifier that accumulates radiation oncology datasets of planning CT and FDG-PET/CT images.

Our method [143, 144] is to feed GTV contours determined by radiation oncologists into a machine learning classifier during the training step, after which the classifier produces the "degree of GTV" for each voxel in the testing step. Six voxel-based image features, including voxel values and magnitudes of image gradient vectors, are derived from each voxel using the planning CT and PET/CT image datasets. Initially, lung tumors are extracted using a support vector machine (SVM) that learns six voxel-based features inside and outside each tumor region determined by radiation oncologists. The final tumor regions are determined using the OCS approach that can be used for selection of a global optimum object contour based on multiple delineations with a level set method around the tumor. Figure 4.24 shows an SVM output image and GTV contours that were multiply delineated using the proposed method on the planning CT image at evolution times of 0, 2000, 3337 (optimum contour), 5000, and 6000. The proposed method achieved an average Dice similarity coefficient of 0.84 in six lung cancer patients, while the conventional method output was 0.78.

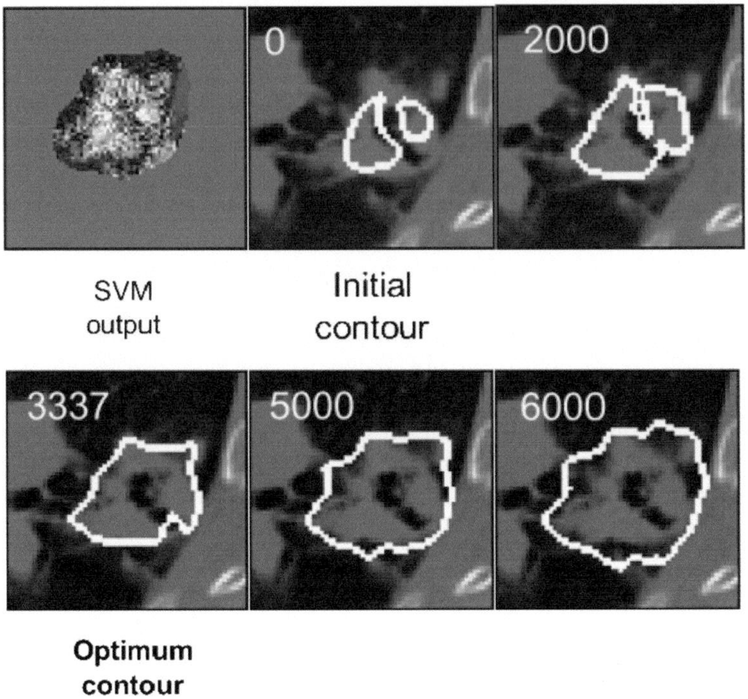

Fig. 4.24 SVM output image and GTV contours that were multiply delineated using the proposed method on the planning CT image at evolution times of 0, 2000, 3337 (optimum contour), 5000, and 6000

Similar-Case-Based Beam Arrangements in Stereotactic Body Radiotherapy

Stereotactic body radiotherapy (SBRT) aims to administer high enough doses to ablate a cancer while minimizing the dose to the surrounding healthy tissues by means of multiple beam arrangements, which generally consist of a large number of coplanar and noncoplanar beams [145]. The determination of beam arrangements is time-consuming, and it is a demanding procedure for less-experienced treatment planners. Treatment planners' skills are generally developed by repeated planning in clinical practice. The planners memorize many planning patterns as an evolving "database," which can be searched for past cases similar to the case under consideration. Several studies have tried to develop computer-assisted methods for this modality [146–149]. Commowick et al. developed a method for selection of a template image, which is the most similar image selected with a distance between transformations in a radiation treatment planning database and can be employed in atlas-based segmentation [146]. Chanyavanich et al. demonstrated the usefulness of prior treatment plans, derived from similar cases, to generate new intensity-modulated radiation therapy (IMRT) plans in prostate neoplasms [147]. Mishra et al. developed a case-based reasoning system, which enables the use of knowledge and experience gained by oncologists in dealing with new patients [148]. Schlaefer et al. reported the feasibility of a framework of case-based beam generation for robotic radiosurgery, which could reduce planning time while maintaining high plan quality for typical clinical cases with similar anatomy [149].

Magome et al. also developed a method for determination of beam arrangements based on similar cases, which proved to be helpful for making new plans in lung cancer SBRT [150]. Beam arrangements were automatically determined based on similar cases using the following two steps: First, the five most similar cases to the current case were retrieved using geometrical features associated with the location, size, and shape of the planning target volume (PTV, the CTV plus a margin allowing for patient movement, position changes, and other variables), lung, and spinal cord. Then, five beam arrangements for the current case were automatically created by aligning five similar cases with the current case in terms of lung regions by use of a linear registration technique. To evaluate the beam arrangements, five treatment plans were manually designed by applying the beam arrangements to the current case. The most useful beam arrangement was chosen by sorting the five treatment plans based on several plan evaluation indices including the D95 (dose that covers 95% of the PTV), mean lung dose, and spinal cord maximum dose. They applied the proposed method to ten test cases by searching in an RTP database of 81 cases of lung cancer and compared the plan evaluation indices between the original treatment plan and the corresponding most useful similar-case-based treatment plan. The method had no statistically significant differences from the original beam arrangements ($p > 0.05$) with respect to the plan evaluation indices. This method could be employed as an educational tool for less-experienced treatment planners. Magome et al. developed a similar-case-based optimization method for beam arrangements in lung cancer SBRT for assisting treatment planners [151]. The local beam direction optimization algorithm, which was developed in their study, improved the quality of treatment plans with significant differences ($p < 0.05$) in the homogeneity index and conformity index for the PTV, V10 (volume receiving $\geq$10 Gy), V20 (volume receiving dose $\geq$20 Gy), mean dose, and NTCP (normal tissue complication probability) for the lung.

The surrounding anatomical environments of tumors, which may affect RTP, were not considered in the study of [151]. We developed a computational framework of retrieving similar cases using a local gradient distribution (LGD) feature for SBRT [152]. We assumed that the LGD feature represents the surrounding anatomical environments of tumors. We adopted a local image descriptor, which was based on scale invariant feature transform [153]. This proposed framework consists of two steps: searching and rearrangement. In the searching step, ten cases most similar to the current case are retrieved from the RTP database based on the shape similarity of a two-dimensional lung region at an isocenter plane. Next, the five most similar cases are selected using geometric features related to the location, size, and shape of the planning target volume, the lung, and the spinal cord. In the rearrangement step, the similarity rank of five selected cases is changed by use of the Euclidean distance between two LGD feature vectors. This is a similarity index based on the magnitudes and orientations of image gradients within a region of interest (ROI) around the isocenter. The gradient magnitude $m(x, y)$ and orientation $\theta(x, y)$ are computed by the following equations based on a gradient vector $(S_x(x, y), S_y(x, y))$ obtained by a Sobel filter:

$$m(x, y) = \sqrt{S_x(x, y)^2 + S_y(x, y)^2} \tag{4.9}$$

and

$$\theta(x, y) = \left(\tan^{-1} \frac{S_y(x, y)}{S_(x, y)} \cdot \frac{180}{\pi}\right) \tag{4.10}$$

The histogram of 36 discrete orientations is constructed by the following equation:

$$h(\theta') = \sum_x \sum_y w(x, y)\delta\left[\theta', \theta(x, y)\right] \tag{4.11}$$

where the weighted magnitude $w(x, y)$ is obtained by multiplying the gradient magnitude by a Gaussian function with the scale estimated at the isocenter and $\delta\left[\theta', \theta(x, y)\right]$ is the Kronecker delta. The orientation with the highest value in the histogram $h(\theta')$ is considered the major orientation.

Gradient magnitudes and orientations of image gradient vectors are recalculated for each pixel in the ROI, which is divided into 4×4 subregions. Figure 4.25 shows illustrations of derivation of the LGD feature. The arrow length corresponds to the sum of the gradient magnitudes as shown in Fig. 4.25b.

An orientation histogram weighted by the vector magnitudes is produced in each subregion using eight bins covering 360°, as shown in Fig. 4.25b. The orientation histogram represents the relationship between eight orientations and the sum of

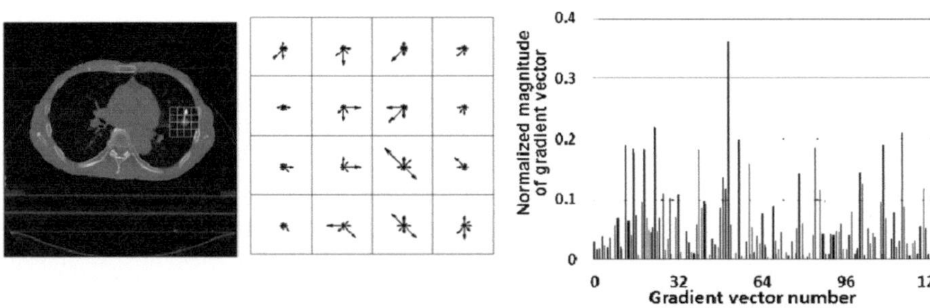

(a) (b) (c)

Fig. 4.25 Illustrations of derivation of the local image descriptor: (**a**) a CT image with an ROI and an arrow showing the major gradient orientation, (**b**) magnitudes of gradient vectors in 16 subregions, (**c**) a local image descriptor. The arrow length corresponds to the sum of the gradient magnitude in (**b**) gradient magnitudes by the following equation:

$$h(\theta') = \sum_{x}\sum_{y} m(x, y)\delta\left[\theta', \theta(x, y)\right] \tag{4.12}$$

Finally, an LGD feature is assembled from an orientation histogram in 16 subregions, which is composed of 128 gradient magnitude features as shown in Fig. 4.25c. As a result, the cases, which are selected as cases similar to the test cases by the proposed method, resemble the test cases more than those selected by the method without the LGD features, in terms of the tumor location. This suggests that the use of the LGD feature is important in providing similar cases to treatment planners.

To evaluate Nonaka's method [152], we applied the similar-case-based beam arrangement method, which was developed by Magome et al. [150]. Figure 4.26 shows a plan generated using the original beam arrangement and five plans determined by similar-case-based beam arrangements, which were generated using the proposed method [152]. The method has the potential to provide superior beam arrangements from the treatment planning point of view.

Quantitative Evaluation of the Robustness of Beam Directions for Charged Particle Therapy

The finely adjusted dose distribution produced in charged particle therapy such as proton or carbon ion beams is vulnerable to setup errors and/or organ motion. We investigated the quantification of the robustness of particle beam directions against patient setup errors in charged particle therapy [154]. Power spectral analysis of target water-equivalent path length (WEPL) images in beam's eye views was

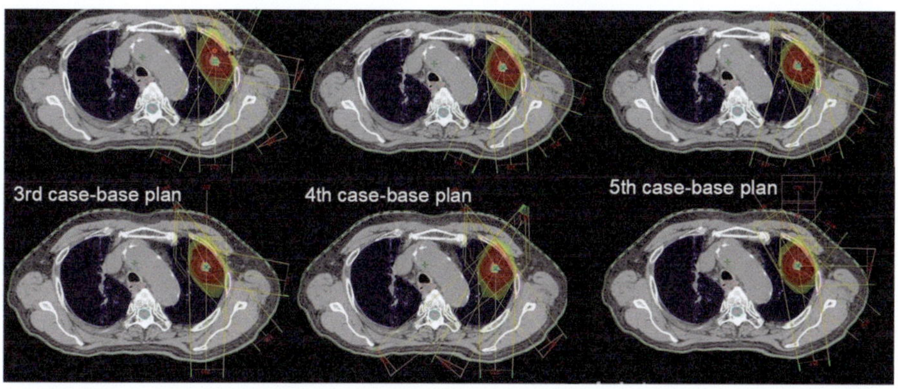

Fig. 4.26 A plan obtained by the original beam arrangement (**a**) and five plans determined by similar-case-based beam arrangements (**b**)–(**f**)

employed for quantifying the robustness of the beam directions. The relationship between the beam direction and the 0th moment of the power spectrum was derived for estimation of the robustness of each beam direction. We applied the proposed evaluation method to seven patients with head and neck cancer. The mean of the 0th moment in the conventional beam directions, which were empirically selected by a manual method, was statistically smaller than that for the avoided beam directions ($p < 0.05$), which means that the conventional beam directions based on planners' experiences and knowledge were appropriate from the theoretical point of view. The results of this preliminary study may lead to an automated selection of beam directions based on the relationships mentioned above.

4.4 Autopsy Imaging (Ai)

4.4.1 Past, Present, and Future of Autopsy Imaging

4.4.1.1 The Nature of Autopsy Imaging (Ai)

Cadavers can be analyzed using diagnostic imaging, which comprises one aspect of medical assessment after death. Imaging of cadavers has also been referred to as postmortem imaging (PMI). Ezawa et al. [155] noted that computed tomographic (CT) imaging before autopsy was valuable, and they referred to the procedure in Japan as autopsy imaging (Ai). Images obtained after death are variously described as "virtopsy" in Switzerland [156] and "virtual autopsy" in France [157, 158] and in Germany [159]. Although the descriptions and concept of Ai somewhat differ, all involve analysis of a cadaver by CT and/or MRI to acquire postmortem medical information.

4.4.1.2 The History of Ai

Early forensic imaging was applied to identify foreign objects in cadavers such as bullets. A murder victim killed by a bullet was assessed using forensic radiography in Lancashire, England in April, 1896 [160, 161]. One report describing a dissection view in 1910 is considered an example of a chest radiographic examination of a cadaver. Although cadavers thereafter were sporadically visualized by radiography and photographed, the information generated in this manner was of little value, and forensic radiology did not develop further.

The appearance and dissemination of computed tomography (CT) during the 1970s generated vast amounts of useful medical information. Imaging of cadavers also began to attract interest, particularly from the viewpoint of crime-related forensics and accidental death. Reports began to describe systematic postmortem imaging during the late 1980s, when Shiotani et al. applied systematic postmortem CT to patients in cardiopulmonary arrest on arrival (CPAOA) [162–164].

A questionnaire sent to hospitals associated with Japanese university medical schools by the Japan Radiological Society between November and December 2010 revealed that 51% of hospitals already employed Ai. Hospitals had begun sporadic CT assessment of patients with CPAOA, and by 2006, about 90% of hospitals in Japan had used CT-based Ai on an emergency basis [165]. Why Ai was implemented in this manner in Japan might be explained by the fact that a medical examiner system had not yet been fully established and the diffusion rate of CT in Japan is the highest in the world.

4.4.1.3 Present Status

Whether Ai can be an alternative to autopsy and determine causes of death without dissection is controversial. This has led to comparisons of causes of death determined by Ai and autopsy in several countries.

Fukayama et al. [166] described the value of Ai in an investigation of deaths associated with medical treatment. That study examined the value of Ai in 165 assessments of causes of death. They judged that autopsy was not necessary in only 3% of these assessments and that the highest correlation of Ai and autopsy findings was 20%. They also proposed classifications for pathological findings determined by Ai alone, compared with the results of both Ai and autopsy findings. They defined obvious pathological findings such as aortic dissection and vague conditions such as systemic infectious disease and embolism (Table 4.3).

Roberts et al. [167] assessed 182 random adult deaths in the UK to compare the accuracy of CT Ai, MRI Ai, and gross autopsy. The major discrepancy rates between causes of death identified by imaging and autopsy were 32% for CT, 43% for MRI, and 30% for the consensus radiology report, which was created by four general radiologists based on CT and MRI images. The most common imaging errors in identification of cause of death occurred in cases of ischemic heart disease, pulmonary embolism, pneumonia, and intra-abdominal lesions, findings

Table 4.3 The classification of disease/pathological findings by the certainty of diagnostic imaging

Classification	Disease/pathological findings
Confident Ai diagnosis	Aortic dissection, aortic aneurysm, end-stage kidney disease, pleural effusion and/or ascites, interstitial pneumonia, pneumothorax
Possible Ai diagnosis	Pericardial effusion, cardiac tamponade, pneumonia/bronchitis (when there is no complication of pulmonary edema), subdural hematoma, advanced liver cirrhosis/liver fibrosis
Uncertain Ai diagnosis	Systemic infection (e.g., miliary tuberculosis), thrombosis, embolism, slight liver cirrhosis/liver fibrosis, meningitis, neurodegenerative disease, acute and old myocardial infarction, carcinoma of unknown origin, diffuse infiltrative lesion

http://humanp.umin.jp/

that concurred with those described by Fukayama et al. Moreover, the major discrepancy rate compared with autopsy was 16% among 88 deaths for which radiologists ruled that autopsy was unnecessary.

The 10% lower discrepancy rate for CT compared with MRI suggests that CT is more accurate than MRI in detecting causes of death. Roberts et al. also discussed an adaptive difference between CT and MRI because CT images scan bone better than MRI because it is density based, not simply because it has slightly higher spatial resolution in most cases. CT is thus effective for visualizing fractures and intracranial hemorrhages. In contrast, MRI provides greater detail of soft tissues. They suggested that forensic situations would be better served by CT, whereas non-forensic and pediatric situations should use MRI [168].

They concluded that the error rate was similar to that for clinical death certificates when radiologists provided a confident cause of death and was therefore acceptable for medicolegal purposes. However, CT and MRI frequently miss common causes of sudden death, and unless these weaknesses are addressed, systemic errors in mortality statistics will result if imaging replaces conventional autopsy.

One German study compared causes of death determined by CT and gross autopsy among 162 (57%) of 285 patients who died in nine intensive care units (ICUs) in Hamburg [159]. Among 47 (16%) autopsies that were also assessed with Ai, the main causes of death overlooked by Ai were cardiovascular disease (12.5%) and cancer (40%). In contrast, gross autopsy overlooked 13 traumatic fractures and two cases of pneumothorax. Ai alone diagnosed new findings in 11 of the remaining 115 deaths. The authors concluded that Ai might be useful for establishing diagnoses that have traditionally been identified by medical autopsy. Ai can also at least in part prove equally instructive as gross autopsy in confirming antemortem clinical diagnoses.

A prospective study of the ability of PMCT to determine causes of death at the Department of Pathology at Aachen University Hospital in Germany analyzed 29 CT studies acquired before autopsy [169]. The accuracy of PMCT for determining causes of death was 68%, with a positive predictive value (PPV) of 75%. However, the accuracy and PPV of CT were 21% and 29%, respectively, for defining pathogenic mechanisms. The authors considered that the combined diagnostic yield of autopsy and CT was excellent compared with autopsy alone. These results were similar to those of Ezawa et al. [155].

A French study compared the abilities of PMCT and autopsy to determine the causes of unexpected death among 47 infants and children [170]. Among 18 (38.3%) causes of death determined by autopsy alone, those also determined by CT agreed with 15 of them. The major discrepancies between CT and autopsy findings were associated with pulmonary analysis, and the authors concluded that the autopsy and CT findings essentially agreed. CT is noninvasive and thus more acceptable to the relatives of deceased children.

4.4.1.4 Categories of Causes of Death

Ai is useful for determining causes of death to some degree. Causes of death are classified as immediate (final disease or injury that caused death), intermediate (disease or condition that preceded death and was responsible for the immediate cause of death), and underlying (extant disease or condition that led to intermediate or immediate cause of death; can exist for years before death) [171]. Most causes of death clarified by Ai are immediate. However, underlying causes of death are very important for medical statistics. Ai has not been evaluated in this capacity. The ability of Ai to determine causes of death accurately needs to be further assessed by comparing large numbers of autopsies with Ai findings.

4.4.1.5 The Future of Ai

Radiologic-pathologic correlation is important to determine the types of pathological lesions that Ai can recognize. To correlate the sites on images with lesions identified in pathological specimens is difficult. Radiologic-pathologic correlation has a longer history in living patients. In cadavers, pinpointing the exact location of a lesion seen on Ai can be challenging. CA is an interdisciplinary field at the intersection of computer science, radiology, anatomy, and pathology. The methods of CA can be grouped into some categories: image segmentation, registration, image-based physiological modeling, and others. Registration in particular, which is a process that searches for the correct alignment of images and sites of the human body, might be useful to overcome this problem. Postmortem changes are evident in Ai. Livor mortis in the lungs is one of several changes found in the decedent. It results in the horizontal line in the lung of Ai image. The postmortem changes in the Ai image may be recognized by using the methods of CA, before we grossly identify the changes in the image.

4.4.1.6 Extension of Ai Applications

Although Ai is an important tool with which to determine causes of death, it also has other applications, as presented at a symposium entitled "Extension of the subject matter of Ai" at the 11th Meeting of the Autopsy Imaging Academic Society in 2013 [172]. Such applications include evaluations of the systemic anatomy of cadavers for medical education and of ancient mummified remains. We advocate a new field of Ai that includes all of these applications, called "postmortem imagiology: autopsy imagiology." The advancement of autopsy imagiology will help to improve the accuracy of causes of death determined by Ai and of evaluations of live patients.

4.4.2 Premortem vs. Postmortem Body Imaging and Computational Anatomy of Liver

4.4.2.1 Introduction

Many medical image interpretation algorithms for different organ systems have been proposed, and some of them are closer to practical use. These algorithms extract organs and/or important radiological findings, e.g., tumors or pneumonia, in a medical image to assist doctors faced with an overwhelming amount of data. A state-of-the-art CT scanner puts out several hundred section images per patient. CT is readily available and easy to use in Japan [173]. Because Ai does not involve radiation dosimetry, a larger number of image sections often result, compared with clinical examinations on living patients. Medical image interpretation algorithms, or computer-aided diagnosis, can help in this instance.

Medical image interpretation algorithms designed for living patients will work to some extent with cadavers, but might fail in some cases. Several types of differences are observed in postmortem CT images. For example, postmortem hypostasis (Fig. 4.27a) that is caused by gravity and increases attenuation or CT value is a significant finding in a CT image of a cadaver. Bronchial branches in a cadaver

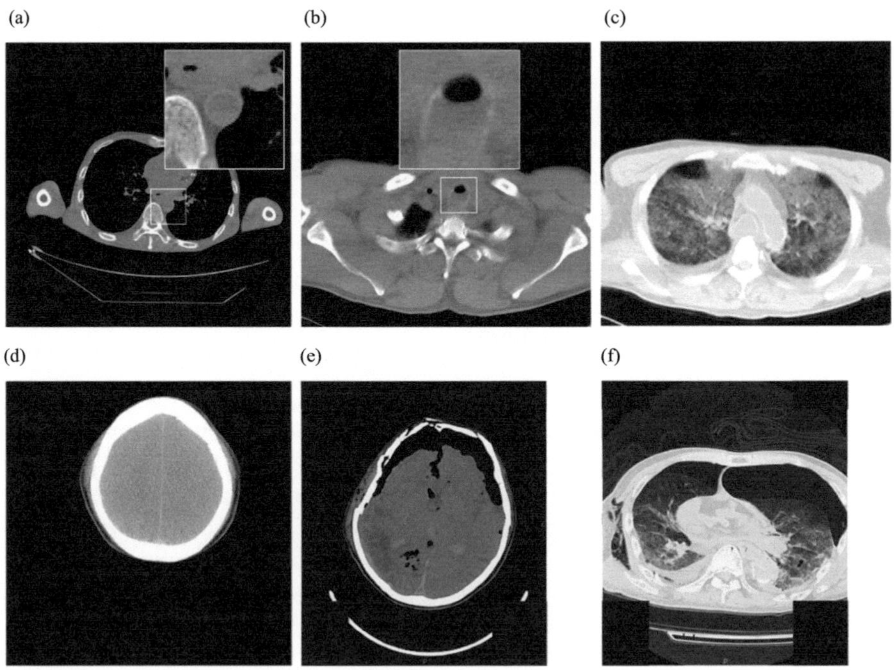

Fig. 4.27 Examples of Ai-specific radiological findings (Original images courtesy of Dr. Yamamoto from Ai Information Center)

are often not filled with air (Fig. 4.27b). Increasing CT values in lung tissue is a typical postmortem change (Fig. 4.27c). As time goes on after death, tissue contrast decreases (Fig. 4.27d). Severe deformation of organs and bone fractures might be observed in an image of a cadaver (Fig. 4.27e, f). Because conventional medical image interpretation algorithms developed for living patients are not designed to deal with these findings, they will fail in segmentation of organs and lesion detection.

This and following sections focus on computational anatomy and segmentation of the liver and lung. As with state-of-the-art segmentation algorithms for a living patient, a CA model or a statistical shape model (SSM)-based approach [174, 175] was employed. The SSM and algorithms are, however, reinforced to deal with postmortem-specific shape variation. This subsection starts with description of differences in shape between an in vivo liver and a postmortem liver.

4.4.2.2 Changes in Liver Morphology

Deformation of the liver from before to after the death is large and might not be accounted for by an SSM trained on in vivo livers. Figure 4.28 shows the typical shapes of an in vivo liver and that of a postmortem liver [176]. Diaphragmatic elevation is caused by respiratory arrest, and enlargement of the right ventricle is caused by circulatory arrest. Such deformation results in the right lobe of the liver

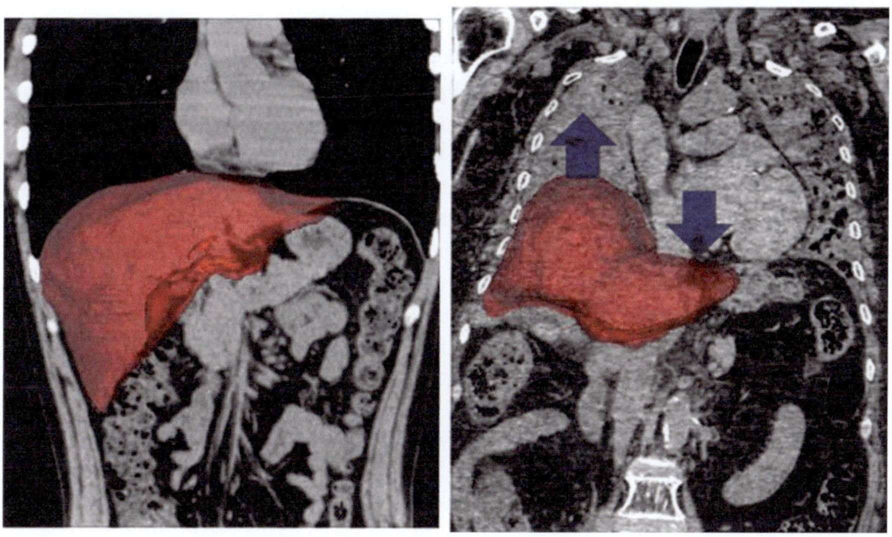

Fig. 4.28 Typical examples from an in vivo liver (*left*) and a postmortem liver (*right*). The *red regions* indicate the true liver regions. In the postmortem liver, the right lobe was elevated owing to respiratory arrest, and cardiac arrest caused enlargement of the right ventricle, resulting in a downward deformation of left lobe

migrating cephalad and the left lobe migrating caudad. The growth of gas-forming bacteria can form an abdominal gas reservoir which pushes the liver cephalad and deforms it.

These are some reasons why a conventional SSM learnt from in vivo liver labels [177–179] may not delineate the postmortem liver shape accurately. One possible solution to the problem would be to collect more postmortem liver labels to construct an SSM. However, because of the shortage of liver labels of postmortem CT volumes used in construction of an SSM and the large diversity of postmortem liver shapes, an SSM learnt only from a small number of postmortem liver labels may display limited ability.

4.4.2.3 Comparisons Between an SSM of an In Vivo Liver and That of a Postmortem Liver

An SSM trained using in vivo liver labels was compared with one trained from postmortem liver labels, in terms of performance in delineating postmortem livers [176]. This study employed a level set distribution model (LSDM) [175, 180] that does not require correspondence between boundaries of shape labels. Materials, performance indices, and SSMs to be compared are given below followed by experimental results:

Materials, Performance Indices, and SSMs

Datasets of 144 in vivo liver labels, L^{144}, as well as 32 postmortem liver labels, D^{32}, were used. The liver labels were manually delineated on unenhanced CT volumes of size $512 \times 512 \times 191 - 3201$ voxels and were reduced to $170 \times 170 \times 170$ voxels for the sake of computational efficacy (voxel size: 2.0 mm isotropic). A spatial standardization of liver labels was carried out before constructing SSMs so that the gravity points of the labels were identical among the training labels.

We computed generalization and specificity of the SSMs as performance indices. Note that generalization is a measure of the ability to describe unknown shapes and specificity a measure of the ability to represent only valid instances of the object. To calculate both indices, in vivo livers and postmortem livers in the database were divided as follows: The L^{144} and D^{32} sets were randomly divided into two equally sized subsets, named L^{72} and D^{16}, respectively. In addition, two L^{16} subsets were randomly selected from the two L^{72} subsets. This study constructed three SSMs, namely, SSM^{L72}, SSM^{L16}, and SSM^{D16}, that were trained from subsets L^{72}, L^{16}, and D^{16}, respectively. Performance indices of the constructed SSMs were calculated using the L^{16} and D^{16} subsets that were not used for training.

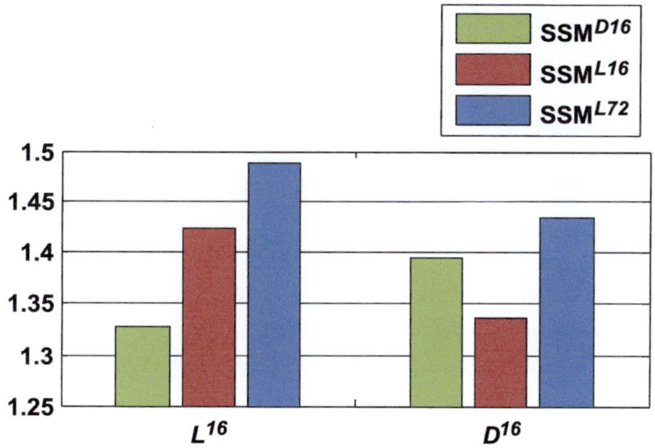

Fig. 4.29 The sum of generalization and specificity (Fig.4c of [176])

Results and Discussion

Figure 4.29 shows the sums of generalization and specificity of SSM. Note that the higher value indicates better performance.

From the experimental results, the following was observed:

1. The performance when applying an SSM to test labels with the same types was always superior to those applying an SSM to test labels with different types. For example, performance of SSM^{D16} evaluated by D^{16} was higher than that of SSM^{L16} evaluated by D^{16}.
2. Another important finding is that the performance of SSM^{L72} was improved by increasing the number of training labels from 16 to 72. Consequently, even when evaluating with D^{16}, the performance of SSM^{L72} was higher than that of SSM^{D16}.

The first observation indicates that the performance of an SSM trained by in vivo liver labels is suboptimal for delineating postmortem liver shapes because of the difference in shape. This finding tells the difficulties in training a postmortem SSM using solely in vivo liver labels. In contrast, the second observation suggests that a larger number of in vivo liver labels would improve the performance of the postmortem SSM to some extent. To solve the problems caused by difference in shape between in vivo liver and postmortem one as well as the shortage of postmortem labels, synthesis-based learning will be introduced in next section.

4.4.3 Computational Anatomy and Segmentation of Postmortem Liver

4.4.3.1 Introduction

As discussed in Sect. 4.4.2.3, increasing training labels that describe the postmortem-specific shape is essential to improve the performance of a postmortem SSM. This section presents a method that solves the abovementioned problem by synthesizing postmortem liver labels, which is inspired by synthesis-based learning [181]. Performance comparisons of SSMs trained using different sets of synthesized postmortem liver labels are presented, followed by a proposal for a postmortem liver segmentation algorithm.

4.4.3.2 Postmortem Liver SSMs Using Synthesized Postmortem Labels

Three transformations are developed to simulate the shape deformation from in vivo livers to postmortem livers. They are categorized into a geometrical transformation F_A and two statistical transformations, F_T and F_{TR}. Details of the methods can be found in [176].

In this study, the transformations yielded five different sets of synthesized postmortem liver labels, or $\mathcal{D}_T$ from F_T, $\mathcal{D}_{TR}$ from F_{TR}, $\mathcal{D}_A$ from F_A, $\mathcal{D}_{AT}$ from F_A followed by F_T, and $\mathcal{D}_{ATR}$ from F_A followed by F_{TR}, respectively. Five postmortem liver SSMs were trained using combinations of the five synthesized liver label sets with original postmortem liver labels, D. The relationships between the five SSMs and the five synthesized label sets are summarized as follows. Note that the training label sets are shown in parentheses:

- SSM_{D+T} model (D and $\mathcal{D}_T$)
- SSM_{D+TR} model (D and $\mathcal{D}_{TR}$)
- SSM_{D+A} model (D and $\mathcal{D}_A$)
- SSM_{D+AT} model (D and $\mathcal{D}_{AT}$)
- SSM_{D+ATR} model (D and $\mathcal{D}_{ATR}$)

In addition, three conventional SSMs constructed solely from original labels were prepared for comparison.

- SSM_D model (D only)
- SSM_L model (L only)
- SSM_{D+L} model (D and L)

Figure 4.30 summarizes the performance of the three conventional and five proposed SSMs in terms of the sum of generalization and specificity explained in the previous section. It was found from the figure that most of the proposed SSMs learnt by synthesized postmortem liver labels outperformed conventional SSMs trained without synthesis-based learning. In particular, the D + T model achieved the highest

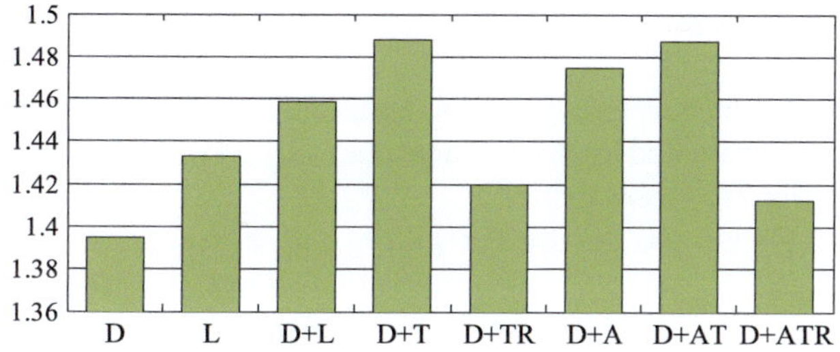

Fig. 4.30 The sum of generalization and specificity (Fig.7c of [176])

score which demonstrated the superiority of the proposed statistical transformation for synthesis-based learning.

4.4.3.3 Postmortem Liver Segmentation Algorithm with an SSM

To develop a segmentation algorithm with an SSM [182] for a postmortem liver, a similar framework for an in vivo liver [183] was employed, but differed in an SSM- and a MAP-based rough segmentation with a probabilistic atlas. A comparative study of algorithms using the aforementioned eight different SSMs is also given in this section.

4.4.3.4 Method

The proposed liver segmentation algorithm consists of three steps: (1) rough segmentation, (2) SSM-based shape estimation, and (3) graph cuts with the estimated shape. The rough segmentation is performed by a probabilistic atlas-guided expectation maximization (EM) algorithm followed by a MAP segmentation [184], in which probability distribution of organs' features is assumed to be a mixture of Gaussians. The difference from the in vivo liver segmentation algorithm [183] is that the atlas-guided EM and MAP are repeated by updating the location of the probabilistic atlas according to the MAP segmentation result of the previous iteration. Such iteration is important to deal with the postmortem-specific deformation of organs. Subsequently, in the shape estimation process, the most similar shape to a MAP segmentation result is selected from an eigenshape space of an SSM. Finally, the graph cut-based segmentation with the estimated shape is performed. The energy function to be minimized is composed of a unary and pairwise terms. The unary term is a negative logarithm of posterior probability of liver given a CT volume in which a patient-specific probabilistic atlas calculated from the estimated shape is used as a prior probability. Pairwise terms consist of a conventional boundary term and a shape term that evaluates the difference in gradients between a segmented shape and the estimated one.

4.4.3.5 Results and Discussion

The same dataset of cadavers in Sect. 4.4.2 was used for validation. The segmentation accuracy was evaluated by the Jaccard index (JI) between the segmented region and corresponding true liver label.

Figure 4.31 shows the JIs of the graph cut segmentation when using the eight different SSMs presented in the previous subsection. The figure tells that the segmentation using SSM_{D+T} achieved the highest performance (average JI = 0.806). Note that SSM_{D+T} was the only model whose segmentation performance is significantly superior to those of all conventional SSMs (SSM_D, SSM_L, and SSM_{D+L}). Here, a Wilcoxon test with a significance level of 0.05 was employed for statistical test. It is worth mentioning that the findings from the figure and statistical test are consistent with the conclusion of the previous subsection, where SSM_{D+T} was proven to be the best model.

Figure 4.32 shows examples of graph cut segmentation when using SSM_L and SSM_{D+T}, in which the contours of the regions are shown in yellow. It was found from the figure that the shape extracted by the algorithm with SSM_{D+T} was obviously superior to that of SSM_L. The failure of SSM_L was explained by failure in shape estimation which was caused by a limited ability to delineate a postmortem liver shape.

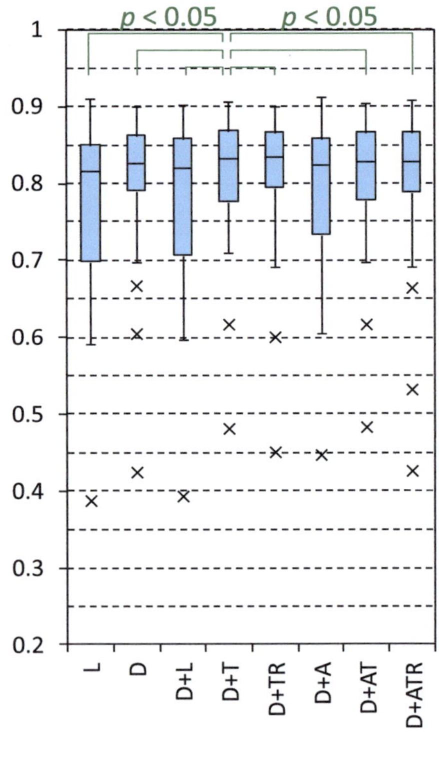

Fig. 4.31 Box plots of the graph cuts segmentation performance of all eight SSMs (Revised version of figure 1 in [182])

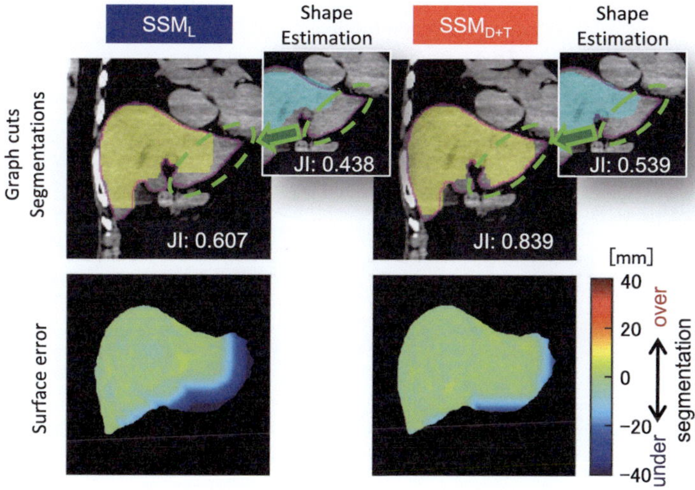

Fig. 4.32 Examples of segmentation results from the conventional SSM$_L$ and from SSM$_{D+T}$, which showed the best performance [182]

4.4.4 Postmortem Lung Segmentation

4.4.4.1 Introduction

Lung segmentation from a postmortem thoracic CT volume is a challenging task owing to large differences not only in shape but also in appearance from a healthy living lung. The change in appearance from in vivo lung to cadaveric one was caused by both postmortem changes and severe pathologies, such as multiple tumors or a large pleural effusion (see Figs. 4.33 and 4.34). One of the state-of-the-art in vivo lung segmentation algorithms is the multi-shape graph cuts with neighbor constraints [185]. It was reported that the algorithm achieved higher accuracy than conventional one, in particular when applying to lung with atypical shape and pathologies. Such in vivo lung segmentation algorithm might still be effective in segmentation of postmortem lung to an extent. Figure 4.33 presents an example of segmentation results for a case with moderate changes in appearance by the previous segmentation algorithm [185]. It was confirmed that the lung boundaries are extracted successfully by the method (JI: 0.963 [left], 0.830 [right]). It might, however, not succeed in extracting a postmortem lung with more severe deformation and/or changes in appearance, as in Fig. 4.34.

This subsection presents a lung segmentation algorithm for a postmortem thoracic CT volume [186]. It is a modified version of the in vivo lung segmentation algorithm [185]. There is twofold contribution: First, to deal with a large difference in appearance of lung that leads to failure in rough location estimation of lung, a robust location estimation algorithm is proposed. It uses the result of postmortem liver segmentation explained in the previous section [182]. Second, a new adaptive

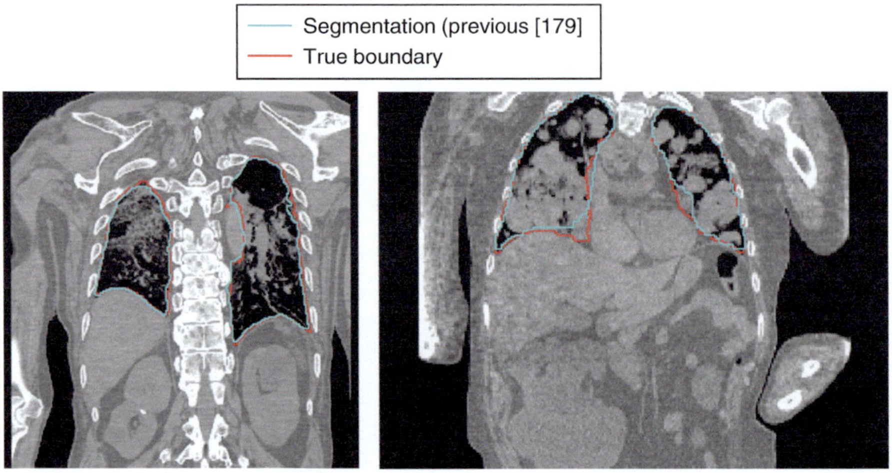

Fig. 4.33 Segmentation results of moderate cases. The previous segmentation algorithm [185] succeeded in extracting the boundaries

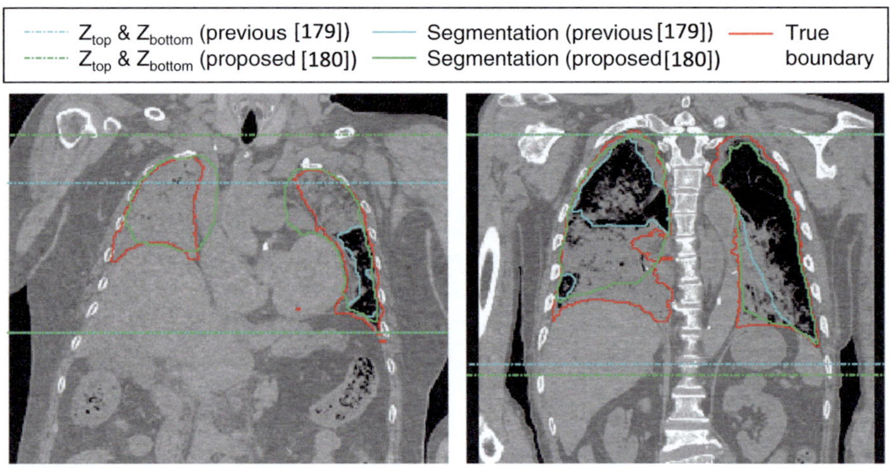

Fig. 4.34 Segmentation results of severe cases with estimated top and bottom sections of the lung. While the previous algorithm significantly failed in segmentation, the proposed algorithm improved the lung boundaries (Revised version of figure 1 in [186])

weight between energy terms in the graph cuts is presented. It adaptively balances between different terms according to the reliability of the rough lung segmentation result. The results of applying the algorithm to 32 postmortem thoracic CT volumes are presented in this subsection.

4.4.4.2 Methodology

A thoracic CT volume is an input to the algorithm. The first step is rough segmentation of the lung, and the second step is body cavity segmentation. Third, a location of lung is roughly estimated, whose results are used to generate the patient-specific shape. Finally, the multi-shape graph cuts with neighbor constraints and an adaptive weight refine the segmentation results. The difference from the algorithm [185] is the third step, or rough location estimation of lung, and the final step, or fine lung segmentation with the proposed adaptive weight.

4.4.4.3 Rough Location Estimation of the Lung

This step estimates top and bottom axial slices of the lung, z_{top} and z_{bottom}, where the z axis is parallel to the craniocaudal axis. Subsequently, landmarks are automatically defined according to the extracted axial slices and bounding box of a body cavity extracted in the previous step. The landmarks are forwarded to a radial basis function-based nonlinear registration between the SSM to an input volume, and the registration results are used to generate the patient-specific shape. In the previous method [185], z_{top} and z_{bottom} slices were determined based on the result of a CT value-dependent rough segmentation only, which frequently failed in segmentation caused by severe pathologies and/or postmortem changes, in particular z_{bottom}. As a result, the estimated positions of z_{bottom} slices would greatly deviate from the true positions. In this study, a linear predictor of z_{bottom} slice was presented based on the z coordinate of the top axial slice of the liver segmentation result [182] explained in the previous subsection. The details can be found in the paper [186].

4.4.4.4 Graph Cut-Based Fine Segmentation of the Lung

The energy function to be minimized has a unary and pairwise terms. The unary term consists of three sub-terms: the likelihood term, the probabilistic atlas term, and the neighbor constraint term. In the postmortem lung segmentation, it is important to balance the probabilistic atlas term with the likelihood one, the latter of which is calculated using the rough segmentation result. So the proposed adaptive weights are defined by the Jaccard index between rough segmentation results and its most similar shape generated by a lung SSM. Note that the lower JI means more irregular shape of rough segmentation, most of which were caused by the failure in segmentation owing to severe pathologies and/or postmortem changes. Specifically, the JI is used as a weight for the probabilistic atlas term and (1-JI) is for the likelihood term. The neighbor constraint term is introduced to force the bottom surface of lung closer to top surface the liver extracted by the algorithm in the previous section.

4.4.4.5 Results and Discussion

The dataset of 32 postmortem CT volumes in Sect. 4.4.3 with the true lung labels were used to demonstrate the effectiveness of the proposed algorithm. A level set distribution model of lung [185] was trained with 40 in vivo studies. Accuracy of the proposed linear predictor of z_{bottom} axial slice was validated by a leave-one-out test of 32 postmortem CT volumes. 26-connectivity was employed in the graph cut-based fine segmentation. The performance was evaluated by JI's between an extracted lung region and the manually delineated true one.

Figure 4.34 shows the extracted top and bottom axial slices of lung as well as segmentation results of the two cases, in which the lungs are affected by severe pathologies and postmortem changes. The proposed algorithm improved the segmentation performance for both cases, but the reasons are different from each other. In the left-hand side case, the estimated positions of z_{bottom} slices were improved, and the better estimation must lead to better segmentation results (JI: 0.048→0.583). In contrast, the estimated slice positions of the right-hand side case are roughly the same as those by the previous algorithm [185], but the boundaries extracted by the graph cuts were improved (JI: 0.676→0.843) because of the proposed term constrained by the extracted liver.

Average JI was improved from 0.784 (previous) to 0.845 (proposed). However, no statistical difference was found by Wilcoxon signed-rank test for 32 cases. When focusing on nine most difficult cases that have lower JI's than the mean by the previous method, statistical difference was observed. It can be concluded that the proposed method must be effective in enhancing the performance of difficult cases.

4.4.5 Perspective

4.4.5.1 Postmortem Liver SSM and Liver Segmentation

Synthesis-based learning was successfully applied to train a postmortem liver SSM in Sect. 4.4.3.2. The advantage of the proposed algorithm lies in the simple simulation of the shape transformation from an in vivo liver to a postmortem one. However, the approach may not simulate all possible shape changes, because of absence of physical point of view. Since the performance of the SSM strongly depends on the quality of synthesized liver labels, physics-based deformation approaches should be applied to synthesize liver labels in the near future, such as the finite element method (FEM)-based transformations [187]. An interesting future topic would be comparison with an FEM-based approach and integration the constructed SSM with a postmortem liver segmentation algorithm.

The current liver segmentation performance for a cadaver is inferior to that for an in vivo patient [177–179, 188] owing to postmortem-specific shape deformation and larger location variation. Extension of the EM algorithm before MAP segmentation to estimate shape and location will be an important endeavor.

4.4.5.2 Postmortem Lung SSM and Lung Segmentation

Unlike postmortem liver, differences in shape of a lung between an in vivo patient and a cadaver are relatively small. Thus, the postmortem SSM was trained by only in vivo lung labels. The shape description ability, however, might be limited, in particular at boundaries neighboring to the liver and heart, because the boundaries are deformed in a cadaver, as explained in Sect. 4.4.2.2. The development of a postmortem lung SSM trained from postmortem lung labels and/or synthesized ones is another important goal.

The most difficult problem in postmortem lung segmentation is changes in appearance, or CT values, caused by severe pathology and/or postmortem changes. A prior knowledge of shape and constraints by surrounding organs is necessary to solve the problem. The proposed method in Sect. 4.4.4 employs an SSM to provide a priori knowledge of shape. In addition, extracted body cavity, aorta, as well as liver were used to set constraints on the shape and location of lung. However, the segmentation performance showed in Sect. 4.4.4.5 is not sufficient for practical use. One possible reason is that the spatial extent of pathologies and postmortem changes are highly different among cases, which means that a priori knowledge and constraints necessary for segmentation are on a case-by-case basis. Thus, development of an algorithm to select appropriate constraints adaptively is vital.

4.4.5.3 Miscellaneous Future Topics

It will be interesting to extend an SSM and a segmentation algorithm to other organs and multiple organs in a cadaver (see Sect. 3.9.6 for in vivo multiple organs). A comparative study of SSMs and segmentation algorithms using more postmortem CT data is also desirable. Modeling both shape and appearance (CT value) changes from an in vivo patient to a cadaver will be essential not only to improve segmentation performance but also to aid in determining the cause of death and to estimate the time after death accurately.

References

1. Doi K (2007) Computer-aided diagnosis in medical imaging: historical review, current status and future potential. Comput Med Imaging Graph 31:198–211
2. Kobatake H (2007) Future CAD in multi-dimensional medical images: project on multi-organ, multi-disease CAD system. Comput Med Imaging Graph 31:258–266
3. Fujita H, Uchiyama Y, Nakagawa T, Fukuoka D, Hatanaka Y, Hara T, Lee GN, Hayashi Y, Ikedo Y, Gao X, Zhou X (2008) Computer-aided diagnosis: the emerging of three CAD systems induced by Japanese health care needs. Comput Methods Programs Biomed 92(3):238–248

4. Lodwick GS, Haun CL, Smith WE et al (1963) Computer diagnosis of primary bone tumor. Radiology 80:273–275
5. Meyers PH, Nice CM, Becker HC, Nettleton WJ, Sweeney JW, Meckstroth GR (1964) Automated computer analysis of radiographic images. Radiology 83:1029–1033
6. Winsberg F, Elkin M, Macy J, Bordaz V, Weymouth W (1967) Detection of radiographic abnormalities in mammograms by means of optical scanning and computer analysis. Radiology 89:211–215
7. Kobatake H (2012) Computational anatomy. In: Fujita H, Ishida T, Katsuragawa S (eds) Handbook of practical image analysis in medicine. Ohmsha Co., Japan, pp 742–748
8. World Health Organization (2013) Latest world cancer statistics. Press Release N223 (pr_223E). http://www.iarc.fr
9. Siegel R, Naishadham D, Jemal A (2013) Cancer statistics. CA Cancer J Clin 63:11–30
10. Kaneko M, Eguchi K, Ohmatsu H et al (1996) Peripheral lung cancer: screening and detection with low-dose spiral CT versus radiography. Radiology 201:798–802
11. Henschke CI, McCauley DI, Yankelevitz DF et al (1999) Early lung cancer action project: overall design and findings from baseline screening. Lancet 354:99–105
12. Swensen SJ, Jett JR, Hartman TE et al (2003) Lung cancer screening with CT: Mayo Clinic experience. Radiology 226:756–761
13. The National Lung Screening Trial Research Team (2011) Reduced lung-cancer mortality with low-dose computed tomographic screening. N Engl J Med 365:395–409
14. Aberle DR, DeMello S, Berg CD et al (2013) Results of the two incidence screenings in the national lung screening trial. N Engl J Med 369:920–931
15. McWilliams A, Tammemagi MC, Mayo JR et al (2013) Probability of cancer in pulmonary nodules detected on first screening CT. N Engl J Med 369:910–919
16. Naidich DP, Bankier AA, MacMahon H et al (2013) Recommendations for the management of subsolid pulmonary nodules detected at CT: a statement from the Fleischner Society. Radiology 266:304–317
17. Doi K (2005) Current status and future potential of computer-aided diagnosis in medical imaging. Br J Radiol 78:S3–S19
18. Sluimer I, Schilham A, Prokop M (2006) Computer analysis of computed tomography scans of the lung: a survey. IEEE Trans Med Imaging 25:385–405
19. Li Q (2007) Recent progress in computer-aided diagnosis of lung nodules on thin-section CT. Comput Med Imaging Graph 31:248–257
20. Goo JM (2011) A computer-aided diagnosis for evaluating lung nodules on chest CT: the current status and perspective. Korean J Radiol 12:145–155
21. Suzuki K (2012) A review of computer-aided diagnosis in thoracic and colonic imaging. Quant Imaging Med Surg 2:163–176
22. Suzuki K (2013) Machine leaning in computer-aided diagnosis of the thorax and colon in CT: a survey. IEICE Trans Inf Syst E96-D:772–783
23. El-Baz A, Beache GM, Gimel'farb G et al (2013) Computer-aided diagnosis systems for lung cancer: challenges and methodologies. Int J Biomed Imaging. Article ID942353. doi: 10.1155/2013/942353
24. van Rikxoort EM, van Ginneken B (2013) Automated segmentation of pulmonary structure in thoracic computed tomography scans: a review. Phys Med Biol 58:R187–R220
25. Firmino M, Morais AH, Mendoça RM et al (2014) Computer-aided detection system for lung cancer in computed tomography scans: review and future prospects. BioMed Eng OnLine 13:41. http://www.biomedical-enginerring-online.com/content/13/1/41
26. Fischback F, Knollmann F, Griesshaber V et al (2003) Detection of pulmonary nodules by multislice computed tomography: improved detection rate with reduced slice thickness. Eur Radiol 13:2378–2383
27. Kim J-S, Kim J-H, Cho G et al (2005) Automated detection of pulmonary nodules on CT images: effect of section thickness and reconstruction interval – initial results. Radiology 236:295–299

28. Sinsuat M, Saita S, Niki N et al (2011) Influence of slice thickness on diagnoses of pulmonary nodules using low-dose CT: potential dependence of detection and diagnostic agreement on features and location of nodule. Acad Radio 18:594–604
29. Tominaga M, Kawata Y, Niki N et al (2011) Measurements of multidetector CT surface dose distributions using a film dosimeter and chest phantom. Med Phys 38:2467–2478
30. Neroladaki A, Botsikas D, Boudabbous S et al (2013) Computed tomography of the chest with model-based iterative reconstruction using a radiation exposure similar to chest X-ray examination: preliminary observations. Eur Radiol 23:360–366
31. Armato SG, McLennan G, McNitt-Gray et al (2004) Lung image database consortium: developing a resource for the medical imaging research community. Radiology 232:739–748
32. van Ginneken B, Armato SG, de Hoop B et al (2010) Comparing and combining algorithms for computer-aided detection of pulmonary nodules in computed tomography scans: The ANODE09 study. Med Image Anal 14:707–722
33. Torres EL, Fiorina E, Pennazio F et al (2015) Large scale validation of the M5L lung CAD on heterogeneous CT datasets. Med Phys 42:1477–1489
34. Jacobs C, van Rikxoort EM, Murphy K et al (2015) Computer-aided detection of pulmonary nodules: a comparative study using the public LIDC/IDRI database. Eur Radiol. doi: 10.1007/s00330-015-4030-7
35. Murphy K, van Ginneken B, Schilham AMR et al (2009) A large-scale evaluation of automatic pulmonary nodule detection in chest CT using local image features and k-nearest-neighbour classification. Med Image Anal 13:757–770
36. Messay T, Hardie RC, Rogers SK (2010) A new computationally efficient CAD system for pulmonary nodule detection in CT imagery. Med Image Anal 14:390–406
37. Tan M, Deklerck R, Jansen B et al (2011) A novel computer-aided lung nodule detection system for CT images. Med Phys 38:5630–5645
38. Meng X, Qiang Y, Zhu S et al (2012) Illustration of the obstacles in computerized lung segmentation using examples. Med Phys 39:4984–4991
39. Sluimer I, Prokop M, van Ginneken B (2005) Toward automated segmentation of the pathological lung in CT. IEEE Trans Med Imaging 24:1025–1038
40. van Rikxoort EM, de Hoop B, Viergever MA et al (2009) Automatic lung segmentation from thoracic computed tomography scans using a hybrid approach with error detection. Med Phys 36:2934–2947
41. Nakagomi K, Shimizu A, Kobatake H et al (2013) Multi-shape graph cuts with neighbor prior constrains and its application to lung segmentation from a chest CT volume. Med Image Anal 17:62–77
42. Niki N (2008) Development of computer aided diagnosis for lung cancer CT screening. IEICE Trans J91-D:1715–1729
43. Matsuhiro M, Suzuki H, Niki N et al (2013) Extraction method of interlobar fissure based on multi-slice CT images. IEICE Trans J96-D:834–843
44. Takahashi E, Suzuki H, Niki N et al (2013) Computer aided diagnosis for osteoporosis using chest multi-slice CT images. IEICE Trans J96-D:892–900
45. Kanazawa K, Kawata Y, Niki N et al (1998) Computer-aided diagnosis for pulmonary nodules based on helical CT images. Comput Med Imaging Graph 22:157–167
46. Armato SG, Giger ML, Moran CJ et al (1999) Computerized detection pulmonary nodules on CT scans. RadioGraphics 19:1303–1311
47. Armato SG, Li F, Giger ML et al (2002) Lung cancer: performance of automated lung nodule detection applied to cancers missed in a CT screening program. Radiology 255:685–692
48. Lee Y, Hara T, Fujita H et al (2001) Automated detection of pulmonary nodules in helical CT images based on improved template-matching technique. IEEE Trans Med Imaging 20:595–604
49. Brown MS, McNitt-Gray MF, Goldin JG et al (2001) Patient-specific models for lung nodule detection and surveillance in CT images. IEEE Trans Med Imaging 20:1242–1250
50. Takizawa H, Yamamoto S, Nakagawa T et al (2002) Recognition of lung nodule shadows from X-ray CT images using 3-D MRF models. IEICE Trans J85-D-II:1401–1412

51. McCulloch CC, Kaucic RA, Mendonça PR et al (2004) Model-based detection of lung nodules in computed tomography exams. Thoracic computer-aided diagnosis. Acad Radio 11:258–266
52. Awai K, Murao K, Ozawa A et al (2004) Pulmonary nodules at chest CT: effect of computer-aided diagnosis on radiologists detection performance. Radiology 230:347–352
53. Bae KT, Kim J-S, Na Y-H et al (2005) Pulmonary nodules: automated detection on CT images with morphologic matching algorithm-preliminary results. Radiology 236:286–294
54. Matsumoto S, Kundel HL, Gee JC et al (2006) Pulmonary nodule detection in CT images with quantized convergence index filter. Med Image Anal 10:343–352
55. Kim KG, Goo JM, Kim JH et al (2005) Computer-aided diagnosis of localized ground-glass opacity in the lung at CT: initial experience. Radiology 237:657–661
56. Dehmeshki J, Ye X, Lin XY et al (2007) Automated detection of lung nodules in CT images using shape-based genetic algorithm. Comput Med Imaging Graph 31:408–417
57. Pu J, Zheng B, Leader JK et al (2008) An automated CT based lung nodule detection scheme using geometric analysis of signed distance field. Med Phys 35:3453–3461
58. Ye X, Lin X, Dehmeshki J et al (2009) Shape based computer-aided detection of lung nodules in thoracic CT images. IEEE Trans Biomed Eng 56:1810–1820
59. Li Q, Li F, Doi K (2008) Computerized detection of lung nodules in thin-section CT images by use of selective enhancement filters and an automated rule-based classifier. Acad Radiol 15:165–175
60. Guo W, Li Q (2012) High performance lung nodule detection schemes in CT using local and global information. Med Phys 39:5157–5168
61. Ge Z, Sahiner B, Chan H-P et al (2005) Computer-aided detection of lung nodules: false positive reduction using a 3D gradient field method and 3D ellipsoid fitting. Med Phys 32:2443–2454
62. Roy AS, Armato SG, Wilson A et al (2006) Automated detection of lung nodules in CT scans: false-positive reduction with the radial-gradient index. Med Phys 33:1133–1140
63. Suzuki K, Armato SG, Li F et al (2003) Massive training artificial neural network (MTANN) for reduction of false positives in computerized detection of lung nodules in low-dose computed tomography. Med Phys 30:1602–1617
64. Das M, Mühlenbruch G, Mahnken AH et al (2006) Small pulmonary nodules: effect of two computer-aided detection systems on radiologist performance. Radiology 241:564–571
65. Yanagawa M, Honda O, Yoshida S et al (2009) Commercially available computer-aided detection system for pulmonary nodules on thin-section images using 64 detectors-row CT: preliminary study of 48 cases. Acad Radiol 16:924–933
66. Zhao Y, de Bock GH, Vliegenthart R et al (2012) Performance of computer-aided detection of pulmonary nodules in low-dose CT: comparison with double reading by nodule. Eur Radiol 22:2076–2084
67. Hansell DM, Bankier AA, HacMahon H et al (2008) Fleischner Society: glossary of terms for thoracic imaging. Radiology 246:697–722
68. Henschke CI, Yankelevitz DF, Mirtcheva R et al (2002) CT screening for lung cancer: frequency and significance of part-solid and nonsolid nodules. Am J Roentgenol 178:1053–1057
69. Yuan R, Vos PM, Cooperberg PL (2006) Computer-aided detection in screening CT for pulmonary nodules. Am J Roentgenol 186:1280–1287
70. Li F, Arimura H, Suzuki K et al (2005) Computer-aided detection of peripheral lung cancers missed at CT: ROC analyses without and with localization. Radiology 237:684–690
71. Brown MS, Goldin JG, Rogers S et al (2005) Computer-aided lung nodule detection in CT: results of large-scale observer test. Acad Radiol 12:681–686
72. Hirose T, Nitta N, Shiraishi J et al (2008) Evaluation of computer-aided diagnosis (CAD) software for the detection of lung nodules on multidetector row computed tomography (MDCT). Acad Radio 15:1505–1512
73. White CS, Pugatch R, Koonce T et al (2008) Lung nodule CAD software as a second reader: a multicenter study. Acad Radiol 15:326–333

74. Sahiner B, Chan H-P, Hadjiiski LM et al (2009) Effect of CAD on radiologists' detection of lung nodules on thoracic CT scans: analysis of an observer performance study by nodule size. Adad Radiol 16:1518–1530
75. Godoy MCB, Kim TJ, White CS et al (2013) Benefit of computer-aided detection analysis for the detection of subsolid and solid lung nodules on thin- and thick-section CT. Am J Roentgenol 200:74–83
76. Metz CE (2006) Receiver operating characteristic analysis: a tool for the quantitative evaluation of observer performance and imaging systems. J Am Coll Radiol 3:413–422
77. Wagner RF, Metz CE, Campbell G (2007) Assessment of medical imaging systems and computer aids: a tutorial review. Acad Radiol 14:723–748
78. He X, Frey E (2009) ROC, LROC, FROC, AFROC: an alphabet soup. J Am Coll Radiol 6:652–655
79. Chakraborty DP, Winter L (1990) Free-response methodology: alternate analysis and new observer-performance experiment. Radiology 174:873–881
80. Chakraborty DP, Berbaum KS (2004) Observer studies involving detection and localization: modeling, analysis and validation. Med Phys 31:2313–2330
81. Thompson JD, Manning DJ, Hogg P (2013) The value of observer performance studies in dose optimization: a focus on free-response receiver operating characteristic methods. J Nucl Med Technol 41:57–64
82. Chakraborty DP (2008) Validation and statistical power comparison of methods for analyzing free-response observer performance studies. Acad Radio 15:1554–1566
83. Ko JP, Betke M (2001) Chest CT: automated nodule detection and assessment of change over time-preliminary experience. Radiology 218:268–273
84. Lee KW, Kim M, Gierada DS et al (2007) Performance of a computer-aided program for automated matching of metastatic pulmonary nodules detected on follow-up chest CT. Am J Roentgenol 189:1077–1081
85. Kubo M, Suzuki H, Saita S et al (2010) Comparative reading support system for lung cancer CT screening. IEICE Trans J93-D:47–58
86. Aoki T, Murakami S, Kim H et al (2014) Temporal subtraction method for lung nodule detection on successive thoracic CT soft-copy images. Radiology 271:255–261
87. Petrick N, Sahiner B, Armato SG et al (2013) Evaluation of computer-aided detection and diagnosis systems. Med Phys 40:087001(17pp.)
88. Huo Z, Summers RM, Paquerault S et al (2013) Quality assurance and training procedures for computer-aided detection and diagnosis systems in clinical use. Med Phys 40:077001(13pp.)
89. Sato Y, Westin C-K, Bhalerao A et al (2000) Tissue classification based on 3D local intensity structures for volume rendering. IEEE Trans Vis Comput Graph 6(2):160–180
90. Frangi AF, Niessen WJ, Vincken KL et al (1998) Multiscale vessel enhancement filtering. Proc MICCAI LNCS 1496:130–137
91. Oda M, Kitasaka T, Mori K et al (2009) Digital bowel cleansing free colonic polyp detection method for fecal tagging CT colonography. Acad Radiol 16(4):486–494
92. Oda M, Kitasaka T, Furukawa K et al (2013) Automated ulcer detection method from CT images for computer aided diagnosis of Crohn's disease. IEICE Trans Inf Syst E96-D(4):808–818
93. Nakamura Y, Nimura Y, Kitasaka T et al (2013) Automatic abdominal lymph node detection method based on local intensity structure analysis from 3-D X-ray CT images. Proc SPIE 8670:86701K-1–7
94. Barbu A, Suehling M, Xu X et al (2010) Automatic detection and segmentation of axillary lymph nodes. Proc MICCAI Part I LNCS 6361:28–36
95. Zhang X, Furukawa T, Zhou X et al (2011) Detection of metastatic liver tumor in multi-phase CT images by using a spherical gray-level differentiation searching filter. Proc SPIE Med Imag 7963:79632K-1–8
96. Welch HG, Black WC (2010) Overdiagnosis in cancer. J Natl Cancer Inst 102:605–613
97. Esserman LJ, Thompson IM, Reid B (2013) Overdiagnosis and overtreatment in cancer: an opportunity for improvement. JAMA 310:797–798

98. Patz EF, Pinsky P, Gatsonis C et al (2014) Overdiagnosis in low-dose computed tomography screening for lung cancer. JAMA Intern Med 174:269–274
99. Noro R, Nonda K, Tsuta K et al (2013) Distinct outcome of stage I lung adenocarcinoma with ACTN4 cell motility gene amplification. Ann Oncol 24:2594–2600
100. Goya T, Asamura H, Yoshimura H et al (2005) Prognosis of 6644 resected non-small cell lung cancers in Japan: a Japanese lung cancer registry study. Lung Cancer 50:227–234
101. Booth CM, Shepherd FA (2006) Adjuvant chemotherapy for resected non-small cell lung cancer. J Thorac Oncol 1:180–187
102. Kawata Y, Niki N, Ohmatsu H et al (2012) Quantitative classification based on CT histogram analysis of non-small cell lung cancer : correlation with histopathological characteristics and recurrence-free survival. Med Phys 39:988–1000
103. Yanagawa M, Tanaka Y, Leung AN et al (2014) Prognostic importance of volumetric measurements in stage I lung adenocarcinoma. Radiology 272:557–567
104. Depeursinge A, Yanagawa M, Leung AN et al (2015) Predicting adenocarcinoma recurrence using computational texture models of nodule components in lung CT. Med Phys 42:2054–2063
105. Kawata Y, Niki N, Ohmatsu H et al (2012) Image-based computer-aided prognosis of lung cancer: predicting patient recurrent-free survival via a variational Bayesian mixture modeling framework for cluster analysis of CT histograms. Proc SPIE Med Imaging 315:83150C-1-8
106. Kawata Y, Niki N, Ohmatsu H et al (2013) Tracking time interval changes of pulmonary nodules on follow-up 3D CT images via image-based risk score of lung cancer. Proc SPIE Med Imaging 8670:86700J-1-6
107. Noguchi M, Morikawa A, Kawasaki M et al (1995) Small adenocarcinoma of the lung: histologic characteristics and prognosis. Cancer 75:2844–2852
108. Aoki T, Tomoda Y, Watanabe H et al (2001) Peripheral lung adenocarcinoma: correlation of thin-section CT findings with histologic prognostic factors and survival. Radiology 220:803–809
109. Kondo T, Yamada K, Noda K et al (2002) Radiologic-prognostic correlation in patients with small pulmonary adenocarcinomas. Lung Cancer 36:49–57
110. Takashima S, Maruyama Y, Hasegawa M et al (2002) Prognostic significance of high-resolution CT findings in small peripheral adenocarcinoma of the lung: a retrospective study on 64 patients. Lung Cancer 36:289–295
111. Nomori H, Ohtsuka T, Naruke T et al (2003) Histogram analysis of computed tomography numbers of clinical T1 N0 M0 lung adenocarcinoma, with special reference to lymph node metastasis and tumor invasiveness. J Thorac Cardiovasc Surg 126:1584–1589
112. Suzuki K, Kusumoto M, Watanabe S et al (2006) Radiologic classification of small adenocarcinoma of the lung: radiologic-pathologic correlation and its prognostic impact. Ann Thorac Surg 81:413–420
113. Seki N, Sawada S, Nakata M et al (2008) Lung cancer with localized ground-glass attenuation represents early-stage adenocarcinoma in nonsmokers. J Thorac Oncol 3:483–490
114. Kakinuma R, Kodama K, Yamada K et al (2008) Performance evaluation of 4 measuring methods of ground-glass opacities for predicting the 5-year relapse-free survival of patients with peripheral non-small cell lung cancer: a multicenter study. J Comput Assist Tomogr 32:792–798
115. Matsuguma H, Oki I, Nakahara R et al (2013) Comparison of three measurements on computed tomography for the prediction of less invasiveness in patients with clinical stage I non-small cell lung cancer. Ann Thorac Surg 95:1878–1884
116. Maldonado F, Boland JM, Raghunath S et al (2013) Noninvasive characterization of the histopathologic features of pulmonary nodules of the lung adenocarcinoma spectrum using computer-aided nodule assessment and risk yield (CANARY)-a pilot study. J Thorac Oncol 8:452–460
117. van Klaveren RJ, Oudkerk M, Prokop M et al (2009) Management of lung nodules detected by volume CT scanning. N Engl J Med 361:2221–2229
118. Cox DR (1972) Regression models and life-table. J R Stat Soc 34:187–220

119. Kaplan EL, Meier P (1958) Nonparametric estimation from incomplete observations. J Am Stat Assoc 53:457–481
120. Camp RL, Dolled-Filhart M, Rimm DL (2004) X-tile: a new bio-informatics tool for biomarker assessment and outcome-based cut-point optimization. Clin Cancer Res 10:7252–7259
121. Wei J, Hagihara Y, Shimizu A, Kobatake H (2002) Optimal image feature set for detecting lung nodules on chest x-ray images. Comp Assist Radiol Surg 18:706–711
122. Kakeda S, Nakamura K, Kamada K, Watanabe H, Nakata H, Katsuragawa S, Doi K (2002) Improved detection of lung nodules by using a temporal subtraction technique. Radiology 224:145–151
123. Ishida T, Ashizawa K, Engelmann R, Katsuragawa S, MacMahon H, Doi K (1999) Application of temporal subtraction for detection of interval changes on chest radiographs: improvement of subtraction images using automated initial image matching. J Digit Imaging 12:77–86
124. Itai Y, Kim H, Ishikawa S, Katsuragawa S, Doi K (2010) Development of a voxel matching technique for substantial reduction of subtraction artifacts in temporal subtraction images obtained from thoracic MDCT. J Digit Imaging 23:31–38
125. Otsu N (1979) A threshold selection method from gray-level histograms. IEEE Trans Syst Man Cybern 9(1):62–66
126. Weiss E, Hess CF (2003) The impact of gross tumor volume (GTV) and clinical target volume (CTV) definition on the total accuracy in radiotherapy theoretical aspects and practical experiences. Strahlenther Onkol 179:21–30
127. ICRU (1999) ICRU report 62, Prescribing, Recording and Reporting Photon Beam Therapy (Supplement to ICRU Report 50)
128. Van de Steene J, Linthout N, de Mey J, Vinh-Hung V, Claassens C, Noppen M, Bel A, Storme G (2002) Definition of gross tumor volume in lung cancer: inter-observer variability. Radiother Oncol 62:37–49
129. Bradley JD, Perez CA, Dehdashti F, Siegel BA (2004) Implementing biologic target volumes in radiation treatment planning for non-small cell lung cancer. J Nucl Med 45:96S–101S
130. Chao KS, Bhide S, Chen H, Asper J, Bush S, Franklin G, Kavadi V, Liengswang-wong V, Gordon W, Raben A, Strasser J, Koprowski C, Frank S, Chronowski G, Ahamad A, Malyapa R, Zhang L, Dong L (2007) Reduce in variation and improve efficiency of target volume delineation by a computer-assisted system using a deformable image registration approach. Int J Radiat Oncol Biol Phys 68:1512–1521
131. Matsushita N, Arimura H, Nakamura K, Yoshikazu K, Shioyama Y, Nakamura Y, Honda H, Hirata H (2013) Computerized production of statistical clinical target volume models in prostate cancer radiation treatment planning. Med Phys 40:89. http://dx.doi.org/10.1118/1.4813953 (Abstract)
132. Biehl KJ, Kong FM, Dehdashti F, Jin JY, Mutic S, El Naqa I, Siegel BA, Bradley JD (2006) 18F-FDG PET definition of gross tumor volume for radiotherapy of non-small cell lung cancer: Is a single standardized uptake value threshold approach appropriate? J Nucl Med 47(11):1808–1812
133. Zhang T, Tachiya Y, Sakaguchi Y, Mitsumoto K, Mitsumoto T, Ohya N, Sasaki M (2010) Phantom study on three-dimensional target volume delineation by PET/CT-based auto-contouring. Fukuoka Acta Media 101(11):238–246
134. Day E, Betler J, Parda D, Reitz B, Kirichenko A, Mohammadi S, Miften M (2009) A region growing method for tumor volume segmentation on PET images for rectal and anal cancer patients. Med Phys 36(10):4349–4358
135. Aristophanous M, Penney BC, Martel MK, Pelizzari CA (2007) Gaussian mixture model for definition of lung tumor volumes in positron emission tomography. Med Phys 34(11):4223–4235
136. Geets X, Lee J A, Bol A et al (2007) A gradient-based method for segmenting FDG-PET images: methodology and validation. Eur J Nucl Med Mol Imaging 34:1427–1438

137. Hatt M, Rest CC, Nidal A et al (2011) PET functional volume delineation: a robustness and repeatability study. Eur J Nucl Med Mol Imaging 38:3663–3672
138. Belhassen S, Zaidi H (2010) A novel fuzzy C-means algorithm for unsupervised heterogeneous tumor quantification in PET. Med Phys 37(3):1309–1324
139. Rousson M, Khamene A, Diallo M et al (2005) Constrained surface evolutions for prostate and bladder segmentation in CT images. In: Liu Y, Jiang T, Zhang C (eds) Computer vision for biomedical image applications. Lecture notes in computer science (LNCS), vol 3765. Springer, New York, pp 251–260
140. Dowling J, Lambert J, Parker J, Greer PB, Fripp J, Denham J, Ourselin S, Salva-do O (2010) Automatic MRI atlas-based external beam radiation therapy treatment planning for prostate cancer. In: Madabhushi A et al (eds) A. Prostate cancer imaging. Computer-aided diagnosis, prognosis, and intervention. Lecture notes in computer science, vol 6367. Springer, Berlin, pp 25–33
141. Acosta O, Dowling J, Cazoulat G, Simon A, Salvado O, de Crevoisier R, Haigron P (2010) Atlas based segmentation and mapping of organs at risk from planning CT for the development of voxel-wise predictive models of toxicity in prostate radiotherapy. In: Madabhushi A et al (eds) Prostate cancer imaging. Computer-aided diagnosis, prognosis, and intervention. Lecture notes in computer science, vol 6367. Springer, Berlin, pp 42–51
142. El Naqa I, Yang D, Apte A, Khullar D, Mutic S, Zheng J, Bradley JD, Grigsby P, Deasy JO (2007) Concurrent multimodality image segmentation by active contours for radiotherapy treatment planning. Med Phys 34(2):4738–4749
143. Arimura H, Jin Z, Shioyama Y, Nakamura K, Magome T, Sasaki M (2013) Auto-mated method for extraction of lung tumors using a machine learning classifier with knowledge of radiation oncologists on data sets of planning CT and FDG-PET/CT images. In: Proceeding of the 35th annual international conference of the IEEE engineering in medicine and biology society (EMBC), Osaka, pp 2988–2991
144. Jin Z, Arimura H, Shioyama Y, Kuwazuru J, Magome T, Nakamura K, Honda H, Toyofuku F, Hirata H, Sasaki M (2013) Computer-aided delineation of lung tumor regions in treatment planning CT images and PET/CT images using localized level set approach. IEICE Tech Rep 112(411):49–51
145. Onishi H, Shirato H, Nagata Y et al (2011) Stereotactic body radiotherapy (SBRT) for operable stage I non-small-cell lung cancer: can SBRT be comparable to surgery? Int J Radiat Oncol Biol Phys 81:1352–1358
146. Commowick O, Malandain G (2007) Efficient selection of the most similar image in a database for critical structures segmentation. Med Image Comput Comput Assist Interv 10(Pt 2):203–210
147. Chanyavanich V, Das SK, Lee WR et al (2011) Knowledge-based IMRT treatment planning for prostate cancer. Med Phys 38(5):2515–2522
148. Mishra N, Petrovic S, Sundar S (2011) A self-adaptive case-based reasoning system for dose planning in prostate cancer radiotherapy. Med Phys 38(12):6528–6538
149. Schlaefer A, Dieterich S (2011) Feasibility of case-based beam generation for robotic radiosurgery. Artif Intell Med 52(2):67–75
150. Magome T, Arimura H, Shioyama Y, Mizoguchi A, Tokunaga C, Nakamura K, Honda H, Ohki M, Toyofuku F, Hirata H (2013) Computer-aided beam arrangement based on similar cases in radiation treatment planning databases for stereotactic lung radiation therapy. J Radiat Res 54(3):569–577
151. Magome T, Arimura H, Shioyama Y, Nakamura K, Honda H, Hirata H (2013) Similar-case-based optimization of beam arrangements in stereotactic body radiotherapy for assisting treatment planners. BioMed Res Int 6:10. Article ID 309534
152. Nonaka A, Arimura H, Nakamura K, Shioyama Y, Soufi M, Magome T, Honda H, Hirata H (2014, accepted) Local image descriptor-based searching framework of usable similar cases in a radiation treatment planning database for stereotactic body radiotherapy. Proc SPIE Med Imaging 9039:10

153. Lowe DG (2004) Distinctive image features from scale-invariant keypoints. Int J Comput Vis 60(2):91–110
154. Arimura H, Kakiuchi G, Shioyama Y, Minohara S, Nakamoto T, Nakamura K, Honda H, Tashiro M, Kanai T, Hirata H (2014) Quantitative evaluation of the robustness of beam directions based on power spectral analysis of water-equivalent path length image in charged particle therapy. Int J Intell Comput Med Sci Image Process 6:1–16
155. Ezawa H, Yoneyama R, Kandatsu S, Yoshikawa K, Tsujii H, Harigaya K (2003) Introduction of autopsy imaging redefines the concept of autopsy: 37 cases of clinical experience. Pathol Int 53:865–873
156. Thali MJ, Yen K, Schweitzer W, Vock P, Boesch C, Ozdoba C, Schroth G, Ith M, Sonnenschein M, Doernhoefer T, Scheurer E, Plattner T, Dirnhofer R (2003) Virtopsy, a new imaging horizon in forensic pathology: virtual autopsy by postmortem multislice computed tomography (MSCT) and magnetic resonance imaging (MRI)-a feasibility study. J Forensic Sci 48:386–403
157. Clarot F, Proust B, Eurin D, Vaz E, Le Dosseur P (2007) Sudden infant death syndrome and virtual autopsy: scalpel or mouse? Arch Pediatr 14:636–639. (Article in French)
158. Dedouit F, Telmon N, Guilbeau-Frugier C, Gainza D, Otal P, Joffre F, Rouge D (2007) Virtual autopsy and forensic identification-practical application: a report of one case. J Forensic Sci 52:960–964
159. Wichmann D, Obbelode F, Vogel H, Hoepker WW, Nierhaus A, Braune S, Sauter G, Pueschel K, Kluge S (2012) Virtual autopsy as an alternative to traditional medical autopsy in the intensive care unit: a prospective cohort study. Ann Intern Med 156:123–130
160. Takazawa Y, Fukayama M (2012) Post-mortem imaging – contrast with Pathology. Pathol Clin Med 30(suppl):123–129 (in Japanese)
161. Brogdon BG (1998) Forensic radiology, Chap 8. CRC Press, Boca Raton, pp 149–187
162. Shiotani S, Kohno M, Ohashi N, Yamazaki K, Nakayama H, Ito Y, Kaga K, Ebashi T, Itai Y (2002) Hyperattenuating aortic wall on postmortem computed tomography (PMCT). Radiat Med 20:201–206
163. Shiotani S, Kohno M, Ohashi N, Yamazaki K, Itai Y (2002) Postmortem intravascular high-density fluid level (hypostasis): CT findings. J Comput Assist Tomogr 26:892–893
164. Shiotani S, Kohno M, Ohashi N, Yamazaki K, Nakayama H, Watanabe K, Itai Y (2003) Dilatation of the heart on postmortem computed tomography (PMCT): comparison with live CT. Radiat Med 21:29–35
165. Hamabe Y (2006) The present condition of Ai on the emergency care field. The 3rd annual academic meeting program, Autopsy imaging academic society (in Japanese)
166. Fukayama M et al (2010) The report of the research on verification of the cause-of-death investigation technique (Post-mortem imaging) which assists the autopsy in "investigation analysis of the death relevant to medical treatment" Grants-in-aid-for-scientific-research in the Ministry of Health, Labour and Welfare (in Japanese)
167. Roberts IS, Benamore RE, Benbow EW, Lee SH, Harris JN, Jackson A, Mallett S, Patankar T, Peebles C, Roobottom C, Traill ZC (2012) Post-mortem imaging as an alternative to autopsy in the diagnosis of adult deaths: a validation study. Lancet 379:136–142
168. Rozzelle CJ, Aarabi B, Dhall SS, Gelb DE, Hurlbert RJ, Ryken TC, Theodore N, Walters BC, Hadley MN (2013) Spinal cord injury without radiographic abnormality (SCIWORA). Neurosurgery 72:227–233
169. Westphal SE, Apitzsch J, Penzkofer T, Mahnken AH, Knuchel R (2012) Virtual CT autopsy in clinical pathology: feasibility in clinical autopsies. Virchows Arch 461:211–219
170. Proisy M, Marchand AJ, Loget P, Bouvet R, Roussey M, Pele F, Rozel C, Treguier C, Darnault P, Bruneau B (2013) Whole-body post-mortem computed tomography compared with autopsy in the investigation of unexpected death in infants and children. Eur Radiol 23:1711–1719
171. Finkbeiner WE, Ursell PC, Davis RL (2004) Autopsy pathology a manual and atlas, Chap 12. Churchill Livingstone, Philadelphia, pp 155–166
172. Takano H, Natsuno Y, Kido S, Inai K, Noriki S (2013) Ai and archaeology, Ai and gross anatomy, Ai and computational anatomy, Ai and post-mortem imagiology. Proc Jpn Soc Autopsy Imaging 11:40–45 (in Japanese)

173. Ezawa E, Shiotani S, Uchigasaki S (2007) Autopsy imaging in Japan. Rechtsmedizin 17:19–20
174. Heimann T, Meinzer HP (2009) Statistical shape models for 3D medical image segmentation: a review. Med Image Anal 13:543–563. doi: 10.1016/j.media.2009.05.004
175. Cremers D, Rousson M, Deriche R (2007) A review of statistical approaches to level set segmentation: integrating color, texture, motion and shape. Int J Comput Vis 72:195–215
176. Saito A, Shimizu A, Watanabe H, Yamamoto S, Nawano S, Kobatake H (2014) Statistical shape model of a liver for autopsy imaging. Int J Comput Assist Radiol Surg 9(2): 269–281. doi: 10.1007/s11548-013-0923-6
177. Okada T, Shimada R, Hori M, Nakamoto M, Chen YW, Nakamura H, Sato Y (2008) Automated segmentation of the liver from 3D CT images using probabilistic atlas and multilevel statistical shape model. Acad Radiol 15:1390–1403. doi: 10.1016/j.acra.2008.07.008
178. Tomoshige S, Oost E, Shimizu A, Watanabe H, Nawano S (2014) A conditional statistical shape model with integrated error estimation of the conditions; application to liver segmentation in non-contrast CT images. Med Image Anal 18:130–143
179. Umetsu S, Shimizu A, Watanabe H, Kobatake H, Nawano S (2014) An automated segmentation algorithm for CT volumes of livers with atypical shapes and large pathological lesions. IEICE Trans Inf Syst E97-D:951–963
180. Uchida Y, Shimizu A, Kobatake H, Nawano S, Shinozaki K (2010) A comparative study of statistical shape models of the pancreas. Int J Comput Assist Radiol Surg 5(supplement 1):385–387
181. Murase H, Nayar SK (1996) Learning by a generation approach to appearance-based object recognition. Int Conf Pattern Recognit 1:25–29. doi: 10.1109/ICPR.1996.545985
182. Saito A, Shimizu A, Watanabe H, Yamamoto S, Kobatake S (2013) Automated liver segmentation from a CT volume of a cadaver using a statistical shape model. Int J Comput Assist Radiol Surg 8(supplement 1):S48–S49
183. Shimizu A, Nakagomi K, Narihira T, Kobatake H, Nawano S, Shinozaki K, Ishizu K, Togashi K (2010) Automated segmentation of 3D CT images based on statistical atlas and graph cuts. In: Proceedings of MICCAI workshop MCV, Beijing, pp 129–138
184. Shimizu A, Ohno R, Ikegami T, Kobatake H, Nawano S, Smutek D (2007) Segmentation of multiple organs in non-contrast 3D abdominal CT images. Int J Comput Assist Radiol Surg 2:135–142
185. Nakagomi K, Shimizu A, Kobatake H, Yakami M, Fujimoto K, Togashi K (2013) Multi-shape graph cuts with neighbor prior constraints and its application to lung segmentation from a chest CT volume. Med Image Anal 17:62–77
186. Saito A, Okata S, Shimizu A, Watanabe H, Yamamoto S, Nawano S (2014) Automated lung segmentation for autopsy imaging using graph cuts with a statistical shape model. Int J Comput Assist Radiol Surg 5(supplement 1):S37–S38
187. Morooka K, Taguchi T, Chen X, Kurazume R, Hashizume M, Hasegawa T (2012) A method for constructing real-time FEM-based simulator of stomach behavior with large-scale deformation by neural networks. SPIE Med Imaging. doi: 10.1117/12.911171
188. Heimann T, Ginneken B, Styner MA et al (2009) Comparison and evaluation of methods for liver segmentation from CT datasets. IEEE Trans Med Imaging 28:1251–1265

Chapter 5
Perspectives

Yoshitaka Masutani

Abstract Computational anatomy (CA) is still a developing discipline. It offers a wide variety of research areas for applications in clinical support and medical science. The editors and the authors hope this book will serve as a guide to students and researchers interested in this exciting new discipline.

Keywords Computational anatomy • Complete medical image understanding

In this book, fundamentals, state-of-the-art techniques, and implementation of computational anatomy (CA) are presented. At this moment, one of the main goals of this discipline, "complete medical image understanding," has yet to be achieved. This is for several reasons, including the need for more data samples, the diversity of pathologies, and so on.

One of the difficulties in establishing theories and techniques for CA is that consistency with conventional anatomy is necessary for physician support. There needs to be a common language and terminology between physicians and engineers. However, ambiguities exist; for example, the explicit boundary between the lung and the hilum is not defined in conventional anatomy, posing difficulties for lung segmentation. This kind of ambiguity directly affects the construction of computational anatomy models in explicit representation of structures. Therefore, a new framework for handling such ambiguity in conventional anatomy is necessary and will need to be worked out in discussions among physicians, engineers, and anatomists.

The new and notable techniques for pattern recognition generally require data samples in the tens of thousands. However, personal medical imaging data are hard to obtain for legal reasons, in addition to the problem of data size. This is not a simple problem solvable by a small research community but should be considered at the national level. In order to increase the availability of data, researchers must show the advantages brought by the computational anatomy approach, such as safer and lower-cost medical practices.

Y. Masutani (✉)
Hiroshima City University, Hiroshima, Japan
e-mail: masutani@hiroshima-cu.ac.jp

H. Kobatake, Y. Masutani (eds.), *Computational Anatomy Based on Whole Body Imaging*, DOI 10.1007/978-4-431-55976-4_5

New imaging modalities promise new opportunities for analyses of anatomical structures from multiple medical specialties. Not only post-processing of given image data but also optimization in the imaging phase should be discussed in the context of medical image understanding.

CA has opened a wide variety of research areas including applications in clinical support and medical science. The editors and the authors hope that this book will help students and researchers garner further understanding of the role of CA in medical education and practice.

MIX
Papier aus verantwortungsvollen Quellen
Paper from responsible sources
FSC® C105338

If you have any concerns about our products,
you can contact us on
ProductSafety@springernature.com

In case Publisher is established outside the EU,
the EU authorized representative is:
Springer Nature Customer Service Center GmbH
Europaplatz 3, 69115 Heidelberg, Germany

Printed by Libri Plureos GmbH
in Hamburg, Germany